MATHEMATICS IN CIVILIZATION

H. L. RESNIKOFF
R. O. WELLS, JR.

DOVER PUBLICATIONS, INC.
NEW YORK

To our students

Published in Canada by General Publishing Company, Ltd., 30 Lesmill Road, Don Mills, Toronto, Ontario.
Published in the United Kingdom by Constable and Company, Ltd., 10 Orange Street, London WC2H 7EG.

This Dover edition, first published in 1984, is an enlarged and corrected republication of the work first published by Holt, Rinehart and Winston, Inc., New York, in 1973. A supplement, "Twentieth-Century Mathematics," an index to the supplement, and solutions to the exercises at the end of chapters have been added to the Dover edition.

Manufactured in the United States of America
Dover Publications, Inc., 31 East 2nd Street, Mineola, N.Y. 11501

Library of Congress Cataloging in Publication Data

Resnikoff, H. L.
Mathematics in civilization.

Bibliography: p.
Includes index.
1. Mathematics—History. I. Wells, R. O. (Raymond O'Neil), 1940–   II. Title.
QA21.R43   1984   510′.9   83-20526
ISBN 0-486-24674-4

PREFACE TO
THE DOVER EDITION

It has been over ten years since the first edition of our excursion into mathematical history appeared, and we are very pleased to have it reappear at this time under the imprint of Dover Publications. The topics covered in the first edition span some 4000 years ending about the turn of the twentieth century. We have added a supplement which discusses some of the major events relating to our theme which have occurred in this century. This supplement was originally prepared for the German edition (*Mathematik im Wandel der Kulturen*, Vieweg, Wiesbaden, 1983) and has been adapted for this volume. In addition, solutions to many of the exercises that follow each chapter now appear at the end of the book. Some of the exercises are of an unusual nature, and these solutions may help to answer some of the questions that were most often asked when we were teaching the course at Rice University.

We want to thank the Dover editors for their help and exceptional patience in bringing out this new edition.

H. L. RESNIKOFF
Cambridge, Massachusetts

R.O. WELLS, JR.
Boulder, Colorado

1984

PREFACE TO
THE FIRST EDITION

This book is an expansion of lectures delivered to freshmen from 1969 to 1972 at Rice University. The Contents indicate the topics treated. In no case have we striven for completeness of presentation but rather have attempted to include sufficient material concerning each topic to enable the reader to form a reasonably accurate idea of the essential similarity of mathematical problems throughout historic times. We hope he will also understand why mathematics is useful as well as beautiful.

Material, other than quotations, that appears in small type in the text is of a subsidiary nature; the reader is expected to be familiar with but not to master this matter, which usually consists of detailed verifications of important statements in the text.

References to the bibliography at the end of the book are enclosed in square brackets; thus [59] refers to the fifty-ninth reference.

Exercises are collected at the end of each chapter; those preceded by an asterisk are of greater difficulty than the rest.

We use the terms *Akkadian* and *Babylonian* interchangeably to refer to the post-Sumerian inhabitants of the Mesopotamian region between 2000 and 6000 years ago.

Dates are written with algebraic sign: thus 1970 for 1970 A.D. and -1970 for 1970 B.C.

We cannot attempt to identify all the colleagues whose comments and criticisms have helped us, but we are nevertheless pleased to have the opportunity to thank them here. Our special appreciation is due David Rector, who taught this material in the 1969 academic year.

We would also like to express our appreciation to Nancy Singleton and Barbara Markwardt for their splendid job of typing the numerous drafts of the manuscript, and to Holly Massey and the staff of Holt, Rinehart and Winston, Inc. for their attention to detail and design in converting our raw manuscript into this finished book. Last, as well as first, it is our pleasure to thank our students whose enthusiastic and patient response to our crude lecture notes encouraged us to transform them into this book.

H. L. RESNIKOFF
R.O. WELLS, JR.

Houston, Texas
December, 1972

My distinguished teachers, J. J. Finkelstein and D. J. Struik, laid the foundation for this book many years ago in their inspiring lectures and conversations which showed me how strong are the bonds between the sciences and the humanities, between the present and the past.

H. L. RESNIKOFF

CONTENTS

INTRODUCTION

§**0.1.** Young people entering college today should attempt to extract two things, at least, from their college years. First is a certain competence in an area or specialized field by means of which they can hope to find a place for themselves in society. Second, they should become aware of and knowledgeable in regard to the many other aspects of society that will engage their attention and to some extent determine the course and quality of the rest of their lives. It is in response to the latter purpose that in recent years college courses have been designed to convey an understanding of the nature and role of fields other than the area of a student's primary specialization. This book is intended for students and others who desire to understand the role that mathematics plays in science and society. It does not teach mathematical technique, nor will it prepare the reader to use mathematics as a tool in any serious way. It is sparing in its demands on the reader's mathematical knowledge; competence in the arithmetic of fractions and decimal expansions, elementary plane geometry, and the rudiments of algebraic manipulation and trigonometry are the only mathematical prerequisites.

One way to introduce mathematics to the uninitiated is to sample attractive topics of current or recent interest to mathematicians and by studying them in some detail attempt to teach the student what mathematics is about and how mathematicians think. Learning about mathematics this way is time consuming and difficult, much as it would be to acquire a knowledge of music by trying to learn to play a number of different musical instruments, but not very well. Just as it is possible to appreciate music and to understand its role in civilization in a more than superficial way without being able to play even one instrument nor even to understand the technicalities of how one is played, so also is it possible to attain a serious comprehension of the nature of mathematical achievements and the impact they have on civilization.

The structure and purpose of this book can be illuminated in terms of another analogy. Mathematics is a growing subject that can be likened to a tree: think of the height of a place on the tree as a measurement of *time,* with early mathematics located near the roots and the most recent advances flowering at the tips of the highest limbs. Those books that sample topics drawn from modern mathematics can be said to exhibit a horizontal section of the mathematical tree near its top, whereas a study of the historical development of one topic corresponds to tracing a vertical path that starts somewhere near the roots and continues up through its limbs. In this book we have concentrated on two major topics and traced their paths up through the tree from antiquity to modern times. These topics are fundamental and pervasive; we think they lie close to the essential nature of mathematics. Moreover, although each preserves its own identity, they have become inextricably intertwined throughout the centuries. By concentrating on them and following their development we hope to provide the reader with a perspective of the process of mathematical development and its symbiotic interaction with the corresponding development of civilization that is impossible to obtain from a study of a sectional selection of recent mathematical topics.

This evolutionary standpoint has another, pedagogical, advantage. The student who lacks technical proficiency in areas of current mathematical interest nevertheless has accumulated a store of mathematical knowledge that is elementary by current standards but once represented the research frontier. By tracing our way from the past to the present, although not always in strict chronological order, and by limiting ourselves to basic mathematical problems of ancient lineage we hope to be able to build on the reader's available technical knowledge to propel him to an understanding of the modern, more sophisticated, forms of these problems and therewith to an understanding of the value and implications of mathematical progress.

We have, as we have said, selected two paths. The first might be termed

the ability to compute.

It is directly related to the growth of technology and to the ability to organize increasingly complex forms of society. The second path can be called the

geometrical nature of space,

that is, the geometrical nature of the physical world in which we find ourselves. This path deals with the evolution of conceptions of the physical universe and their relation to abstract forms of geometry invented and studied by mathematicians through the centuries. In its practical applications this path is a determinant of our ability to control physical reality. It also lies at the foundation of some of the most profound philosophical speculations. The paths are interdependent because the ability to compute underlies the advance from simple and simplistic geometrical considerations to complex ones that can accurately describe portions of reality. On the other hand, the complexity of some geometrical models challenges the available computational capabilities and encourages their further development.

§0.2. The purpose of this book is, as already asserted, to study the role of mathematics in the development and maintenance of civilization. It is part of our thesis that, although the techniques and personalities of mathematics are not and indeed should not be of much interest to the nonmathematician, the purposes and consequences of mathematics are of serious concern for the growth and health of society and therefore are a proper and necessary part of the workaday intellectual baggage that must be carried about by every educated and effective participant in civilized life.

Mathematics occupies a peculiar and unique role in that it is neither a science nor an art but partakes of both disciplines. Art provides the motivation for most pure mathematicians but science (this term understood in its broadest sense) reaps the harvest. That there are important differences between mathematics and science is not simply a matter of personal opinion; they show up in quantitative as well as qualitative ways, some of which are considered below.

Nevertheless, mathematics is usually considered to be a science, and in this guise it participates in the general increase of federal support for research activities. The competition for the taxpayers' dollars having now become quite keen, a serious inquiry into the role played by mathematics in society is well justified. Is the emphasis on mathematics in contemporary American society sufficient, too great, or just right? We have tried to provide you, the reader, with some tools that you can use to answer this difficult question yourself. After all, it is, or shortly will be, your tax dollar that will help to determine the future.

One way to evaluate the importance to society of an activity that it undertakes and supports is to estimate the fraction of its human and other resources this activity consumes. If this fraction is now large but was small in the past, we can reasonably assert that the activity has recently become more important; if the fraction has remained sensibly constant throughout long historical periods, then we should conclude that the activity has always had about the same relative importance as it now has.

Let us examine the size of the mathematics establishment today and compare it with the situation in the past as far back as we can. It will turn out that "memorable" mathematicians have always—at least for the last 2000 years—accounted for a virtually fixed fraction of the population: about one memorable mathematician for every 4 million people. First consider the current situation. The 1966 edition of the *World Directory of Mathematicians* lists about 11,000 persons. In 1965 the population of the world was about 3.3 billion—about 300,000 people for each mathematician. Not all mathematicians produce mathematics that will be memorable in years to come; perhaps only one in 50 will, which means that about 220 or so will be likely to pass into history as memorable. This estimate is, of course, just an opinion, but there is some independent evidence that suggests it cannot be far wrong.

To understand this evidence and to be able to compare the present size of the mathematics establishment with the past let us turn to the history books. Dirk J. Struik, himself a notable mathematician, has written on the history of science and mathematics. His *A Concise History of Mathematics* [60], now in its third edition, is a learned and sophisticated work that has been well received. Let us agree, for the sake of argument, that a mathematician is memorable if he is listed (with his date of birth) in Struik's index. This "defines" what we mean by a "memorable mathematician," and although it is a definition that is open to question with regard to any particular mathematician it certainly will reflect in general what is intuitively meant when it is said that any historical figure is memorable — he occurs in the history books. When counting the number of memorable mathematicians, it makes sense to accumulate all those who were born before a given date, since important mathematical contributions do not tarnish with age.[1] Figure 0.1 displays the graph of this function. Evidently there has been an enormous growth in the number of memorable mathematicians in recent times. On the other hand, it is clear that there must have been some memorable mathematicians who were born before -700, although no names have come down to us. There can be many reasons for this: the ravages of time acting on records, Struik's possible idiosyncrasies, social anonymity in early civilizations,

FIGURE 0.1

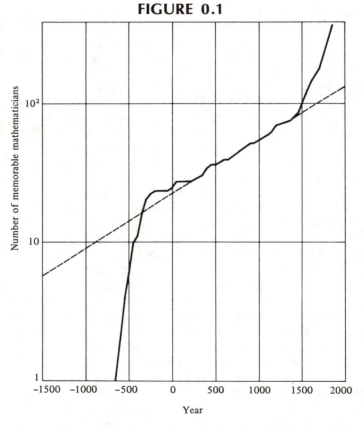

Memorable mathematicians (cumulative, by birthdate). Data from Struik [60].

[1] *For example, many modern textbooks on number theory and recreational mathematics still give Euclid's proof (Euclid: fl. c. -300) that there are infinitely many primes.*

etc. Be that as it may, Figure 0.1 shows a steady growth [2] in the number of memorable mathematicians from about -100 until $+1400$, after which there is a dramatic and still continuing increase in the rate of growth of the curve.

Notice that there were no "Dark Ages" for memorable mathematicians. The "darkness" of medieval Europe, from 600 until about 1200 (depending on the authority quoted), was compensated by the "lightness" of Arab civilization.

Figure 0.2 illustrates the cumulative growth of memorable mathematicians with the portion corresponding to steady exponential growth from -100 to $+1400$ projected back in time; this projection suggests that there should have been "one" memorable mathematician about -3700. We will return to this speculation later. The same figure shows estimates of the world's population at various times. Observe that the population curve and the cumulative memorable mathematicians curve are approximately parallel. This means that memorable mathematicians have constituted about the same fraction of the population in the distant past as they have in more recent times. Inspection of the figure shows that the ratio of population to the cumulative number of memorable mathematicians has varied with time as shown in Table 0.1. This table cannot be brought more closely up to date because Struik considers only pre-twentieth century mathematics, and we have classified memorable mathematicians according to

FIGURE 0.2

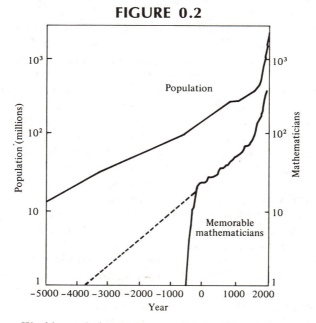

World population and memorable mathematicians.

[2] *The curve is plotted on* semilogarithmic *graph paper; that is, the ordinates are the* logarithms *of the cumulative number of mathematicians. A straight line segment corresponds to* exponential growth. *See Chapter 7 for a discussion of logarithmic and exponential functions.*

TABLE 0.1

Ratio of Population to Cumulative Number of Memorable Mathematicians

Date	Number of Mathematicians Born Before	World Population	Ratio: Population per Number of Mathematicians (in millions)
700	39	2.70×10^8	6.9
1049	54	2.85×10^8	5.3
1449	80	3.75×10^8	4.7
1549	97	4.20×10^8	4.3
1649	143	5.45×10^8	3.8
1749	180	7.28×10^8	4.0
1799	223	9.06×10^8	4.1
1849	296	1.17×10^9	4.0

their date of *birth,* which is at least 20 years before their active intellectual life begins. It is remarkable that the ratios are so close in value: they all lie between 3.8 millions of population per memorable mathematician and 6.9 millions. The apparent trend seems to indicate that the number of mathematicians is growing more rapidly than population. This may be due to an increasing underenumeration of mathematicians the longer ago they lived, or to an overestimation of world population in early times, or it may simply be that the number of memorable mathematicians and world population are just not proportional. The last possibility is perhaps right, but the fact that the population growth and memorable mathematician growth curves are both so far from being simple, yet are almost parallel, suggests that the processes they represent may be closely related.

As a speculative possibility, we propose that memorable mathematician growth is actually proportional to *gross world product* (or some other index of world economic growth) rather than to world population. Gross world product is analogous to gross national product, which measures the annual value of all goods and services produced by a nation and is the best known and probably most reliable indicator of its state of wealth. An expanding economy increases per capita wealth with time, and the economy of the world as a whole has certainly been expanding rapidly since the industrial revolution. Because mathematicians are supported by that fraction of world wealth that remains after food, clothing, shelter, and other "necessities" have been paid for, it seems reasonable to conclude that the number of mathematicians should increase as "excess" wealth increases. This is a possible explanation of the relatively more rapid growth of memorable mathematicians compared with population, but it is also a difficult hypothesis to verify, since there is little direct data available that would permit the calculation of gross world product for past centuries.

In any event, we assume provisionally that there will continue to be about 4 million people for every memorable mathematician, the ratio characteristic for those memorable mathematicians born before 1850. For a world population of

1.6 billion, which was the situation in 1900, there ought to have been about 400 cumulative memorable mathematicians; for a population of about 3 billion (the situation in 1960), about 750. Therefore the number of memorable mathematicians born between 1900 and 1960 should be $750 - 400 = 350$; those born since 1940 have not, for the most part, been heard from yet, although they must constitute at least $(60 - 40)/60 = \frac{1}{3}$ of the total. This means that there should be (very approximately) $(\frac{2}{3})350 \cong 233$ recognized memorable mathematicians alive today, which is in reasonable agreement with our earlier intuitive estimate that perhaps one in 50 working mathematicians is memorable.

Figure 0.2 shows that world population experienced the same kind of dramatic — one might aptly say "explosive" — growth after 1400 that was experienced by the number of memorable mathematicians. This period overlaps the humanistic Renaissance, and all three events are to a large degree responses to one critical development: the *invention of movable type* and the use of the *printing press* about 1450 by Gutenberg and Fust in Mainz, Germany. The importance of this event cannot be too greatly stressed. With a means of large-scale and rapid dissemination of information in permanent form and at low cost, it became possible to accumulate library archives for effective reference and to concentrate on the discovery of new knowledge without the need of continually reproducing what had already been done but had not been communicated to others. Again and again historical indicators point to the late fifteenth century as the most critical period in the last two millenia for the development of civilization.

The growth of the number of "memorable technologists" (Figure 0.3) shows one effect of the printing press in a clear way. In this case we have agreed that a technologist is memorable if he appears in the index to Derry and Williams' *Short History of Technology* [26], a standard work. Observe that technologists grew in number at about the same rate as mathematicians in the earliest times, from about -700 to $+400$, but their growth remained completely stagnant throughout the Dark Ages (is this a possible definition of the Dark Ages?) until a phoenix-like resurrection after 1400. Since 1400 the number of technologists has grown much more rapidly than the number of mathematicians. In the past it was much easier to find mathematicians among the scientists than it is today.

Technology depends on the timely dissemination of knowledge, even more than pure science and much more than mathematics; therefore it should come as no surprise that for hundreds of years before the invention of movable type technology was stagnant, as Figure 0.3 so clearly shows. How, then, can we explain the growth of the number of technologists that apparently took place from -700 on? Before we turn to this question let us look at one measure of the effect of printing that makes clear how rapidly human knowledge has accumulated since its invention.

The Library of Congress of the United States is the largest in the world. Although founded only in the early nineteenth century, it has grown with remarkable rapidity and now attempts to acquire a copy of almost every significant printed work, regardless of language; in 1966 it held nearly 14 million books [56]. Figure 0.4 shows the number of books held since 1865 displayed on semi-

FIGURE 0.3

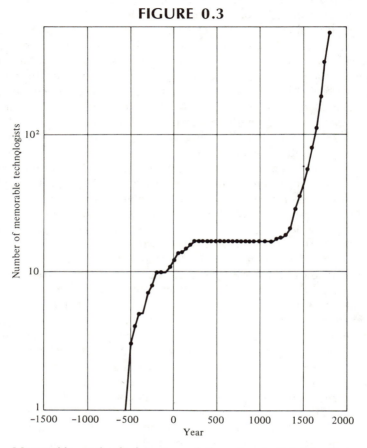

Memorable technologists (cumulative, by birthdate). Data from Derry and Williams [26].

logarithmic graph paper. The points fall nearly on a straight line; if this line were extended back in time, we would discover that the "first" book ought to have been printed about 1500, which is reasonably consistent with the facts considering the quality of the data we have used. We can therefore provisionally conclude that the sum of human knowledge, as represented in printed books, has been growing exponentially with time since the invention of printing.

With regard to the fifteenth century we have argued that it is rapid and inexpensive communication facilities, coupled with the ability to preserve information in permanent but easily retrieved form, that really set civilization going. During the period from -4000 to -700 two major advances were made which established ancient civilizations that we easily recognize as forerunners of our own, cast in the same basic pattern. These were the *invention of writing* and the much later *invention of the alphabet*. Figure 0.5, taken from Gelb's *A Study of*

FIGURE 0.4

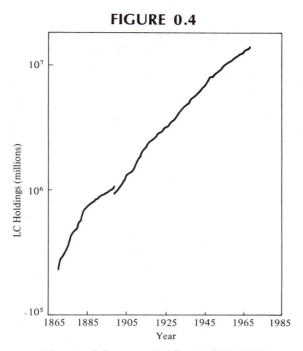

Library of Congress holdings, 1865–1966.

Writing [37], shows that pictographic writing systems were in existence in −3100 but probably not much before that. These systems gave rise to the great near-eastern languages of antiquity, *Egyptian hieroglyphs* and *Akkadian cuneiform*,[3] which lasted for more than 2000 years before they were displaced by the much more efficient alphabetic systems. Returning to Figure 0.2, we see that the straight line projection of the cumulative number of mathematicians indicates that the "first" memorable mathematician should have been born in −3743. This date is consistent with the first postneolithic developments of civilization and is some hundreds of years before the earliest known pictographic writing systems.

It is conceivable that mathematical needs for notational symbolism were later developed into full-fledged pictographic writing systems; that is, that mathematics preceded and was the catalytic agent for the formation of writing systems. Certainly primitive mathematical records are ancient. We quote from Struik [60]:

> *Numerical records were kept by means of . . . strokes on a stick, ... The oldest example of the use of a tally stick dates back to paleolithic times and was found in 1937 in Vestonice (Moravia). It is the radius of a young wolf, 7 inches long, engraved with 55 deeply incised notches, of which the first 25 are arranged in groups*

[3] *See Chapters 1 and 2 for a description of these writing systems.*

FIGURE 0.5

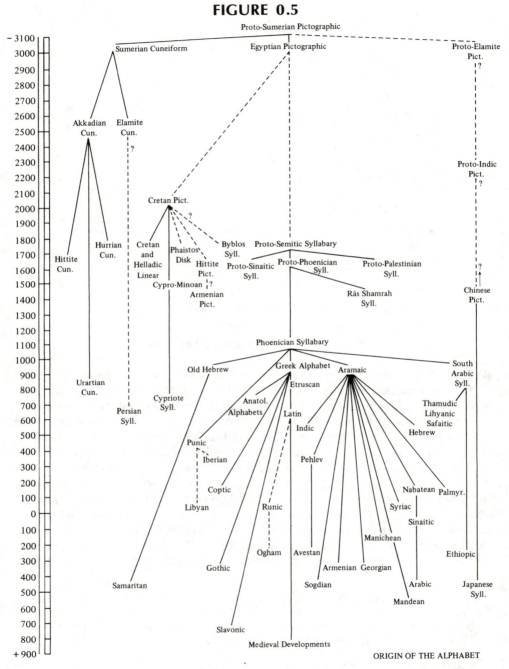

Origin of the alphabet. (Gelb [37], reprinted with permission of the publisher).

of 5. They are followed by a simple notch twice as long which terminates the series; then starting from the next notch, also twice as long, a new series runs up to 30.

This tally stick has been dated at about −30,000. If numbers as large as 55 were already necessary in the primitive hunting societies of the Paleolithic, it should be no surprise that settled agrarian societies would soon find it necessary to introduce much larger numbers and an efficient means for denoting them. We think it quite likely that mathematical notations are anterior to writing systems.

The invention of writing systems led to the production of massive quantities of records, some of which have survived the ravages of time and have been deciphered to reveal the nature of the civilizations that created them. Thus we know quite a bit about the Egyptian and Akkadian people, although somewhat less about the Sumerian forerunners of the Akkadians. The Sumerian and Egyptian writing systems were originally pictographic; the Akkadian was a modification of the Sumerian and ultimately was simplified to a still quite complex syllabic system that utilized hundreds of different cuneiform signs to express the different syllables of the language. Pictographic systems are extremely inflexible and inefficient ways to write; large syllabary-based writing systems are only slightly more efficient. The major improvement in writing systems, which made them vastly more efficient as well as much simpler to learn, was the invention of the alphabet. Although the Phoenicians invented a small efficient syllabary about −1000, it was the Greeks who by −700 modified it to form a real alphabet. Shortly thereafter the Greeks began their phenomenal rise to political power and intellectual eminence, the latter quality persisting in its influence to this day. The chronological scheme of Figure 0.6 connects linguistic developments with the civilizations that produced them.

Return to Figures 0.2 and 0.3, and observe the sudden growth of both memorable mathematicians and memorable technologists that began between −800 and −700. Could this effect be due to the invention of the alphabet, an efficient tool for recording and retrieving information?

We have seen striking changes in the nature of civilization which occurred shortly after or contemporaneously with three advances in recording and communicating knowledge: the invention of *writing systems* before −3100, the invention of the *alphabet* before −700, and the invention of *movable type* about +1450. During the last two decades we have witnessed the invention and explosive development of a similar fourth advance: the *digital computer*. Would it be unreasonable to posit a corresponding change in the fabric of our society in response to this novel and almost unbelievably efficient means for recording and retrieving knowledge?

§**0.3.** Our two main themes, mentioned in §0.1, recur throughout the book in various guises. In Part I we concentrate on early examples related to the ability to compute and the geometrical nature of space in the context of three ancient civilizations: Babylonian, Egyptian, and Greek. For instance, the positional notation of the Babylonian civilization is contrasted with the nonpositional and

FIGURE 0.6

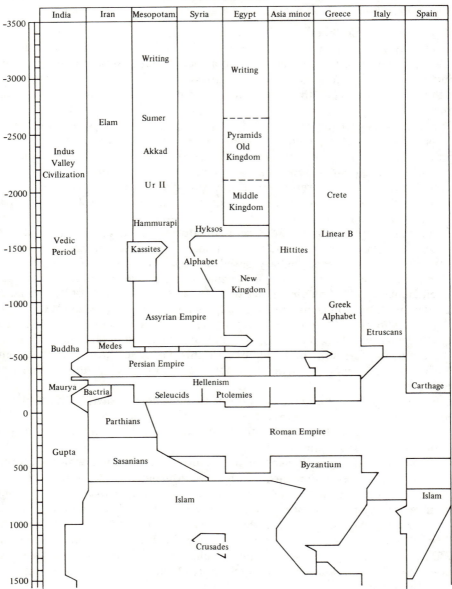

Chronological scheme (Neugebauer [50]).

much more cumbersome notation of the Egyptians. Positional notation was one of the major conceptual advances in the computational ability of mankind. The geometry of the solar system, as exemplified by our development in Chapter 4 of the Greek mathematician Aristarchus' estimate of the radius of the moon and its distance from the earth, is a significant conceptual development in man's understanding of the geometry of the space surrounding him, which supplanted the more primitive notion that all the stars and celestial bodies lay on a large transparent sphere that rotated about its center in the heavens. The concept that the moon is much closer to the earth than is the sun would have made no sense at all from earlier viewpoints.

In Part II we discuss the development of our themes during the Middle Ages and post-Renaissance period, and, in particular, in the chapters on navigation and cartography we find that the geometrical model of the universe built by the Greeks can be used for exploring the surface of the earth. A contemporaneous computational development was the invention of logarithms, which became an indispensable tool for the astronomer of fifteenth- and sixteenth-century western Europe, just as the invention of positional notation had been for the Babylonian astronomers in more ancient times. The invention of analytic geometry by Descartes in 1637 welded the computational tools of algebra based on Babylonian achievements to the geometry of the Greeks, thus forming an effective bridge between the two disciplines. This was a critical development on the road to the calculus as well as an important tool of mathematics in its own right.

In Part III we discuss the increased computational sophistication provided by the calculus. Its invention by Newton and Leibniz in the 1680s, based on a long sequence of advances originating with the first limiting processes in mathematical history introduced by the Greeks (and discussed in Chapter 4), marked the advent of what is now considered the period of "modern mathematics." Its importance, practical, theoretical, and philosophical, cannot be overestimated. Its complexity should not be underestimated.

In the hands of Newton, and utilizing the brilliant work of Kepler, calculus provided a new model of the universe that supplanted the older synthetic geometrical Greek model which depended on the properties of synthetically defined curves (principally circles). The application of these limiting processes to the geometry of surfaces, initiated by Gauss in 1824, provided a new perspective for considerations of the type and nature of the geometry that might possibly pertain to celestial space. These notions inevitably led, in a process still unfolding, to the view of the universe as an instance of a geometry of a four-dimensional curved "space" incorporating both perceptual space and perceptual time. Within its grand framework it is possible to question the shape, size, and age of the universe and to calculate answers that, although still incomplete, combine beauty and philosophical satisfaction in a way that seems not far from the truth.

PART ONE

Mathematics in Antiquity

CHAPTER 1

Number Systems and the Invention of Positional Notation

This chapter discusses the differences between counting systems *of notation for* numbers and positional systems *and introduces the* real numbers. *The invention of positional notation was the first profound mathematical advance. It made accurate and efficient calculations possible. Positional notations with respect to a general base are described. The Egyptian and Greek nonpositional notations are contrasted with the ·much more powerful Babylonian positional notation which uses the base 60.*

§**1.1.** The tally notches incised 30,000 years ago in the wolf's bone described in the introduction are examples of symbols used to denote the cardinal numbers 1, 2, 3, . . . ; each notch corresponds to one unit. A similar primitive system is in common use today for counting a number of objects; strokes are grouped in fives like this:

$$\text{册}$$

Repetition of a fundamental symbol to record a tally is cumbersome and space consuming. The number represented cannot be read off easily; it must be *counted* off, and it does not permit the development of arithmetic. For instance, in order to add tallies, say ||||||| and ||||, it is necessary only to group both sets together; thus |||||||||||. But what is the sum? Each complex civilization invented a notation or symbolism to represent large numbers and developed methods of performing elementary arithmetic operations with their symbols, although different civilizations attained varying degrees of effectiveness and sophistication. A well-known example of a notational system based on counting is the Roman, employed extensively in Western civilization until about 1400. Their symbols I, V, X, L, C, D stand for 1, 5, 10, 50, 100 and 500, respectively, and are now used to denote chapter numbers and positions of the hours on clock faces and for other decorative purposes. This system was supplanted by the present system of Arabic numerals, which is superior in many ways, as we shall see.

Measurement of distances immediately demands that the system of notation for numbers permits the expression of fractions corresponding to distances shorter than the basic unit of measure. For instance, if that unit is the meter, there must be a means of expressing centimeters; if it is the centimeter, there must be a means of expressing millimeters, and so on. However small the unit chosen, there will always be distances to be measured that are not integral multiples of it and that will therefore have to be represented as a fractional part. Systems of notation based on tally counting are obviously poorly suited to measuring distance and similar quantities.

Some of the systems of notation developed by early civilizations had a few special symbols for common fractions lumped together with a counting system notation; others, particularly the one introduced by the Akkadians, were almost as complete as our own decimal system but somewhat less efficient and convenient.

§1.2. We discuss the methods used for representing numbers by the early Egyptian, Akkadian, Greek, and Roman civilizations in order to show how well adapted each was to the measurement of geometrical quantities. First consider the problem of representing whole numbers.

Consider a positive integer, the number 36,521, for instance. Recalling that its rightmost digit corresponds to the number of units, the next digit to tens, the third to hundreds, and so on with higher powers of the base 10, we see that 36,521 is shorthand notation for

$$3 \times 10^4 + 6 \times 10^3 + 5 \times 10^2 + 2 \times 10^1 + 1 \times 10^0$$

where $10^4 = 10,000$, $10^3 = 1000$, $10^2 = 100$, $10^1 = 10$, and $10^0 = 1$. More generally, we know today that any positive whole number n can be represented in *decimal notation* as

$$n = a_r a_{r-1} \cdot \cdot \cdot a_0$$

where $a_0, \ldots, a_r$ are symbols, each of which stands for one of the numbers $0, 1, \ldots, 9$. This decimal shorthand stands for the *positional representation*

$$n = a_r \cdot 10^r + a_{r-1} \cdot 10^{r-1} + \cdot \cdot \cdot + a_0$$

Moreover, we can take any positive integer $g > 1$, write

$$n = b_s \cdot g^s + b_{s-1} \cdot g^{s-1} + \cdot \cdot \cdot + b_0$$

with each of $b_0, \ldots, b_i, \ldots, b_s$, standing for one of the numbers $0, 1, \ldots, g - 1$, and call the sequence of symbols

$$b_s b_{s-1} \cdot \cdot \cdot b_0$$

the *positional representation for n in the base g*. The position of a symbol determines the power of g it multiplies, hence its *value*.

Various positional number systems of different bases have been used by past civilizations, although the base 10 has been dominant. For instance, 5, 12, 20, and 60 have been used as bases; our present-day clock shows the influence of both base 12 and base 60. There is no universally *best* base, but for different purposes different bases will serve best. Historically there were various reasons for each system's use.

Modern computers employ combinations of the *binary system* (base 2), the *octal system* (base 8), the *decimal system* (base 10), the *duodecimal system* (base 12), and the *hexadecimal system* (base 16). The binary system is used to simplify computation, the hexadecimal to save storage space (the number of *places* needed to represent a given number is smaller, the larger the base, but correspondingly the number of *symbols* needed is larger), and the decimal to communicate with people. Hexadecimal digits are often expressed by pairs of decimal digits, but sometimes symbols drawn from the alphabet are used.

§**1.3.**   Among the ancient cultures the only ones that used a positional system to represent numbers were the Babylonians and their predecessors, the Sumerians (of whom no written mathematical works survive). The Egyptians, Greeks, and Romans had more primitive, only partly positional, decimal systems that utilized counting by powers of 10 but required the introduction of a new symbol for each power of 10 rather than a new position. For example, the Egyptians used the hieroglyphic symbols shown in Figure 1.1

FIGURE 1.1

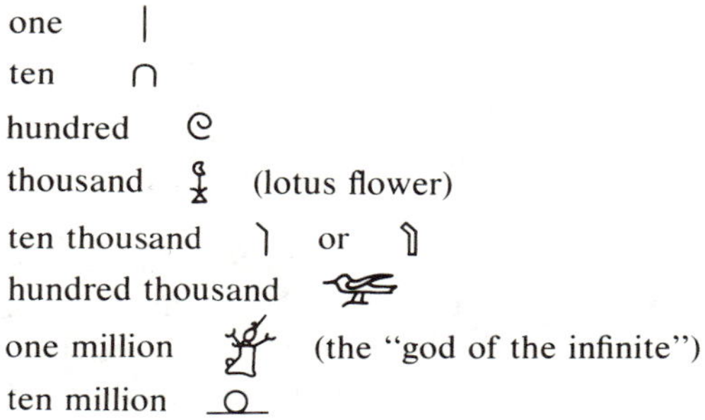

Egyptian hieroglyphic numerals.

and expressed numbers by their juxtaposition (representing addition of the values represented by the symbols). Thus

$$23 = ∩∩||| 59 = \begin{matrix} ∩∩|||| \\ ∩∩∩||||| \end{matrix}   213 = ℮℮∩|||$$

The analogy with Roman numerals is clear; the Romans used additional symbols to represent 5, 50, etc, (V, L . . .) so that there would be less writing involved, but the principle is the same.

The Greeks used their alphabetic characters to represent numbers in a manner similar to the Egyptians and the Romans but with an enormous saving of space (Table 1.1).

Note that three letters (Ϛ.ϙ.ϡ) in this table are not in the usual Greek alphabet. According to Heath's *Greek Mathematics* [40], a multitude of Greek alphabets was derived from the earlier Phoenician syllabary, each with its own variations (*cp*. the variation in the modern Russian and Ukrainian alphabets). The first two extra letters were kept in their original places for use as numerals, even though they had fallen out of literary use. The last letter,ϡ, already discarded, was tacked on at the end, since it no longer had a natural place at the time of the invention of the numeration system (about −700).

This system of numeration had the advantage over the Roman (and a previous

TABLE 1.1

1	α	alpha	10	ι	iota	100	ρ	rho
2	β	beta	20	κ	kappa	200	σ	sigma
3	γ	gamma	30	λ	lambda	300	τ	tau
4	δ	delta	40	μ	mu	400	υ	upsilon
5	ϵ	epsilon	50	ν	nu	500	ϕ	phi
6	ς	vau	60	ξ	xi	600	χ	chi
7	ζ	zeta	70	o	omicron	700	ψ	psi
8	η	eta	80	π	pi	800	ω	omega
9	θ	theta	90	$\koppa$	koppa	900	λ	sampi

$,\alpha = 1000, ,\beta = 2000$, etc.

Greek numerals.

Greek system, similar to the Roman, called *Attic*) that much less space was needed to represent a given number; for example,

$$849 \quad = \quad \omega\mu\theta \quad = \lceil^{\mathsf{H}} \mathsf{HHH} \triangle\triangle\triangle\triangle \lceil \mathsf{IIII}$$
(decimal)   (alphabetic)   (Attic)

We can guess what the various Attic symbols must mean from the context. The new numbers had the political advantage that they could be stamped on coins. The great Greek mathematicians, such as Archimedes, developed remarkable skill in computing with the alphabetic numeration system, as we show in Chapter 4. The Attic system was primarily used to denote ordinal numbers, much as we use Roman numerals today to denote the number of a given chapter in a book, a volume in a serial publication, or an hour on the face of a clock.

For large numerals the Greeks wrote the symbol M (Attic for 10,000) with alphabetic numerals above it; for example,

$$\overset{\beta}{M} = 20,000$$

$$\overset{,\zeta\rho o\epsilon}{M} {}_{,\epsilon\omega o\epsilon} = 71,755,875$$

The M serves as a place system on which a number system of base 10,000 could be built up by inventing symbols (or *positions*) for higher order powers of M = 10,000 (a *myriad*). Something similar was done by Archimedes in his *Sandreckoner*, wherein he estimated the number of grains of sand in the universe by making certain "astronomical assumptions" about its size. He "computed" with numbers to base 10^8 (the *second myriad* $= 10^4 \cdot 10^4 = M \cdot M$).

In fact Archimedes considered all numbers from 1 to 10^8 to be of *first order* and took the last number 10^8 as the unit of numbers of the *second order* (10^8 to 10^{16}), up to numbers of the 10^8-th *order* [all numbers from $10^{8(10^8-1)}$ to $10^{8\cdot10^8}$]. All numbers from 1 to the 10^8-th order form the *first period*; that is, if $P = (100,000,000)^{10^8}$, then the first period consists of the numbers between 1 and P. P is the unit of the first order of the *second period*, that is, the numbers from P to $10^8 \cdot P$; continue in this way to construct $10^8 \cdot P$ to $10^{16} \cdot P$, etc. Archimedes ends with *The Period*, which is given by

$$\text{The Period} = P^{10^8} = [\,(10^8)^{10^8}]^{10^8}$$

which Archimedes calls "a myriad-myriadth unit of the myriad-myriadth order of the myriad-myriadth period" (see Heath's *Greek Mathematics* [40]). Archimedes did not use symbolic notation for these large numbers but described them in words. This work exhibits the playful speculations of a brilliant mathematician but was far removed from the practical computational problems, general numerical ability, or burning theoretical issues of the day.

The next basic conceptual step, which occurred much earlier chronologically, was taken by the Babylonians who invented the *sexagesimal* system, that is, a base 60 positional number system. This system was used strictly for scientific purposes and in a very sophisticated manner. First we describe their notation and then employ a transliteration device invented by O. Neugebauer so that we can later analyze Babylonian mathematics in context. Basically they had two numerical symbols, Υ and $\langle$, which correspond to the *one* and *ten* of a primitive decimal system, such as the Egyptian | and ∩ . Numbers smaller than 60 were formed in a straightforward fashion,

$$3 = \Upsilon\Upsilon\Upsilon$$

$$25 = \langle\langle\ \overset{\Upsilon\Upsilon\Upsilon}{\Upsilon\Upsilon}$$

$$49 = \overset{\langle\langle}{\langle\langle}\ \overset{\Upsilon\Upsilon\Upsilon}{\underset{\Upsilon\Upsilon\Upsilon}{\Upsilon\Upsilon\Upsilon}}$$

which is as primitive and cumbersome as the Egyptian-Greco-Roman notations. Later on they introduced shorthand versions, for example,

$$9 = {}^{\Upsilon}\Upsilon{}_{\Upsilon}$$

The basic wedge (*cuneiform*) symbols Υ and $\langle$ formed the characters in the Babylonian writing system as well as their numbers.

These symbols are difficult to *write;* they were originally impressed on soft clay by a wedge-shaped instrument, as shown schematically in Figure 1.2. This was an advance over earlier methods of Babylonian writing which corresponded to engraving (hieroglyphic style characters taken from daily life) on a clay surface with a sharp stylus. Gradually these forms developed into a syllabary for writing and keeping records, which principally concerned economic matters. There were two fortunate finds of libraries in which thousands of clay tablets were assembled, some of which correspond to the mathematics sections of our own libraries. An important selection of the collection has been deciphered, analyzed, and published (with pictures or sketches of the actual clay tablets) by Neugebauer and Sachs, *Mathematical Cuneiform Texts* [51]. Analyses of these and similar tablets are our principal and most reliable source of information about the state of mathematics and astronomy in Babylonia from about −3000 to −200.

With this system of writing, based on the symbols Υ and $\langle$, it is clear that the Babylonians could easily record numbers up to, say, 100, without requiring any

FIGURE 1.2

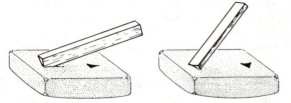

Impression of cuneiform symbols on clay tablets. (Redrawn from Neugebauer [81] with permission of the publisher.)

new symbols. Their remarkable achievement, one of the most distinguished in antiquity, was that with these two symbols they were able to systematically represent arbitrarily large numbers *and arbitrarily small fractions.* For instance, they wrote 𐎀𐎀 (in which the first 𐎀 has the value 60) and read it as $60 + 11 = 71$. Similarly,

$$\text{𐎀𐎀}\langle\langle\text{𐎀𐎀𐎀} = 2 \cdot 60 + 23 = 143$$

$$\langle\text{𐎀}\langle = 11 \cdot 60 + 10 = 670$$

This process could be carried on to three or more places, and each time a new power of 60 would appear:

$$\text{𐎀 𐎀 𐎀𐎀𐎀} = 1 \cdot 60^2 + 1 \cdot 60 + 3$$
$$= 3663$$

This example shows how (and why!) an individual number such as 3 was written with vertices touching to distinguish the symbols from the notation for two distinct numbers, each with a different value. The analogy to our decimal system is quite clear; this system is called the *sexagesimal number system.*

As used by the early Babylonians, the sexagesimal number system lacked two basic features of our modern decimal system. There was no zero and there was no "sexagesimal point." Later on the special symbol 𐎐 was introduced to denote an unfilled position, as in

$$\text{𐎀 𐎐 𐎀} = 1 \cdot 60^2 + 0 \cdot 60 + 1$$
$$= 3601$$

but initially there was no set way to tell which power of 60 a given 𐎀 represented, except from the context and the fact that the powers decreased to the right. The spacing between the basic numerals was not uniform. Presumably most of their practical work was done in context, and there was no confusion, just as when we ask "how much is a loaf of bread" and the answer given is "thirty," it is clear that we do not mean $30.

Because their system lacked a decimal point, the symbol 𐎀 could also mean 3600, 1/60, or 1/3600, depending on the context, but as we know from our system, as far as computations are concerned, we need only make sure that we add (or subtract) the corresponding coefficients of a given power of 60, whereas for multiplication and division we operate with the coefficients and place the decimal point correctly after the computation. In practice, we often know ahead of time where the decimal point will finally be and only need to carry out the computation on the various place entries. This was, as we shall see, the secret of Babylonian computational ability and why their astronomy and algebra were far superior to that of their Egyptian contemporaries.

Following Neugebauer, we represent a Babylonian sexagesimal number by the following scheme:

$$...,a,b,c;d,e,...$$

where $a,b,c,...$ are (decimal) numbers between 0 and 59 and the semicolon represents the *sexagesimal point*, which separates the integral part of the number from the nonintegral part just as the *decimal point* does in the decimal system. For instance, 12,20;21 represents the number

$$12 \cdot 60 + 20 + 21 \cdot \frac{1}{60} = 740 \frac{7}{20}$$

A system of numerical notations must be capable of expressing parts of a whole as well as whole (integral) quantities. All positional notations have this capability. The following table, which shows how reciprocals of small integers are represented in Neugebauer's sexagesimal notation, is easy to construct:

$$\frac{1}{2} = 0;30$$
$$\frac{1}{3} = 0;20$$
$$\frac{1}{4} = 0;15$$
$$\frac{1}{5} = 0;12$$
$$\frac{1}{6} = 0;10$$
$$\frac{1}{7} = 0;8,34,17,8,34,17,8,34,17, \ldots$$
$$\frac{1}{8} = 0;7,30$$

.

.

.

The decimal (base 10) representations of these fractions are familiar:

$$\frac{1}{2} = 0.5$$
$$\frac{1}{3} = 0.3333 \ldots$$
$$\frac{1}{4} = 0.25$$
$$\frac{1}{5} = 0.2$$
$$\frac{1}{6} = 0.16666 \ldots$$
$$\frac{1}{7} = 0.142857142857 \ldots$$
$$\frac{1}{8} = 0.125$$

.

.

.

In the binary system the same numbers appear in a different guise:

$$\tfrac{1}{2} = (0.1)_2$$
$$\tfrac{1}{3} = (0.010101 \ . \ . \ .)_2$$
$$\tfrac{1}{4} = (0.01)_2$$
$$\text{etc.,}$$

where ( )$_2$ denotes the base 2 representation.

By examining these expressions it is easy to convince ourselves that some fractions have *finite* expressions, whereas others appear to have unending but ultimately repetitive expressions. Moreover, no matter what base is chosen, *some* fractions will always have unending expressions relative to that base.

These observations lead us to inquire about the nature of numbers expressed by infinite expansions in, say, the decimal system of notation, and this in turn leads to a characterization of the *real number system* which underlies all the mathematical accomplishments described in this book.

§1.4. Much of mathematics is, and was, concerned with properties of the *real number system*. The *integers* (the set of positive and negative whole numbers, together with 0: . . . , $-4,-3,-2,-1,0,1,2,3, \ . \ . \ .$) are included amongst the real numbers. The *rational numbers,* that is, the *ratios of integers* (with denominator different from 0, of course), are also contained in the embracing collection of real numbers. This means that all numbers of the form p/q, where p and q are integers and q is different from 0, are real numbers. You already know that there are some "real" numbers that are *not* rational — these are called, unimaginatively enough, *irrational numbers*. Here are three examples: $1 + \sqrt{3}$, $\sqrt[3]{2}$, and π.

It is not hard to prove that $1 + \sqrt{3}$ and $\sqrt[3]{2}$ are irrational — we do so later on in this chapter — but the proof that π (the ratio of the circumference of a circle to its diameter) is irrational is much more difficult. In any case, how often do numbers like $1 + \sqrt{3}$ or $\sqrt[3]{2}$ occur in the normal course of human affairs? And, although π is more popular — perhaps, you may think, because wheels are useful in daily life — it nevertheless seems to be a unique type of number, the sole example of its species. Can you think of another number as strange as π? Perhaps you have heard of one: e, the base for *natural logarithms* (see Chapter 7).

Your experience may lead you to believe that most real numbers must be rational or at least that all of the "important" numbers, with perhaps a few exceptions admitted to spice the pie, are rational and that the irrationals for all practical and most other purposes could well be ignored; that the "real numbers" are really the friendly fractions of old, dressed up in a new name. If you so believe, gentle reader, you're in for a surprise.

Consider, for instance, why it is that when we refer to a particular rational number, say "six-sevenths," it is described in a straightforward way, namely, $\tfrac{6}{7}$; but the description of some of those irrational numbers recalls to mind the practice of certain cults which forbear the explicit mention of the name of a certain deed or god or edible. For instance, we write π or e, which conceals as much as it reveals. These letters from alphabets, foreign and domestic, are certainly poor substitutes for a concrete description of a real number, like $\tfrac{6}{7}$. We want to understand why it is that some numbers *must* be expressed with help from outside the ordinary apparatus of arithmetic. It must already be clear to you that if there really are such numbers we would be wise to avoid using them if possible, especially in daily life, for otherwise what a muddle would result from the inexplicit nature of their descriptions. One of our state legislatures decreed that $\pi = 3$, but not even a village council will try to assert that $1 = 2$. Why?

What is a real number? We can give you one perfectly good definition which has the advantage of using a body of knowledge that you already possess. A *real number* is a number expressed by a decimal expansion. You are all familiar with *decimal expansions*. A number expressed in this form can be written as

$$(1.1) \qquad x = a_n a_{n-1} a_{n-2} \cdots a_1 a_0 . a_{-1} a_{-2} a_{-3} a_{-4} \cdots$$

where each of the numbers a_i is a nonnegative integer less than 10 and the overemphasized dot "." denotes the decimal point. The sequence of a's to the right of the decimal point may or may not terminate, and here of course is the difficult and important point; when the decimal expansion does not terminate, we are no longer considering a simple finite process. Intuition is easily led astray, and notions that in normal finite circumstances are of the most elementary character must in the infinite situation be redefined and used with utmost care.

The sequence of integers $a_n \cdots a_1 a_0$ is called the *integral part* of x; it may be 0. The sequence of integers to the right of the decimal point is sometimes *mis*-named the *fractional part* of x; we shall see that it cannot always be presented as a fraction.

Here are a few examples of real numbers to illustrate what the three dots to the right of a_{-4} mean in the above expression. A real number is given when each term of the expansion (1.1) is well defined. Consider the expressions

        (a) 0.50000 . . .
        (b) 0.6666 . . .
        (c) 0.857142857142857 . . .
        (d) 0.212112111211112 . . .
        (e) 3.141596 . . .
        (f) 1.414213 . . .
        (g) 0.156231 . . .

Which of these represent real numbers? This is a matter of interpretation. In (a) and (b) the obvious rule is that 0 in the first expression and 6 in the second should be indefinitely repeated; these expansions then represent the real numbers $\frac{1}{2}$ and $\frac{2}{3}$, respectively. Similarly, in the next expression (c) we see that the block of six numbers "857142" is to be repeated, and this represents a *rule* for determining the *value* of any term arbitrarily far out in the expansion. (This expansion represents the fraction $\frac{6}{7}$.) For (d) we see that the rule is clear from the context: a "2" followed by one more "1" than the preceding time around. This rule will allow anyone to write out the decimal expansion to as many decimal places as desired; hence it is well determined and well defined. In (e), (f), and (g) we have expressions in which the rule for forming the next terms in the sequence is not clear from what is written down. Therefore with only this information the last three expressions given do not represent real numbers unless we give more information. We can specify (e) by requiring that the expansion for (e) be the ratio of the circumference to the diameter of any circle (usually denoted by π). This will uniquely specify the remainder of the expansion, not an easy fact to understand completely. In (f) we require that the expansion represent a positive number whose square is 2 (the expansion should represent $\sqrt{2}$). Again this uniquely specifies some positive real number and will uniquely determine the rest of the terms; that is, we can find an algorithm that will give increasingly accurate approximations to the decimal expansion (see Chapter 3). In (g) we have a sequence of six numbers which has been written down with no apparent rule in mind. Hence the remaining terms (. . .) are not well determined; for example, one person might write 0.1562313 and another might write 0.1562314444 . . . , both of which are now well determined and obviously distinct real numbers.

We have given examples of real numbers expressed as decimal expansions. Since we want to add, subtract, multiply, and divide such numbers, we may well ask how to proceed. Let us illustrate the simple answer by some examples. Suppose we wanted to add the expressions in (b) and (c). We could proceed as follows: (b) is $\frac{2}{3}$, (c) is $\frac{6}{7}$, so (b) + (c) is $\frac{2}{3} + \frac{6}{7} = \frac{32}{21}$, and by long division we can see that $\frac{32}{21}$ has the decimal expansion 1.52380 . . . . Thus we have added the two expansions in (b) and (c). This procedure, however, really avoids the real problem because we used the rules for adding *fractions,* which are not applicable to those decimal expansions that do not correspond to fractions. Suppose we try to add (c) to (d). What do we do? Since (d) is not a fraction, none of the usual rules applies. The correct idea is to construct a sequence of finite decimal expansions which approaches, that is, approximates the value of, the desired sum:

0.8	0.85	0.857	0.8571	0.85714
+0.2	+0.21	+0.212	+0.2121	+0.21211
1.0	1.06	1.069	1.0692	1.06925

Thus the sum (c) + (d) is given by 1.06925 . . . , and the rule for determining the rest of the terms is clear. Similarly, for (b) + (c), we have

0.6	0.66	0.666	0.6666	0.66666
+0.8	+0.85	+0.857	+0.8571	+0.85714
1.4	1.51	1.523	1.5237	1.52380

We note three things here:

1. A finite approximating expansion may differ in its *last* digit from the real expansion at that same level but nowhere else; *e.g.* 1.51 but 1.523. The rule is still well defined.

2. We get precisely the same expansion as we did above by the long division $\frac{32}{21}$.

3. It is not necessary to know that the sum represents the fraction $\frac{32}{21}$ to be able to compute it.

In a similar manner we can easily subtract, multiply, and divide decimal expansions of numbers.

Is it ever the case that two decimal expansions represent the *same* number? What should we mean by "same"? Let us agree that two real numbers are the *same* (although possibly expressed in different external forms) if their difference is smaller than any number we might choose. Then you will see immediately (if you do not, stick with it for a few moments anyway) that the two decimal expansions

$$y = b_n \cdots b_0 . b_{-1}b_{-2} \cdots b_{-k}000000 \ldots$$

and

$$x = b_n \cdots b_0 . b_{-1}b_{-2} \cdots (b_{-k}-1)9999 \ldots$$

represent the same number if b_{-k} is greater than 0; for example,

$$y = 2.5630000 \ldots$$

and

$$x = 2.5629999 \ldots$$

represent the same real number.

If one decimal expansion continues with all 9's from some point on, it represents the same number as the corresponding expansion with the 9's replaced by 0 and the digit preceding the 9's increased by 1. This is the most general case possible; in every other instance different decimal expansions represent different real numbers.

Now let us see what the decimal expansions that correspond to *rational* numbers look like. Examples (a) through (c) above exhibit decimal expansions of rational numbers; they each ultimately repeat some sequence of digits indefinitely. In general, if $x = p/q$ is a rational number, its decimal expansion can be found by long division. Let us show that there is an integer t such that $p = tq + r$, where the remainder r is less than q. (It can always be assumed that r and q are not negative; if x is negative, let p be negative.) It is always possible to find just one t with this property. Indeed, consider the integral multiples of q: q, $2q$, $3q$, ... After some time we will find a first (that is, smallest) integer t such that $(1 + t)q$ is larger than p if p is positive (if p is negative, read "larger than $-p$"). This is the t we want. Then

$$\frac{p}{q} = \frac{tq + r}{q} = t + \frac{r}{q}$$

Since r is less than q, t is the integral part of the decimal expansion of $x = p/q$. Now the process of dividing r by q to determine the "fractional" part of the expansion is a repetitive one involving division of certain remainders by q. Since there are only q possible remainders, namely 0, 1, 2, 3, . . . , $q - 1$, it is obvious that after we have divided $(q + 1)$ times in the repetitive process at least one of the remainders will have appeared twice. Once a remainder appears the second time the division process becomes a true and exact repetition of what has already occurred since the first appearance of that remainder, and therefore the sequence of digits in the quotient must repeat periodically; for example,

$$\frac{6}{7} = : 7 \overline{) 6.00000000000} \quad 0.857142857142857142 \ldots$$

$$
\begin{array}{r}
5\,6 \\ \hline
④0 \\
3\,5 \\ \hline
⑤0 \\
4\,9 \\ \hline
①0 \\
7 \\ \hline
③0 \\
2\,8 \\ \hline
②0 \\
1\,4 \\ \hline
⑥0
\end{array}
$$

Remainders are circled. Notice that once the remainder 6 is attained the calculation must be a copy of what went before and the answer can be written out with no further effort. Find the expansions of $\frac{1}{11}$ and $\frac{3}{13}$ yourself for comparison. Similar repetitions will occur when a rational number is expressed relative to any base.

We have just proved the important fact that the decimal expansion of a rational number has the form

$$(1.2) \quad x = a_n \cdots a_1 a_0 \, . \, a_{-1} a_{-2} \cdots a_{-k} c_1 c_2 \cdots c_m c_1 c_2 \cdots c_m c_1 c_2 \cdots$$

where the block $c_1 \cdots c_m$ is repeated indefinitely. We now ask the question: does every real number (decimal expansion) of the form (1.2) represent a rational number? In other words, given (1.2), can we find integers p and q so that p/q will give (1.2) when we carry out the long division?

In order to determine that the answer is *yes*, we shall look at an example. Consider the simple repeating decimal

$$r = 0.726666 \ldots$$

How do we write r as the quotient of two integers? Observe that

$$10r = 7.26|66 \ldots$$
$$r = 0.72|66 \ldots$$

On subtracting r from $10r$ the repeating parts to the right of the vertical bar will cancel and

$$10r - r = 7.26 - 0.72 = 6.54$$
$$= \frac{654}{10^2}$$

This gives $9r = 654/100$ whence r is the rational number

$$r = \frac{654}{900}$$

Now apply the same procedure to the general ultimately repeating decimal in (1.2):

$$x = a_n \cdots a_1 a_0 . a_{-1} a_{-2} \cdots a_{-k} c_1 c_2 \cdots c_m c_1 c_2 \cdots c_m \cdots$$

in which the repeating part is $c_1 c_2 \cdots c_m$. Multiply x by 10^m; this is the same as shifting the decimal point m places to the right. Thus

$$10^m x = a_n \cdots a_1 a_0 a_{-1} \cdots a_{-m} . a_{-m-1} \cdots a_{-k} c_1 c_2 \cdots c_m | c_1 c_2 \cdots$$
$$c_m \cdots$$

$$x = \qquad a_n \cdots \cdots a_0 . a_{-1} \cdots \qquad \cdots a_{-k} | c_1 c_2 \cdots$$
$$c_m \cdots$$

On subtraction the repeating parts to the right of the vertical bar will cancel and

$$10^m x - x = (10^m - 1)x$$

is a number of the form, say, $b_n \cdots b_0 . b_{-1} \cdots b_{-k}$ which is rational and equal to $(b_n \cdots b_0 b_{-1} \cdots b_{-k})/10^k$; so x must also be rational:

$$x = \frac{(b_n \cdots b_0 b_{-1} \cdots b_{-k})}{10^k(10^m - 1)}$$

An alternative approach to the same problem is given by the use of *geometric series,* which we introduce here as an example of an infinite summation process that appears in later chapters. Consider, for instance, the expression

$$1 + \tfrac{1}{2} + (\tfrac{1}{2})^2 + (\tfrac{1}{2})^3 + (\tfrac{1}{2})^4 + \cdots +$$

This is an example of a geometric series. It has the *partial sums*

$$
\begin{aligned}
1 &= 1\\
1 + \tfrac{1}{2} &= 1.5\\
1 + \tfrac{1}{2} + (\tfrac{1}{2})^2 &= 1.75\\
1 + \tfrac{1}{2} + (\tfrac{1}{2})^2 + (\tfrac{1}{2})^3 &= 1.875
\end{aligned}
$$

It is not hard to convince ourselves geometrically that these partial sums get closer and closer to the number 2 as we take more and more terms in them. The limiting value of such partial sums is what is meant by "summing an infinite series."

In general, a *geometric series* is of the form

$$a + ar + ar^2 + ar^3 + \cdots$$

and can be summed if $-1 < r < 1$; in this case the sum is given by

(1.3)
$$a + ar + ar^2 + ar^3 + \cdots = \frac{a}{1 - r}$$

In the example above, $a = 1$ and $r = \frac{1}{2}$. This formula is in a simple sense the *limit* of similar finite formulas (*cp.* Chapter 9). Note that

(1.4)
$$(1 + r + \cdots + r^{n-1})(1 - r) = 1 - r^n$$

Indeed, the left side equals

$$(1 + r + \cdots + r^{n-1})(1) - (1 + r + \cdots + r^{n-1})r$$
$$= (1 + r + \cdots + r^{n-1}) - (r + r^2 + \cdots + r^n)$$

and all terms cancel except $1 - r^n$; for any r and any positive integer n the middle terms in the multiplication cancel. If $-1 < r < 1$, we see that r^n will tend to zero as n gets very large; for example,

$$\left(\frac{1}{2}\right)^n = \frac{1}{2^n},$$

which clearly becomes as close to zero as we want if n is taken large enough. Indeed,

$$\frac{1}{2} = \frac{1}{2} = 0.50000$$
$$\left(\frac{1}{2}\right)^2 = \frac{1}{4} = 0.25000$$
$$\left(\frac{1}{2}\right)^3 = \frac{1}{8} = 0.12500$$
$$\left(\frac{1}{2}\right)^4 = \frac{1}{16} = 0.06250$$
$$\left(\frac{1}{2}\right)^5 = \frac{1}{32} = 0.03125$$

Repetition of this process evidently will produce as many zeros after the decimal point as desired; that is $(1/2)^n$ tends to zero. So by using (1.4) we have the following sequence of formulas:

$$a + ar = \frac{a - ar^2}{1 - r} = \frac{a}{1 - r} - \left(\frac{a}{1 - r}\right) \cdot r^2$$
$$a + ar + ar^2 = \frac{a - ar^3}{1 - r} = \frac{a}{1 - r} - \left(\frac{a}{1 - r}\right) \cdot r^3$$
$$\vdots \qquad\qquad \vdots$$
$$a + ar + ar^2 + \cdots + ar^n = \frac{a}{1 - r} - \left(\frac{a}{1 - r}\right) \cdot r^{n+1}$$

On the right-hand side of each equation stands $a/(1 - r)$, which does not depend on n, added to a term multiplied by r^{n+1}, which we know approaches zero as n gets large. Therefore we say that the infinite series on the left has the *sum* $a/(1 - r)$, since the "error term" for any finite partial sum tends to zero as n gets large. Notice that not every geometric series can have a sum. For instance, if $a + ar + ar^2 + \cdots$ had a sum with $a > 0$ and $r > 1$, then $s = a/(1 - r)$ would be negative, which is absurd; for example, uncritical use of the formula suggests the silly result

$$1 + 2 + 2^2 + 2^3 + 2^4 + \cdots = 1 + 2 + 4 + 8 + 16 + \cdots = \frac{1}{1 - 2} = \frac{1}{-1} = -1$$

Similarly, if $a \neq 0$ but $r = 1$, the series has no sum. It has a sum, given by (1.3), only if $-1 < r < +1$.

Now we show that the expansion (1.2) corresponds to a rational number by using the formula for the sum of a geometric series. The expression in (1.2) is equal, by straight-forward factorization, to the number

$$(a_n a_{n-1} \cdots a_1 a_0) + \left(\frac{a_{-1}}{10} + \frac{a_{-2}}{10^2} + \cdots + \frac{a_{-k}}{10^k}\right) +$$
$$\left(\frac{c_1}{10^{k+1}} + \frac{c_2}{10^{k+2}} + \cdots + \frac{c_m}{10^{k+m}}\right) \times \left(1 + \frac{1}{10^m} + \frac{1}{10^{2m}} + \cdots\right)$$

The infinite sum in the last set of parentheses

$$\left(1 + \frac{1}{10^m} + \frac{1}{10^{2m}} + \cdots\right)$$

is a geometric series with $r = 1/10^m$ and $a = 1$; using (1.3), it sums to

$$\left(\frac{1}{1 - \frac{1}{10^m}}\right) = \left(\frac{10^m}{10^m - 1}\right)$$

which is a rational number. Therefore the original number x must also be rational, since it is the sum of the integer $(a_n \cdots a_0)$ and the rational $(0 . a_{-1} a_{-2} \cdots a_{-k})$ added to the product of the two rationals

$$\left(\frac{c_1}{10^{k+1}} + \cdots + \frac{c_m}{10^{k+m}}\right) \quad \text{and} \quad \left(\frac{10^m}{10^m - 1}\right)$$

We have shown that every decimal expansion with a periodic (repeating) sequence of digits represents a rational number and that, conversely, every rational number can be expressed as a periodic decimal.

We have also just learned how to construct decimal expansions for an infinite number of different irrational numbers. All that is necessary is to ensure that the decimal expansion *never* can become periodic. So, for instance, the number

$$0.212112111211112 \ldots$$

in example (d) above, where each group of 1's has one more member than the group of 1's directly to its left, must be an *ir*rational number. Construct several irrational numbers yourself.

§1.5. Since we are back on the subject of irrationals, let us prove that $1 + \sqrt{3}$ and $\sqrt[3]{2}$ are irrational. Note that $1 + \sqrt{3}$ and $\sqrt[3]{2}$ both have decimal expansions, hence "qualify" as real numbers according to our definition. What is needed is an *algorithm* (rule) for writing down approximating finite decimal expansions which *converge* in the same sense that the finite expressions that approximated the sum of two decimal expansions "converged" to the desired decimal expansion. The early Babylonians had such an algorithm for $\sqrt{x}$, for any positive integer x (see Chapter 3), and, similarly, we can find an algorithm for higher order roots of numbers. It is simpler, however, to work (algebraically, *not* computationally) with the formal expressions $1 + \sqrt{3}$ and $\sqrt[3]{2}$. What we shall do now is prove (algebraically) that it is impossible to find integers p and q so that

$$\frac{p}{q} = 1 + \sqrt{3}$$

The proof we give goes back to Euclid and is a "proof by contradiction" or indirect proof. The idea is this: we suppose $1 + \sqrt{3}$ to be rational and show that this logically must imply a contradiction. Since we agree that we do not want any contradictions in our mathematical system, the conclusion is that the *assumption* we made cannot be true, hence is false, which is the fact (that $1 + \sqrt{3}$ is *not* rational) that we wanted to prove.

Carrying out this program, we first note that $1 + \sqrt{3}$ is rational if and only if $\sqrt{3}$ is rational. So we shall prove that $\sqrt{3}$ is irrational. Suppose $\sqrt{3}$ were rational, that is, $\sqrt{3} = m/n$, where m and n are integers and where we can also assume that the fraction m/n is in *lowest terms;* that is, m and n have no common factor other than 1. Squaring both sides shows that

$$3n^2 = m^2$$

so 3 divides m^2. Now either m is divisible by 3 or it leaves a remainder of 1 or of 2 on division by 3; that is,

$$m = 3k \qquad m = 3k + 1 \qquad \text{or} \qquad m = 3k + 2$$

for some integer k. Then

$$m^2 = 9k^2 \qquad m^2 = 9k^2 + 6k + 1 \qquad \text{or} \qquad m^2 = 9k^2 + 12k + 4$$

Since only the first of these squares can be divided by 3, it follows that the divisibility of m^2 by 3 implies that 3 divides m itself. So $m = 3k$ and therefore from $3n^2 = m^2$ we find $3n^2 = 9k^2$; that is,

$$3k^2 = n^2$$

Then 3 divides n^2 and by the same argument we have already used 3 divides n. We have shown, on the assumption that $\sqrt{3}$ is rational, that 3 divides both m and n. This is a contradiction, since we assumed that m/n was in lowest terms. Hence $\sqrt{3}$ cannot·be rational; it must be *irrational*.

We can proceed with the proof that $\sqrt[3]{2}$ is irrational in a similar·way. Assume that $\sqrt[3]{2} = m/n$. Then $2n^3 = m^3$ and 2 divides m^3. Show that this implies that 2 divides m; that is, $m = 2k$. Then $2n^3 = 8k^3$, so $n^3 = 4k^3$ and 2 divides n^3. Then both m and n are even and m/n is not in lowest terms, and so on.

You can understand that the proof of the irrationality of a particular number can be difficult despite the fact that we already know how to *construct* infinitely many different irrationals. This is an example of the sad but universal fact that quite different methods are generally needed to solve problems of the *particular* as opposed to problems of the *general* and not only in mathematics.

§1.6. How many rationals are there? An infinite number, of course. Still, it would be nice to have some idea, be it intuitive or imprecise, of the rough proportion among the infinite collection of all real numbers that is filled out by the rationals. In other words, do the periodic decimal expansions account for most of the decimal expansions or for a small portion? To make some sense of these questions, it is imperative that we state clearly how we will decide when one infinite set has more members than another. You might be tempted to argue, as is so easy in the finite case: there are 713 men and 4 women in a box. Are there more men than women? Since 713 is greater than 4, there *are* more men than women. There is another way to answer the question, however, which extends to the case of infinite sets; *pair* the men and women and remove the paired people from the box. Then, if the box is empty, we say that there is an equal number of men and women; if the box is not empty, it will contain only men or only women. We say that there were *more* men than women if men are left in the box, *more* women than men if women are left in the box.

Pleasant though it may be to consider men and/or women in a box, it is clear that we are really merely pairing the elements of two sets in order to compare their size; the names of the elements or of the sets do not matter at all. This technique of pairing works nearly as well for infinite sets as it does for finite collections of people. It will lead to some startling results because you (probably) have a fixation on finite things and either have *no* intuition or *wrong* intuition about what is true of things infinite.

The first place in which intuition may let you down concerns subsets of a given set. If a set S is finite, and the set T is a proper subset of S (that is, everything in T is also in S but there are things in S that are not in T), then you can be sure that there are *more* things in S than in T. This is not true for infinite sets. Indeed, let N denote the set of positive integers 1,2,3,4,5, . . . and let 2N denote the set of *even* positive integers 2,4,6,8, ... Then 2N is a proper subset of N, but now pair each integer n in N with the even integer $2n$ in 2N. It is clear that every integer in 2N belongs to just one of these pairs and the same is true of every integer in N. Therefore, according to our agreement about pairing, we must say that there are *just as many integers in* N *as in* 2N. Difficult to countenance? The choice is between this, and deciding that pairing of objects does not work for infinite sets. It has been agreed for some time (since the late 1890's) that the pairing process leads to useful (as well as startling) results and that it represents perhaps a more elementary and fundamental mode of reasoning than that involved in rejecting the notion that N and 2N have the same number of elements. There are certain ways of describing this process that may make it appear more reasonable to you; unfortunately, we have no time to explore these byways.

Once this method of pairing sets to determine whether they are the "same size" is accepted, it is easy to test some of the more common subsets of the real numbers to find out how large they are. We have already seen that 2N is as large as N; in the same way we can discover that the integers that are multiples of any fixed nonzero integer are as numerous as the entire set of integers.

A more striking observation is that the set of rational numbers is no larger than the set of integers. We will content ourselves with showing that the set of positive rationals is the same size as the set of positive integers, for by the remarks above the set of negative integers can be put in a one-to-one correspondence with the positive integers, the set of negative rationals with the set of positive rationals, and it is really enough to prove that these two sets of positive numbers are the same size. Consider the diagram of rational numbers in Figure 1.3. Every rational finds a place in this diagram; p/q lies at the intersection of the pth row and qth column, but some rationals are repeated because $kp/kq = p/q$ for any positive integer k. Therefore strike out all multiples of rationals that are expressed in reduced form, numerator and denominator having no common divisors. Then each rational occurs exactly once in the modified diagram. Now trace a path through the diagram as shown and number the rational stepping-stones as they occur along the path.

FIGURE 1.3

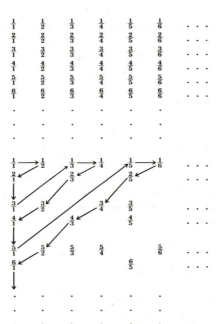

The rational numbers are countable.

No matter how large the integer n, there is a rational number (stepping-stone) corresponding to it in the diagram and to every rational there corresponds an integer n. Hence the rationals and the integers are equally numerous.

Let us say that a number x is *algebraic* if it is the root of a polynomial equation whose coefficients are integers, that is, if

$$a_0x^n + a_1x^{n-1} + \cdots + a_{n-1}x + a_n = 0$$

where the a_i are integers. For instance, every rational number p/q is algebraic because it is the root of $qx - p = 0$; $\sqrt{2}$ is irrational but algebraic because it is a root of $x^2 - 2 = 0$; π is *not* algebraic, but it is difficult to prove that this is so. Almost all numbers we ever come in contact with are algebraic, but an application of the same diagonal path procedure that we used above shows that *there are just as many integers as there are algebraic numbers.*

The wary reader may suspect that we are wasting his time by not simply stating that the set of real numbers is no larger than the set of integers, but this is false. There are *more* real numbers (that is, decimal expansions) than there are integers (or rationals or algebraic numbers).

Suppose that it were possible to put the integers (say positive integers for convenience) into correspondence with the decimal expansions. There would be a *first* decimal, a *second*, and so on, as indicated in Table 1.2 where a_k^n stands for the kth digit after the decimal point in the expansion of the nth number in the list. We have taken the liberty of not writing the integer parts of the decimal expansions because the notation is already cluttered; you can imagine that they are there.

TABLE 1.2

Ordinal Numbers	Decimal Expansion
1	$.a_1^1 a_2^1 a_3^1 \ldots$
2	$.a_1^2 a_2^2 a_3^2 \ldots$
3	$.a_1^3 a_2^3 a_3^3 \ldots$
.	.
.	.
.	.
n	$.a_1^n a_2^n a_3^n \ldots$
.	.
.	.
.	.

Now we show that such a list is impossible. Recall that the list is asserted to contain every positive real number; to show that such a list cannot exist all we must do is produce a number—a decimal expansion—that is not in the list. Here is one:

$$a = .a_1 a_2 a_3 \ldots$$

where

$$a_1 = 1 \text{ if } a_1^1 \neq 1 \text{ but } a_1 = 2 \text{ otherwise}$$
$$a_2 = 1 \text{ if } a_2^2 \neq 1 \text{ but } a_2 = 2 \text{ otherwise}$$
$$a_3 = 1 \text{ if } a_3^3 \neq 1 \text{ but } a_3 = 2 \text{ otherwise}$$

$$a_n = 1 \text{ if } a_n^n \neq 1 \text{ but } a_n = 2 \text{ otherwise, and so on.}$$

It is clear that the real number a with the decimal expansion just defined is different from the first number in the table in the first decimal place, from the second number in the second decimal place, from the third in the third, and so on. In fact a is different from every number in the list, but it is itself a real number, that is, a decimal expansion, and so the list must not be complete. Therefore there cannot be a one-to-one correspondence between the integers and the real numbers. There will always be a real number (indeed many of them) left over, so the set of real numbers is larger than the set of integers or rationals or algebraic numbers. This elegant proof was created by the German mathematician Georg Cantor at the end of the nineteenth century and has revolutionized our understanding of the system of real numbers.

§1.7. One of the principal reasons that the branch of mathematics known as *analysis* (which includes *calculus;* see Chapter 9) is particularly difficult is that it is concerned with the properties of the real numbers rather than those of just the rational or algebraic numbers.

It is remarkable that man has been able to master the set of real numbers, to use it for his science and technology and sometimes for his art, when he himself is finite and has at his disposal only the integers and other "countable" sets of numbers and, as a tool, only the finitary principle of mathematical induction. This is in large measure why mathematics is difficult; its difficulty should be expected, but man's achievements in the face of the Uncountable Infinite are, we think, awesome and inspiring.

Since there are more real numbers than algebraic numbers and since the symbols that we must use to denote them are all drawn from a finite collection (including the integers, signs such as "+", "−", "×", and "÷", and letters from various alphabets), it follows that we can express only countably many different numbers by using finite combinations of these symbols. This means that *there must be numbers for which we can have no general and systematic means of expression as a finite combination of symbols drawn from a fixed finite inventory.* By using the symbols for the integers and for division we can express any rational number in a systematic way as p/q; similarly, any algebraic number can be expressed in finite form, although it is not customary to do so. For instance, if the algebraic number x is a root of the polynomial

$$a_0x^n + a_1x^{n-1} + \cdots + a_n = 0$$

in which the a_i are all integers, this equation is uniquely determined by the ordered sequence of integers

$$(a_0, a_1, \ldots, a_n)$$

Since an algebraic equation of degree n has exactly n (complex) roots, which are all distinct if we assume that the polynomial has no repeated factors, the particular root x will be determined if we state which of the n roots it is. Therefore a finite symbol such as

$$(a_0, a_1, \ldots, a_n \,|\, k)$$

completely specifies an algebraic number, where k is an integer specifying which root of the polynomial is meant. Since it is clear that each algebraic number can be associated with an expression of this kind, we have constructed a *notation* that enables us to express any algebraic number in a systematic way.

No such scheme can be constructed for expressing *all* real numbers, as we have already seen. This means that we will have to provide *special* and *distinctive* expressions for any *transcendental numbers* that crop up in the course of our pursuits (a *transcendental number* is a real number which is not algebraic — it "transcends" the realm of algebraic methods). When a transcendental number is important enough to require a name, we give it one. The ratio of the circumference of a circle to its diameter (in a rather awkward way that expression defines a number! Is it larger than 3? Not easy to tell from this description) is the most famous transcendental number in history; its conventional name is π. Two other transcendentals that have appeared often enough in mathematics to deserve a name along with verbal "descriptions" of what the names are intended to "remind" you, are

e, base of the system of natural logarithms (see Chapter 7)

and

γ, the Euler-Mascheroni constant

Many of the other transcendentals that actually appear in a natural way are logarithms of algebraic numbers and algebraic powers of e. Notice that none of these symbolic notations tells you anything at all about how to compute the value of the number represented and that this differs in an essential way from the usual notation for rational numbers or even from the notation for algebraic numbers introduced above; letter names for transcendental numbers are just *names,* whereas the *notations* for rational and algebraic numbers contain information about the nature of the number that is sufficient to tell you how to calculate its value. This is one of the main reasons why in normal life we deal primarily with rational numbers; we have a systematic way of writing them that dovetails with their mathematical properties and, most important of all, with their *size* (that is, *value*). Imagine the difficulties that would occur in everyday calculation of bills, football scores, odds on betting, and income taxes if each number had a special symbolic name entirely unrelated to its size or arithmetical properties. That is precisely the situation in which the ancient Egyptians, Greeks, and Romans found themselves.

EXERCISES

1.1. Express the following decimal numbers in base 2 and base 8:
    (a) 346   (b) 64
    (c) 512   (d) 1696

1.2. (a) Express the binary number 10011 (base 2) in decimal form.
    (b) Express the octal number 1071 (base 8) in decimal form.
    (c) Does the expression 1278 represent a number in the octal number
    system? Why?

1.3. The expression of numbers in the base 12 system requires 12 different
    symbols to represent the numbers 0, 1,...,9, 10, 11. Use the notation 0,
    1, 2,...,9, T, E to represent zero and the first eleven positive inte-
    gers. (a) Express 1512 (base 10) in the duodecimal system. (b) Express
    the following base 12 numbers in base 10: (i) TEE, (ii) TOE, (iii) 111,
    (iv) $21E3$.

1.4. Carry out the indicated operations in the base indicated.
    (a) 110×101 (base 2)   (b) $123 + 457$ (base 8)
    (c) $310E - TEE$ (base 12)   (d) $55 \div 17$ (base 8).

1.5. Express the following numbers (in base 10) in the Egyptian, Greek, and
    Roman systems:
    (a) 251   (b) 2462
    (c) 28   (d) 10,270

1.6. Carry out the indicated operations:
    (a) [Egyptian numeral] + [Egyptian numeral] = _____
    (b) $\chi\lambda\alpha + \omega\iota\beta =$ _____
    (c) $\phi\kappa\beta + \sigma\xi =$ _____
    (d) [Egyptian numeral] − [Egyptian numeral] = _____

1.7. Express the following in decimal notation:

    (a) [cuneiform] (b) [cuneiform] (c) [cuneiform] (d) [cuneiform] (e) [cuneiform] (f) [cuneiform]

    (g) [cuneiform] (h) [cuneiform] (i) [cuneiform]

1.8. Express the following (base 10) numbers in Babylonian notation:
(a) 1   (b) 12   (c) 123
(d) 1234   (e) 12345   (f) 123,456
(g) 1,234,567

1.9. Using the Neugebauer system (see p. 27) to represent sexagesimal numbers (base 60), convert the following numbers (base 10) to sexagesimal form:
(a) 621   (b) 3600   (c) .24
(d) 120.50   (e) $\frac{1}{3}$ (f) $\frac{1}{7}$ (exact answer please!)

1.10. What is different about part (f) in Problem 1.9?

1.11. Express the following in decimal notation:
(a) 1,1   (b) ;1   (c) 1;0,1
(d) ;1,50   (e) 12,20;21   (f) 1;24,51,10
                                (we shall see this
                                number again!)

1.12. Carry out the indicated operations in base 60:
(a) 12,2 + 15,4   = _____
(b) 12,45 + 13,55   = _____
(c) 25,6;37 × 1,2;3   = _____
(d) 12,20 ÷ 2,1   = _____
(e) 1 ÷ 36   = _____

1.13. Show that any rational number expressed in base 2 has an ultimately *repeating* binary expansion. (Consider the example

$$\frac{1}{11} = 0.01010101 \ . \ . \ .$$

which is repeating the pair of symbols (01).)

*1.14. Show that any rational number expressed in *any* base has a repeating expansion with respect to that base.

1.15. (a) State at least one advantage of the:
Egyptian number system compared with the Roman number system;
Greek number system compared with the Egyptian number system;
Babylonian number system compared with the Greek number system;
decimal number system compared with the Babylonian number system;
binary number system compared with the decimal number system;
decimal number system compared with the binary number system.

    (b) Which of the above-mentioned number systems used positional notation?

    (c) Which number system (not necessarily one of the above) is best? Why?

1.16. Show that the series

$$2 + \tfrac{4}{3} + \tfrac{8}{9} + \tfrac{16}{27} + \cdots$$

is a geometric series and find its sum.

1.17. Express the following repeating decimal expansions as fractions:

    (a) 1.21212 . . .   (b) 0.12121 . . .

    (c) 0.185918591859 . . .   (d) 21.9011111 . . .

1.18. (a) Does the decimal expansion of $\sqrt{5}$ contain a sequence of digits that ultimately repeats?

    (b) Prove that the answer you gave in (a) is the correct one.

1.19. Define a decimal expansion that represents an irrational number.

1.20. Show that there are as many positive integers as there are integers (positive and negative).

1.21. Show that there are as many squares of integers (1, 4, 9, 25, . . . ,) as there are positive integers.

*1.22. Consider a quadratic polynomial of the form $P(x) = x^2 + ax + b$, where a and b are integers. The roots of $P(x) = 0$ are called *algebraic integers of degree* 2. Show that there are as many algebraic integers of degree 2 as there are integers. (*Hint:* Follow the proof in the chapter that there are as many integers as rational numbers. Note that any integer d satisfies the equation $x^2 - d^2 = 0$ and in particular, that every integer *is also* an algebraic integer of degree 2.).

*1.23. Suppose that a, b, c, d, and D are integers and that D is fixed throughout the problem.

    (a) Show that $a + b\sqrt{D}$ is an algebraic integer of degree 2.

    (b) If $a + b\sqrt{D}$ and $c + d\sqrt{D}$ are algebraic integers, define their *sum* and *product by* $(a + b\sqrt{D}) + (c + d\sqrt{D}) = (a + c) + (b + d)\sqrt{D}$ and $(a + b\sqrt{D})(c + d\sqrt{D}) = (ac + bdD) + (ad + bc)\sqrt{D}$, respectively. Are the sum and product of these algebraic integers also algebraic integers of degree 2?

    (c) Show that every solution of $x^2 - 2ax + (a^2 + b^2) = 0$ can be written in the form $x = a \pm b\sqrt{-1}$. Algebraic integers of this type are called *complex*, or *Gaussian*, integers.

CHAPTER 2

Egyptian Arithmetic

The Egyptians were the first to invent a method of expressing the result of division of one integer by another when the quotient is a fraction. Their solution was terribly cumbersome because their number system was not positional. This chapter studies the Egyptian number system and their arithmetical techniques. Its principal purpose is to demonstrate how difficult elementary arithmetical calculations are when expressed in an inefficient notation and to suggest that the Egyptians' further advance in mathematics and science was inhibited by their lack of positional notation. To give the reader a closer understanding of the nature of Egyptian mathematics and thought a brief sketch of the decipherment and structure of hieroglyphic Egyptian is included.

§2.1.　Our knowledge of Egyptian mathematics rests on two important papyri: the *Rhind Papyrus,* named for Mr. A. H. Rhind who donated it to the British Museum, and the *Moscow Papyrus,* which now resides in the Museum of Fine Arts in Moscow. Both of these major texts as well as a number of less important papyri were written between -2060 and -1580 in a cursive script (called *hieratic*). In modern works it is common to transliterate this script into the *hieroglyphic* ideograms that were used to record information on monuments and tombs. The signs used to denote Egyptian numbers in the previous section were in hieroglyphic writing; their equivalents in the hieratic script are shown in Figure 2.1. The problem of deciphering and interpreting the mathematical texts involves the general problem of the decipherment of hieroglyphic Egyptian, which is the oldest form of their written language. The key to decipherment was the

FIGURE 2.1

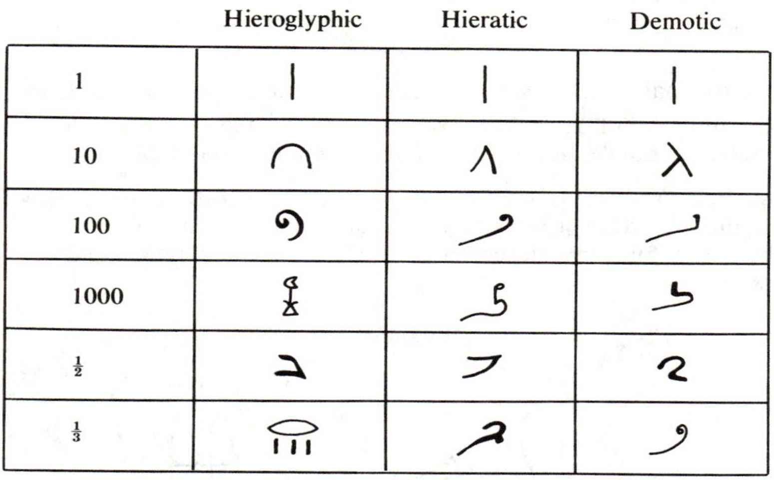

	Hieroglyphic	Hieratic	Demotic
1						
10				
100				
1000				
$\frac{1}{2}$				
$\frac{1}{3}$				

Typical Egyptian notations.

Rosetta Stone, which contains 14 lines of hieroglyphics, 32 lines of a late cursive form of ancient Egyptian writing called *demotic,* and 54 lines of Greek. It was found by the French in 1799 near the mouth of the Nile, subsequently passed into British hands, and now is housed in the British Museum. Although many attempted to decipher it and other hieroglyphic remains, success was not achieved until 1822 when Champollion published an "alphabet" and some decipherments. Today a street near the Sorbonne in Paris is named for him.

The principle of decipherment depended on the supposition that each of the three inscriptions appearing on the Rosetta Stone represented the same message. It had been noted that certain collections of hieroglyphic characters were sometimes enclosed in a cartouche (the French word for "cartridge"), and it was generally supposed that they represented the names of important persons. In the Greek inscription on the Rosetta Stone only the name *Ptolemy* occurs, and the hieroglyphic message, supposedly a translation of the Greek, contains a sequence of hieroglyphs enclosed in a cartouche:

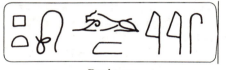

Ptolemy

Champollion assumed that this represented the name *Ptolemy.* On another multilingual monument he found two names in cartouches, one of which was *Ptolemy,* and the other, as he determined from the Greek translation, *Kleopatra;* in hieroglyphics it appeared this way:

Kleopatra

Since these two names have some letters in common, it is reasonable to suppose that the common hieroglyphs will also correspond. Together with a small amount of "extraneous" knowledge, this is enough to effect the decipherment.

Following E. A. Wallis Budge's discussion in his *Egyptian Language* [13], we will indicate how this was done. Let us rearrange the symbols occurring in the two names and number them for convenience, after recalling that in Greek *Ptolemy* is written *Ptolemaios:*

Ptolemaios

1   2   3   4   5   6   7

Kleopatra

| 1 | 2 | 3 | 4 | 5 | 6 | 7 | 8 | 9 | 10 | 11 |

Symbol 1 of the first name is identical with symbol 5 of the second and so should represent the letter *P*. Similarly the fourth and second are identical and, according to their positions, ought to represent *L*. Thus we have so far

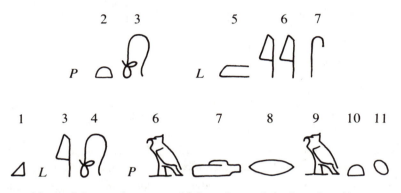

From the positions of the two letters we think we know, it is clear that the top name must
be *Ptolemy* and the other *Kleopatra*. Since the third and fourth symbols are the same and
the first name is *Ptolemy*, this symbol must represent a vowel sound something like *O* to
give

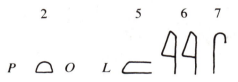

Between the *P* and *O* in *Ptolemy* must be *T*, and the Greek form of that name ends with
S; consequently two more hieroglyphs are in our possession and we have

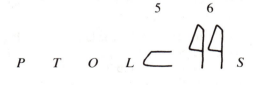

We already have a symbol for *T*, but there seems to be a *T* missing in *Kleopatra*, at least where we would expect to find it, and an extra *T* is stuck almost at the end. At this point we need a piece of extraneous information. As Champollion had observed in other inscriptions, *Kleopatra* was sometimes spelled with the sign ⌐⌐ instead of ◠ ; this means that ⌐⌐ must also stand for *T*; the sign preceding it ought to denote a vowel like *A*, and the first symbol evidently denotes *K*. At this point we have

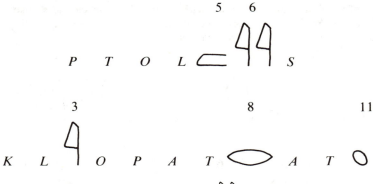

Now ⌐ must stand for a vowel sound like *E*, so 𝄇𝄇 should stand for *I* pronounced in continental fashion (as *EE* in *NEED*); therefore we can assume that doubled ⌐ stands for *I*, and ⊏ must be *M*. Since the hieroglyphs in *Ptolemaios* have been exhausted, ancient Egyptian must not have had a sound for the *O* preceding the *S*. Turning to *Kleopatra*, the Englishman Young, a contemporary of Champollion, had observed that the symbols ◠ always occurred at the ends of names of goddesses; since the kings and queens of Egypt (including the Greek monarchs) were revered as deities, the symbols ◠ in *Kleopatra*'s cartouche must be the terminative sign for *goddess*. Therefore the remaining unknown symbol in her cartouche must stand for *R*, that is, ⌾ = *R*. Finally, then, the two inscriptions can be translated as

PTOLMIS

and

KLEOPATRA   goddess

Extending this technique and using their knowledge of Coptic, which is Egyptian written in Greek letters, Champollion and his followers were able to decipher complete texts and build up a reasonably complete vocabulary and grammar.

   The hieroglyphic signs stand for entire words as well as for alphabetic sounds. This is one of the remarkable peculiarities of the writing systems of ancient civilizations in general and the Egyptian in particular. They used an awkward and inefficient combination of *ideographic* ("picture") writing, a *syllabary*[1], and an *alphabet* and on this complicated structure erected a superstructure of special determinative signs to indicate the type of reading to be used. For instance, the goddess indicator terminating the cartouche containing Kleopatra's name was intended not only as a title but also to tell the reader that the foregoing symbols

---

[1] *In a syllabary the symbols represent syllables, whereas with an alphabet several letters, in general, are used to denote a syllable.*

represented the name of a diety and should be read alphabetically or syllabically but not ideogrammatically. For instance, the verb *write* could be denoted by the sequence of symbols

a reed for writing, an inkpot, and a palette with red and black ink, which were materials used for writing, followed by the determinative sign ⌒ which always denotes an abstract concept and actually represents a rolled papyrus. A *writer* would be indicated this way:

Only the determinative has changed from the "abstract concept" indicator to the determinative for "man," to indicate a person. By throwing away the ideogrammatic, syllabic, and determinative components of their writing system, the Egyptians would have been left with an efficient and convenient alphabet, 2000 years before its introduction by the Greeks.

Mathematical terminology was subject to the same awkward conventions. The addition of two numbers was thought of in terms of one number "going into" or "joining" the second; the word for it was *per,* which also means "house," and the symbolic designation could be either ⌒ , a pair of walking legs "going into" the house, or the ground plan symbol that denotes a house followed by a pair of walking legs "going into" to act as a determinative to ensure that the ideogram for house was read in the sense of the concept of addition; thus ⌷ ⌒ . For subtraction, walking legs "going out" of the house were used. This raises an interesting question: which direction of motion corresponds to "in" and which to "out"? The Egyptians did not write from left to right or right to left consistently. Their hieroglyphic writing system carries built-in information to give the reader the correct direction. Many of the hieroglyphic symbols are animals or birds, and the reading rule (for text written in a row) is simply this: read from head to tail. Then feet walking against the direction of reading are "going in" and denote addition, whereas feet walking in the direction of reading are "going out" and denote subtraction.

As we saw while deciphering the hieroglyphic representation of *Ptolemy* and *Kleopatra,* the Egyptians were not consistent in their use of a single symbol to denote a single sound or concept. This casual attitude carried over to their mathematical works. Thus the Moscow Papyrus has a pair of legs walking "into" (the addition sign), but there it stands for the square of the associated number. The meaning must be determined from the context of the problem. Other notations for addition were

$$= demedsh = \text{"unite"}$$

and

$$\text{[hieroglyphs]} \quad \begin{array}{l} = wah = \text{``put together''} \\ \text{or ``collect''} \end{array}$$

Observe that both notations use the "abstract" determinative sign. Subtraction is also represented by the notion of "breaking up" the minuend; thus

$$\text{[hieroglyphs]} \quad \begin{array}{l} = hebet = \d{h} + b + t + \text{determinative} \\ \text{for ``splitting''} \end{array}$$

What *remains* after a calculation has been performed which we would call the *result*, is denoted by the symbols

$$\text{[hieroglyphs]} \quad \begin{array}{l} = tat = ta + t + \text{plural sign } (=|\,|\,|) \\ = \text{``result,'' ``that which remains''} \end{array}$$

The printing press and movable type provided a vital impetus to consistent and permanent symbolism which had been lacking. Before the fifteenth century the ambiguity and parochialism of the Egyptian notational system were the norm; there were almost as many different notations for basic concepts as there were mathematicians. For instance, the oldest known use of the unambiguous plus sign "+" for addition occurs in a Latin manuscript, now in the Dresden Library, written in +1486.[2] Nor were such inconsistencies peculiar to the mathematical discipline.

Figure 2.2 shows a hieratic transcription and a hieroglyphic transliteration of Problem 28 of the Rhind Papyrus. It is evidently to be read from *right* to *left*. The translation (from T. Eric Peet's *The Rhind Mathematical Papyrus* [78]) follows:

> *Two-thirds added and one-third [of this sum] taken away, 10 remains. Make one-tenth of this 10: the result is 1, remainder 9. Two thirds of it, namely, 6, are added to it; total 15. A third of it is 5. It was 5 that was taken away: remainder 10.*

We recognize the walking feet, which denote both addition and subtraction, on the first line and later on also. The integers are clear in the hieroglyphic text. $\frac{2}{3}$ is denoted by the special symbol [glyph] and $\frac{1}{3}$ by writing the denominator 3 and placing a special sign over it which means "one divided by"; thus

$$\tfrac{1}{3} = \text{[glyph]} \qquad \tfrac{2}{3} = \text{[glyph]}$$

Notice that $|$ is half of $|\,|\,|$; therefore [glyph] should be 1 divided by half of 3; that is, $\frac{2}{3}$.

§2.2. The hieroglyphs that denote numbers and the signs for addition and subtraction are clear in Figure 2.2. The recurrent figure of the owl, which corresponds to the word *em*, is a preposition that translates as "from," "out of," "in," "into," "on," "among," "as," "comformably to," "with," "in the state of," "if,"

---

[2] *See Chapter 8 for a discussion of the development of mathematical notation.*

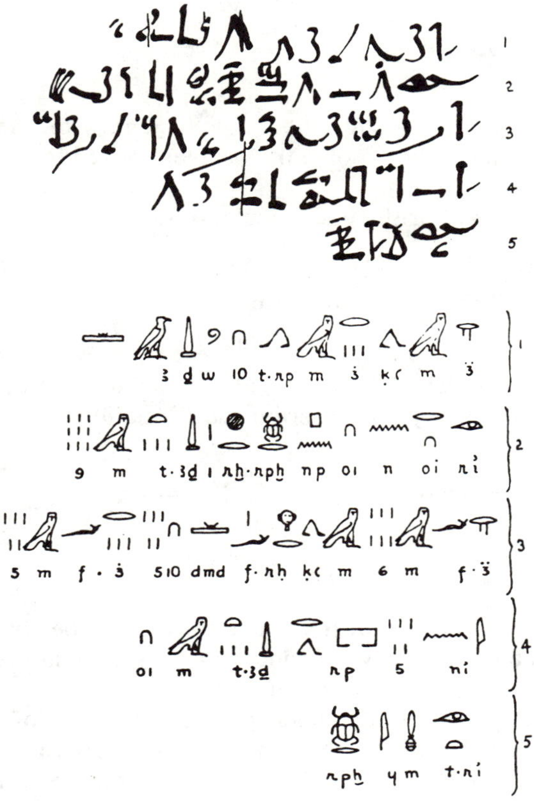

3 ḏw 10 t·n̄p m ṡ kꜥ m ṡ   ⟩1

9 m t·ṡḏ i n̄ḥ·n̄pḫ n̄p oi n oi n̄i   ⟩2

ṡ m f·ṡ 510 dmd f·n̄ḥ kꜥ m 6 m f·ṡ   ⟩3

oi m t·ṡḏ n̄p 5 n̄i   ⟩4

n̄pḫ ꞽ m t·n̄i   ⟩5

FIGURE 2.2

Rhind mathematical papyrus, Problem 28. (From Vogel [72], p. 56, originally in A. B. Chace, *The Rhind Mathematical Papyrus,* Oberlin, 1929)

or "when." In the context of Figure 2.2 the owl should be read as "from," or perhaps better as "from it," although the "it" is not indicated. Peet's translation might be improved to read

> *Two-thirds from it added and one-third [of this sum] taken away, 10 remains*

and so on. This reinterpretation is important, for without it the problem appears to be concerned with a numerical calculation and with this interpretation it does not make sense. For instance, the first line of Peet's translation seems to describe the equation

$$\tfrac{2}{3} - \tfrac{1}{3} = 10$$

which is absurd. What is actually meant and what Peet suggested by his use of the phrase "of this sum" is that $\tfrac{2}{3}$ of some unknown quantity should be added to the quantity itself, and then $\tfrac{1}{3}$ of this sum subtracted to yield 10; thus

(2.1) $$(\tfrac{2}{3}x + x) - \tfrac{1}{3}(\tfrac{2}{3}x + x) = 10$$

The problem concerns the solution of a linear equation in one unknown. Collecting all multiples of x, we find that the equation reduces to

$$\tfrac{10}{9}x = 10$$

with the solution $x = 9$.

Our interpretation must be the correct one, for the remaining lines of the hieroglyphic inscription in Figure 2.2 tell how to solve the problem posed in the first line.

They describe the following sequence of arithmetic calculations:

$$10 \rightarrow 10/10 = 1 \rightarrow 10 - 1 = 9 \qquad \text{(line 2)}$$
$$\rightarrow \tfrac{2}{3}9 = 6 \rightarrow 6 + 9 = 15 \rightarrow \tfrac{1}{3}15 = 5 \qquad \text{(line 3)}$$
$$\rightarrow 15 - 5 = 10 \qquad \text{(line 4)}$$

When they are written out in full, these calculations provide the solution to the original equation, for the calculations just given describe the formation of the arithmetic expression

$$[\tfrac{2}{3}(10 - \tfrac{10}{10}) + (10 - \tfrac{10}{10})] - \tfrac{1}{3}[\tfrac{2}{3}(10 - \tfrac{10}{10}) + (10 - \tfrac{10}{10})] = 10$$

Comparison of this expression with

$$(\tfrac{2}{3}x + x) - \tfrac{1}{3}(\tfrac{2}{3}x + x) = 10$$

makes it clear that $x = 10 - \tfrac{10}{10} = 9$ is a solution of the equation. We can be confident that the Egyptians never troubled themselves about the possible existence of other solutions; it was difficult enough for them to find one.

This method is really a way of verifying that a number, perhaps the result of a guess or trial and error, is in fact a solution. It does *not*, however, provide a prescription for solving other equations of the same kind. In fact, consider the following:

$$(ax + x) - b(ax + x) = c$$

This equation simplifies to Rhind Papyrus Problem 28 if $a = \tfrac{2}{3}$, $b = \tfrac{1}{3}$, and $c = 10$. Factoring x and then $(a + 1)$ shows that $x = c/(a + 1)(1 - b)$. Now the prescription for solving Rhind 28 instructs us to calculate

$$\left[a \left(c - \frac{c}{c} \right) + \left(c - \frac{c}{c} \right) \right] - b \left[a \left(c - \frac{c}{c} \right) + \left(c - \frac{c}{c} \right) \right]$$

which, if it equals c, will show that $x = (c - c/c) = c - 1$ is the solution. However, the expression can be simplified to $(a + 1)(1 - b)(c - 1)$, which is equal to c only for very special choices of a and b. With $a = \tfrac{2}{3}$ and $b = \tfrac{1}{3}$, this expression is in fact $\tfrac{10}{9}(c - 1)$, which equals c only if $c = 10$.

This example suggests that the Egyptians lacked a systematic method for solving linear equations with one unknown, which is true in the same sense that they lacked an alphabetic writing system. Just as their cumbersome system of writing included all the ingredients for a successful method of alphabetic writing so too did their collection of mathematical techniques include a general procedure for solving linear equations in one unknown (which we study next); but in neither case did they recognize that their systems included special or superfluous techniques.

In more than a dozen problems in various papyri a technical word appears for an unknown quantity similar to our use of x; its hieroglyphic symbols are

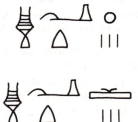

or

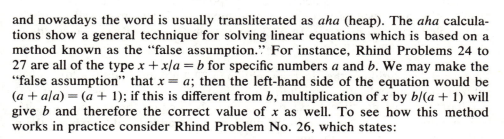

and nowadays the word is usually transliterated as *aha* (heap). The *aha* calculations show a general technique for solving linear equations which is based on a method known as the "false assumption." For instance, Rhind Problems 24 to 27 are all of the type $x + x/a = b$ for specific numbers a and b. We may make the "false assumption" that $x = a$; then the left-hand side of the equation would be $(a + a/a) = (a + 1)$; if this is different from b, multiplication of x by $b/(a + 1)$ will give b and therefore the correct value of x as well. To see how this method works in practice consider Rhind Problem No. 26, which states:

A quantity whose fourth part is added to it becomes 15.

This corresponds to the equation $x + x/4 = 15$. Now assume (falsely) that $x = 4$ (because it is easy to calculate $x/4$); the left-hand side of the equation would be $4 + \frac{4}{4} = 5$; which is only one-third of the right-hand side. Therefore multiply the false assumption 4 by 3 to get $x = 12$, which will make the left-hand side of the equation agree with the right and solve the problem.

Since this technique will work in general, but was not always used by the Egyptians, it seems clear that they did not know that it would always work, or possibly believed that other methods, such as the one used in Rhind Problem No. 28, discussed above, were also general methods. In any event, it cannot be said that they really understood linear equations.

It may be worthwhile to show how the general method of the *aha* calculations can be used to solve Rhind Problem No. 28. Recall that the problem is equivalent to solving equation (2.1):

(2.1) $$(\tfrac{2}{3}x + x) - \tfrac{1}{3}(\tfrac{2}{3}x + x) = 10$$

Considering the expression $(\tfrac{2}{3}x + x)$ as a new unknown, say

(2.2) $$u = \tfrac{2}{3}x + x$$

(2.1) can be rewritten as

$$u - \tfrac{1}{3}u = 10$$

Apply the method of the false assumption: if $u = 3$, then the left side is $3 - \frac{3}{3} = 2$, which must be multiplied by 5 to equal the right side. Hence the correct value of u must be $u = (3)(5) = 15$. But u stands for $\frac{2}{3}x + x$ [by (2.2)], and we are led to a new equation of the same type:

$$x + \tfrac{2}{3}x = u = 15$$

Again applying the method of the false assumption, we suppose that $x = 3$. Then $x + \frac{2}{3}x = 3 + 2 = 5$, which must be multiplied by 3 to equal the right side. Hence $x = (3)(3) = 9$, as the scribe of Problem No. 28 found by his special method.

The *aha* calculations represent the pinnacle of Egyptian algebra. They could not advance for several reasons: first, their cumbersome notation and lack of a systematic way of representing unknown quantities made it difficult for them to express more complex problems; compare the relative effectiveness of later systems of algebraic notation described in §8.2. More important, their technique of calculation was so complicated and so poorly adapted to the solution of problems whose answers were not whole numbers that it presented an insurmountable barrier to progress. In §2.3 we examine these arithmetic methods in order to understand their deficiencies.

§2.3. The solution of the algebra problem in §2.2 involved addition (including subtraction), multiplication (including division), and operations with fractions. Addition presents no difficulties, for it is enough to count the number of units, tens, hundreds, *etc.* that appear in the numbers to be added and to express them in the usual notation. For instance,

$$\cap \cap \; {}^{||||}_{||||} \; + \; \cap \; {}^{||}_{||} \; = \; {}^{\cap \cap}_{\cap \cap} \; ||$$

Multiplication introduces complications of a much higher order and requires a special technique. The Egyptians introduced a method based on successive doublings, related to the way digital computers multiply, which works because every positive whole number can be expressed as a sum of powers of 2; that is, every positive integer has a base 2 representation.

Consider an example from Rhind Papyrus No. 32, which describes the multiplication 12×12. In Figure 2.3 the multiplication is written out with hieroglyphics on the left in the manner of the Egyptians (read from right to left) and in

FIGURE 2.3

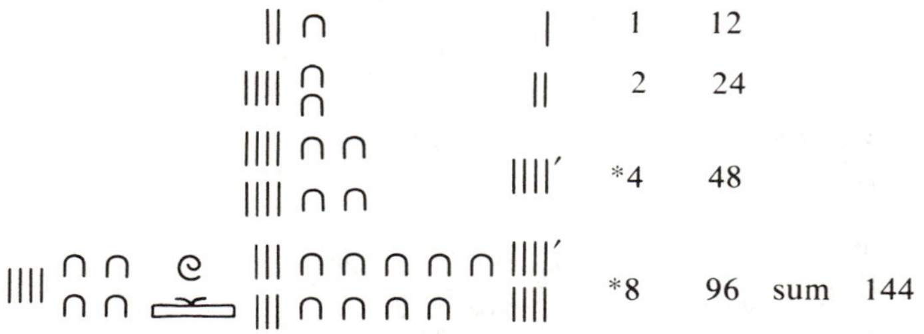

$12 \times 12 = 144$ from Rhind mathematical papyrus, Problem 32.

modern notation on the right. Each successive row represents a doubling of the preceding row and therefore 12, 24, 48, and 96 appear as successive doubles of 12. Also 1, 2, 4, and 8 appear as successive doubles of 1, and the 4 sitting next to the 48 indicates that 48 is 4 times the original quantity 12. We want to multiply 12×12, which is the same as $(4 + 8) \times 12 = 4 \times 12 + 8 \times 12$, the sum of the rows marked by 4 and 8. These rows are indicated by a * which tells the scribe to add the corresponding quantities to obtain the product 12×12. To obtain, for example, 13×12, write 13 as a sum of powers of 2, thus $13 = 8 + 4 + 1$; * the corresponding rows in Figure 2.3 and add the starred quantities:

$$
\begin{array}{lll}
*1 & 12 & \\
2 & 24 & \\
*4 & 48 & \\
*8 & 96 & \text{sum} \quad 156
\end{array}
$$

More generally, suppose an ancient Egyptian wanted to multiply 72 by 7. He would first (usually) represent the smaller number as a sum of powers of 2 (perhaps mentally if the number were small enough); thus $7 = 2^2 + 2^1 + 2^0$. Next, double the larger number repeatedly; thus

$$
\begin{array}{lll}
1 & (=2^0) & 72 \\
2 & (=2^1) & 144 \\
3 & (=2^2) & 288
\end{array}
$$

Now add the results of the doubling operation that correspond to the powers of 2 occurring in the expression of the smaller multiplier; in our example all the powers occur and the sum of the numbers yields the product 7(72). To compute 9(72) observe that $9 = 2^3 + 2^0$. Continuing the doubling table for 72, find

$$
\begin{array}{llll}
* & 1 & (=2^0) & 72 \\
& 2 & (=2^1) & 144 \\
& 4 & (=2^2) & 288 \\
* & 8 & (=2^3) & 576
\end{array}
$$

and add the doubled numbers appearing on the lines corresponding to 0 and 3 doublings (that is, to 2^0 and to 2^3) shown by the asterisks listed. This will produce the desired product.

It must have been clear to the Egyptians that this procedure would always work, for it is based on the fact that every positive integer is a sum of powers of 2, which they probably assumed without question. Construction of a proof of this fact is easy, but no one before the Greek civilization ever wanted to try to construct a proof of any assertion.

It may be worthwhile to point out the relation between this method of multiplication, which is after all not so convenient for people, and the way most digital computers multiply. If N is a positive integer, it can be written as

$$
N = a_r 2^r + a_{r-1} 2^{r-1} + \cdots + a_2 2^2 + a_1 2^1 + a_0
$$

where $a_0 = 0$ or 1, $a_1 = 0$ or 1, and so forth; the expression $a_r a_{r-1} \cdots a_2 a_1 a_0$, which consists of a sequence of zeros and ones, is the base 2 representation of N. For instance, the base 2 representations of 2, 10, and 123 are 10 ($= 1 \cdot 2^1 + 0 \cdot 2^0$), 1010 ($= 1 \cdot 2^3 + 0 \cdot 2^2 + 1 \cdot 2^1 + 0 \cdot 2^0$), and 1111011 ($= 1 \cdot 2^6 + 1 \cdot 2^5 + 1 \cdot 2^4 + 1 \cdot 2^3 + 0 \cdot 2^2 + 1 \cdot 2^1 + 1 \cdot 2^0$), respectively (*cf.* Chapter 1).

Suppose that N and M, both whole numbers, are to be multiplied. We can express N as a sum of powers of 2 as above and the product NM is the same as

$$NM = (a_r 2^r + a_{r-1} 2^{r-1} + \cdots a_1 2^1 + a_0) M$$

where all the a_i are either zero or one. Then the product is equal to

$$\sum 2^i M$$

where the sum[3] includes only those terms $2^i M$ for which $a_i = 1$ and i runs through 0, 1, 2, 3, . . . , r. Now it is easy to calculate $2^i M$; just double M, then double the result, and continue doubling until $2^i M$ is reached, which will happen after i doublings. Therefore, if we can *add* and *double* numbers, we can also multiply them. If *both* N and M are expressed as sums of powers of 2 and we are willing to accept the answer as a number expressed the same way, then all the calculations will be performed in the *base 2* representation, and instead of writing N as $a_r 2^r + \cdots + a_1 2^1 + a_0 2^0$, as we did above, we might just as well write the base 2 representation for N, which will be $a_r a_{r-1} \cdots a_2 a_1 a_0$, a sequence of ones and zeros. Similarly for M.

Egyptian multiplication of N by M amounts to successive doublings of M and addition of certain of the results of this doubling; this raises the question how numbers expressed in base 2 notation are doubled.

If the number is $b_s 2^s + b_{s-1} 2^{s-1} + \cdots + b_1 2^1 + b_0 2^0$, its double is $b_s 2^{s+1} + b_{s-1} 2^s + \cdots + b_1 2^2 + b_0 2^1 + 0 \cdot 2^0$; each exponent of 2 has been increased by one. The base 2 representation of the original number is just $b_s b_{s-1} \cdots b_2 b_1 b_0$ and the base 2 representation of its double is $b_s b_{s-1} \cdots b_2 b_1 b_0 0$. Therefore to double a number expressed in base 2 notation all that is necessary is to *shift the expression of the number one place to the left.* Thus to double 3, whose base 2 representation is 11, shift to the left one place to get 110, which represents $1 \cdot 2^2 + 1 \cdot 2^1 + 0 \cdot 2^0 = 6$; similarly, the expression for 9 in base 2 is 1001; its double, 18, is expressed as 10010. This is analogous to the fact that multiplication by 10 in our everyday base 10 system simply amounts to shifting one place to the left — the decuple of 3 is 30 — but you cannot multiply arbitrary numbers if you know only how to add and multiply by 10. The reason that digital computers utilize the doubling scheme for multiplication is that the process of shifting to the left can be implemented electronically in an inexpensive and extremely fast device called a *shift register;* together with the *adder,* it forms the heart of the arithmetic part of a computer.

Egyptian division proceeds along the pattern set out by their multiplication. To divide 15 by 3 a number x is required such that $3x = 15$; therefore double 3 repeatedly and try to see if some of the rows of the doubling table can be made to sum to 15; if they can, the sum of the powers of 2 corresponding to those

---

[3] *The symbol Σ, a capital Greek sigma, stands for "sum"; we shall see more of it in Chapter 4 and thereafter.*

rows is the desired quotient $\frac{15}{3}$. It works this way:

* 1	3
2	6
* 4 $(=2^2)$	12
8 $(=2^3)$	24
.	.
.	.
.	.

Addition of the lines preceded by asterisks gives

5	15

whence $\frac{15}{3} = 5$.

If we wanted to divide 16 by 3, we would be unable to find a collection of rows in the table above that would sum exactly to 16. This tells us that 3 does not divide 16 with an integral quotient and demands that we introduce some way of dealing with *fractions*. The Egyptian could quickly observe that summing the starred rows leads to 15, which, since it is less than 16, means that 5 is smaller than $\frac{16}{3}$; addition of the second and fourth rows yields 18, which means that 6 is bigger than $\frac{16}{3}$.

As we know, there is a special symbol to denote the fraction $\frac{2}{3}$, and any unit fraction, $1/N$, can be written by placing the sign $\frown$ over the hieroglyphs for N. In general the Egyptians had no notation for any other fraction, so a fraction that we would write M/N would have had to be expressed by them in terms of $\frac{2}{3}$ and the unit fractions. For instance, $\frac{3}{7}$ could not be written directly; of course, we could write $\frac{1}{7} + \frac{1}{7} + \frac{1}{7}$, which is reasonable but not handy. But expressing $\frac{99}{100}$ as a sum of 99 instances of $\frac{1}{100}$ is pointless. There are other more practical representations of $\frac{99}{100}$ as a sum of unit fractions.

The fundamental problem in Egyptian arithmetic was to provide a method for finding such "simple" expressions so that fractions could be used in a systematic way.

We could proceed as follows: $\frac{99}{100}$ is greater than $\frac{1}{2}$, and there is no unit fraction larger than $\frac{1}{2}$. Subtraction of $\frac{1}{2}$ from $\frac{99}{100}$ shows that $\frac{99}{100} = \frac{1}{2} + \frac{49}{100}$. Since $\frac{49}{100}$ is greater than $\frac{1}{3}$, which is the next largest unit fraction, we calculate $\frac{49}{100} = \frac{1}{3} + \frac{47}{300}$, which is not so useful because the denominator 300 is rather large and the calculations will become increasingly difficult. The Egyptians would have appreciated this point more than we. The next largest unit fraction is $\frac{1}{4}$, and this will work better because 4 divides 100; we find $\frac{49}{100} = \frac{1}{4} + \frac{24}{100}$, which is fine because we have not increased the denominators beyond 100, the denominator of the original fraction. Now $\frac{24}{100}$ is slightly larger than $\frac{20}{100}$, which we recognize as the unit fraction $\frac{1}{5}$, leaving $\frac{4}{100}$ remaining; the latter is $\frac{1}{25}$, and we have found the decomposition

$$\tfrac{99}{100} = \tfrac{1}{2} + \tfrac{1}{4} + \tfrac{1}{5} + \tfrac{1}{25}$$

This is certainly a more wieldy expression than 99 repetitions of $\tfrac{1}{100}$ would be.

Decompositions into sums of unit fractions are obviously not unique; we have just given two for $\tfrac{99}{100}$, one involving four summands and the other requiring 99. Had we continued by expressing $\tfrac{49}{100}$ as $\tfrac{1}{3} + \tfrac{47}{300}$ we would have found

$$\tfrac{99}{100} = \tfrac{1}{2} + \tfrac{1}{3} + \tfrac{1}{7} + \tfrac{1}{71} + \tfrac{41}{149,100}$$

and have been forced to continue before finding a unit fraction expression. The point of this observation is that there is no obvious rule governing the selection of a "best" unit-fraction representation. Picking fractions with the smallest denominators, as we have done above, does not always lead to the smallest number of summands and may not even result in a process that terminates after a finite number of steps.

Modern historians and mathematicians have tried to discover the rules that the Egyptians actually followed in performing calculations with fractions, and several different opinions have been published. We think that the most closely reasoned argument is Neugebauer's, which is reproduced here from his book, *The Exact Sciences in Antiquity* [50] (Neugebauer denotes $1/N$ by $\bar{N}$, and classes $\tfrac{2}{3}$ with the unit fractions, denoting it by $\bar{\bar{3}}$):

> . . . *a few of the main features [of Egyptian calculation with fractions] must be described in order to characterize this peculiar level of arithmetic. If, e.g., $\bar{\bar{3}}$ and $\overline{15}$ should be added, one would simply leave $\bar{\bar{3}}\ \overline{15}$ as the result and never replace it by any symbol like $\tfrac{4}{5}$. . . . $\bar{\bar{3}}$ forms an exception in so far as the equivalence of $\bar{2}\ \bar{6}$ and $\bar{\bar{3}}$ is often utilized.*

> *Every multiplication and division which involves fractions leads to the problem of how to double unit fractions. Here we find that twice $\bar{2}, \bar{4}, \bar{6}, \bar{8}$, etc., are always directly replaced by $1, \bar{2}, \bar{3}, \bar{4}$, respectively. For twice $\bar{3}$ one has the special symbol $\bar{\bar{3}}$. For the doubling of $\bar{5}, \bar{7}, \bar{9}, \ldots$ however, special rules are followed which are explicitly summarized in one of our main sources, the mathematical Papyrus Rhind. One can represent these rules in the form of a table which gives for every odd integer n the expression for twice $\bar{n}$.*

> *This table has often been reproduced and we may restrict ourselves to a few lines at the beginning:*

n	twice $\bar{n}$
3	$\bar{2} + \bar{6}$
5	$\bar{3} + \overline{15}$
7	$\bar{4} + \overline{28}$
9	$\bar{6} + \overline{18}$
	etc.

> *The question arises why just these combinations were chosen among the infinitely many possibilities of representing 2/n as the sum of unit fractions.*

I think the key to the solution of this problem lies in the separation of all unit fractions into two classes, "natural" fractions and "algorithmic" fractions, combined with the previously described technique of consecutive doubling and its counterpart, consistent halving. As "natural" fractions I consider the small group of fractional parts which are singled out by special signs or special expressions from the very beginning, like $\bar{\bar{3}}, \bar{3}, \bar{2},$ and $\bar{4}$. These parts are individual units which are considered basic concepts on an equal level with the integers. They occur everywhere in daily life, in counting and measuring.[4] The remaining fractions, however, are the unavoidable consequence of numerical operations, of an "algorism," but less deeply rooted in the elementary concept of numerical entities. Nevertheless there are "algorithmic" fractions which easily present themselves, namely, those parts which originate from consistent halving. This process is the simple analogue to consistent duplication upon which all operations with integers are built. Thus we obtain two series of fractions, both directly derived from the "natural" fractions by consecutive halving. One sequence is $\bar{\bar{3}}, \bar{3}, \bar{6}, \overline{12},$ etc., the other $\bar{2}, \bar{4}, \bar{8}, \overline{16},$ etc. The importance of these two series is apparent everywhere in Egyptian arithmetic. A drastic example . . . [is] . . . that $\bar{\bar{3}}$ of $\bar{3}$ was found by stating first that $\bar{\bar{3}}$ of 3 is 2 and only as a second step 3 of 3 is 1. This arrangement $\bar{\bar{3}} \to \bar{3}$ is standard even if it seems perfectly absurd to us. It emphasizes the completeness of the first sequence and its origin from the "natural" fraction $\bar{\bar{3}}$.

If one now wishes to express twice a unit fraction, say $\bar{5}$, as a combination of other fractional parts, then it seems natural again to have recourse to these two main sequences of fractions. Thus one tries to represent twice $\bar{5}$ as the sum of a natural fraction of $\bar{5}$ and some other fraction which must be found in one way or another. At this early stage, some trials were doubtless made until the proper solution was found. I think one may reconstruct the essential steps as follows. We operate with the natural fraction $\bar{\bar{3}}$, after other experiments (e.g., with $\bar{2}$) have failed. Two times $\bar{5}$ may thus be represented as $\bar{\bar{3}}$ of $\bar{5}$ or $\overline{15}$ plus a remainder which must complete the factor 2 and which is $1 \bar{\bar{3}}$. The question of finding $1 \bar{\bar{3}}$ of $\bar{5}$ now arises. This is done in Egyptian mathematics by counting the thirds and writing their number in red ink below and higher units, in our case

$$1 \ \bar{\bar{3}} \quad \text{(written in black)}$$
$$3 \ 2 \quad \text{(written in red)}$$

This means that 1 contains 3 thirds and $\bar{\bar{3}}$ two thirds. Thus the remaining factor contains a total of 5 thirds. This is the amount of which $\bar{5}$ has to be taken. But 5 fifths are one complete unit and this was a third of the original higher unit. Thus we obtain for the second part simply $\bar{3}$ and thus twice $\bar{5}$ is represented as $\bar{3} + \overline{15}$. This is exactly what we find in the table.

For the modern reader it is more convenient to repeat these clumsy conclusions with modern symbols, although we must remember that this form of expression is totally unhistorical. In order to represent $\frac{2}{5}$ in the form $1/m + 1/x$ we choose $1/m$ as a natural fraction of $\frac{1}{5}$, in this case $(\frac{1}{3})(\frac{1}{5}) = \frac{1}{15}$. For the remaining fraction we have

---

[4] *There are, in most languages, familiar special words to denote the "natural fractions." In English, for instance, we use "half" or "one-half" to denote $\frac{1}{2}$ which is linguistically quite different from the way we say $\frac{1}{8}, \frac{1}{9},$ etc. ("eighth," "ninth").*

$$\frac{1}{x} = \left(1 + \frac{2}{3}\right)\frac{1}{5} = \frac{5}{3} \cdot \frac{1}{5} = \frac{1}{3}$$

Thus we have the representation

$$\frac{2}{5} = \frac{1}{15} + \frac{1}{3}$$

of the table. In general we have

$$\frac{2}{n} = \frac{1}{3} \cdot \frac{1}{n} + \frac{5}{3} \cdot \frac{1}{n}$$

and the second term on the right-hand side will be a unit fraction when and only when n is a multiple of 5. In other words a trial with the natural fraction $\frac{1}{5}$ will work only if n is a multiple of 5. This is indeed confirmed in all cases available in the table of the Papyrus Rhind which covers all expressions for 2/n from n = 3 to n = 101.

We may operate similarly with the natural fraction $\frac{1}{2}$. Then we have

$$\frac{2}{n} = \frac{1}{2} \cdot \frac{1}{n} + \frac{3}{2} \cdot \frac{1}{n}$$

which shows that we obtain a unit fraction on the right-hand side if n is divisible by 3. For n = 3 we obtain

$$\frac{2}{3} = \frac{1}{6} + \frac{1}{2}$$

The main Egyptian mathematical accomplishment was the invention of a system of notation that made it *possible* to express the result of an arbitrary division of integers. Unfortunately the Egyptian process is cumbersome and does not result in a unique expression for a fraction, which in turn makes it difficult to compare the size of numbers.

With this understanding of their calculational capabilities we can think of one reason why the Egyptians did not know how to solve quadratic equations. This is usually thought of as a problem in algebra, but suppose for a moment that the "quadratic formula" were made known to an Egyptian mathematician. There would be little use that he could make of it, for the Egyptian notation for numbers and computational techniques that can be derived from it are not adapted to the calculation of square roots. Indeed, it is not clear how one could proceed to approximate the square root of a sum of unit fractions in a systematic way. How, for instance, could an Egyptian mathematician evaluate the square root of so simple an expression as $\overline{3} + \overline{9}$ by using a systematic procedure? We, with the advances of three millenia behind us, easily recognize that $\overline{3} + \overline{9} = \frac{4}{9}$, which has $\pm\frac{2}{3}$ as its square roots. Even this elementary example shows how advances in algebra are dependent on an efficient notation for numbers and on systematic methods for computing with them.

EXERCISES

2.1. Consider Figures E2.1a and b.

(a) Locate and translate all mathematical symbols and numbers in the text of Figure E2.1a.

FIGURE E 2.1A

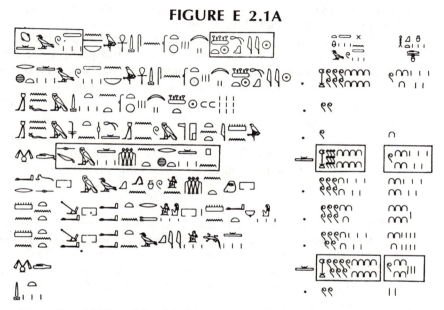

(From Vogel [72], p. 46, original in A. H. Gardiner, *Egyptian Grammar,* Oxford, 1927)

FIGURE E 2.1B

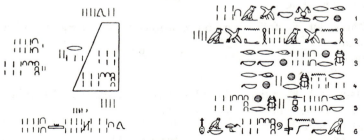

(From Vogel [72], p. 71, original in W. W. Struve, *Mathem. Papyrus d. staatl. Museums der schönen Künste in Moskau,* Quellen und Studien zur Geschichte der Mathem. A 1, Berlin, 1930).

(b) The two rightmost columns contain certain calculations. Decipher the calculations.

(*Note:* this papyrus represents daily bookkeeping for bread and beer for the royal palace [72, Vol. 1], p. 46.)

2.2. In Figure E2.1*b* a problem from the Moscow papyrus is shown ([72, Vol. 1], p. 71). It describes the computation of the volume of the frustum of a pyramid [the Egyptians always drew ▷ or ▢ to represent areas (or two dimensional objects) and △ or ▱ to represent volumes (or three dimensional objects)]. In this case ▱ represents a vertical cross section of the volume to be computed. To solve this problem the Egyptian mathematician first squared the numbers 4 and 2 and then doubled the number 4; the text at the right continues (see Figure E2.1*b*)

"*Combine this 16 with this 8 with this 4, and one obtains 28. Take $\frac{1}{3}$ of 6, giving 2; take 28 twice, giving 56. You see, it is 56, you have it correctly.*"

(a) Find the numbers described above in the Egyptian text.

(b) The volume of the frustum of the pyramid is given by the formula

$$V = (a^2 + ab + b^2) \cdot \frac{h}{3}$$

Identify and interpret the quantities a, b, h in the diagram and in the Egyptian text and verify the Egyptian scribe's use of this formula to obtain the volume. Also explain the computations near the diagram.

(c) Derive the above formula for the volume of the frustum of the pyramid (recall that the volume of a right pyramid of base area A and height h is $\frac{1}{3} Ah$).

2.3. Using the Egyptian *procedure* (described on pp. 53–54, in the text) for solving certain kinds of linear equations. Solve

$$(2x + x) - \tfrac{1}{2}(2x + x) = 3$$

State two disadvantages of their procedure. (Calculations may be performed in usual base 10 notation). *Prove* that there is no other solution to this equation.

2.4. Carry out the following computations by using Egyptian methods and notation:
(a) $45 \div 15$ (c) $100 \div 10$
(b) 28×17

2.5. Using the Egyptian method (but not necessarily Egyptian notation), perform the following calculations:
(a) 64×32 (b) 64×31
(c) 65×31 (d) 63×31
(e) 10×10 (f) $10 \times 10,000$

Notice the advantages of *positional* notation for parts (e)–(f).

2.6. The addition and multiplication tables for base 2 are

+	0	1
0	0	1
1	1	10

×	0	1
0	0	0
1	0	1

Calculate in base 2.
(a) $(10)_2 \times (1010)_2 =$
(b) $(1010)_2 \times (1010)_2 =$
(c) $(111)_2 \times (10001)_2 =$
(d) $(1010)_2 \times (1.01)_2 =$

2.7. (a) Express $\sqrt{2} = (1.414...)_{10}$ in base 2 correct to 3 (binary) digits. *Hint:* $(0.4)_{10} = \frac{2}{5} = \frac{1}{4} + \frac{1}{8} + \cdots$
(b) Square your answer as a check.
(c) In what sense does part (b) check your work?

2.8. Express the following duplications by using the formula

$$(**) \qquad \frac{2}{n} = \frac{1}{2} \cdot \frac{1}{n} + \frac{3}{2} \cdot \frac{1}{n}$$

(a) $2 \times \overline{3}$
(b) $2 \times \overline{6}$
(c) $2 \times \overline{9}$
(d) $2 \times \overline{12}$
(e) $2 \times \overline{15}$
(f) $2 \times \overline{18}$

2.9. Suppose n is divisible by 7. Find a formula similar to (**) above that will express $2/n$ as a convenient sum of unit fractions.

2.10. Express each of the following numbers as a sum of powers of 2:
(a) 15
(b) 24
(c) 63
(d) 1023
(e) 5113
(f) 341

*2.11. Prove that every positive integer can be expressed as a sum of distinct powers of 2. Can there be more than one expression of this kind for a given integer?

CHAPTER 3

Babylonian Algebra

The present chapter shows how the Babylonians were able to utilize the base 60 positional notation to solve many geometrical and algebraic problems. Several of their discoveries, such as a technique for approximating square roots, are still in use today. Positional notation enabled them to consider algebraic problems whose solution demanded more than the rational operations of addition, subtraction, multiplication, and division. Their astronomical methods are described to illustrate the power that their notation gave them, but astrology was their chief motivation and interest for reasons that are briefly discussed. The reader should compare the accomplishments of the Babylonians and Egyptians and observe that the major apparent relevant difference was the kind of notation for numbers that each society adopted.

§**3.1.** Our knowledge of Babylonian mathematics rests entirely on the contents of a few hundred clay tablets excavated in the Assyrian area, principally from the ruins of Nippur and Kiš; it is somewhat remarkable that no mathematical texts were found at the better known ruins of Nineveh, where some 15,000 tablets of a nonmathematical nature have been cataloged. The mathematical tablets have been translated and published with extensive commentaries on the nature of Babylonian mathematics (see O. Neugebauer, *Mathematische Keilschrifttexts,* 1935–1937, F. Thureau-Dangin, *Textes Mathematiques Babyloniens,* 1938, and O. Neugebauer and A. Sachs, *Mathematical Cuneiform Texts* [51], 1945). The dates of these publications show that our knowledge is quite recent indeed, considering that the authors of the tablets did their work during the period *circa* -2000 till -300 in Mesopotamia, approximately from the reign of Hammurabi till that of Alexander the Great. There are also extensive astronomical texts, but most are from a later Babylonian time known as the *Seleucid* period, after -300.

In van der Waerden's *Science Awakening* [71], Chapter III, there is a concise account of the state of the art of mathematics in Babylonia which draws on the three sources we have mentioned. We shall follow his example and select some of the interesting tablets and their interpretation as representative of the Babylonian mathematician at work.

First we mention that the tablets [see Figure 3.3 for an example (Plimpton 322)] come in two categories. There are *table texts* and *problem texts*. The table texts correspond to the tables at the back of a high school algebra book and contain such things as reciprocals, squares, cubes, and square roots. There are also the ever-necessary multiplication tables, since the base 60 multiplication table was a little too large to be readily memorized. The practical computations and problems that the Babylonians attacked would have been impossible without the availability of these tables, just as trigonometric tables or a slide rule are necessary to the solution of trigonometrical problems as taught in high school today.

§**3.2.** In Chapter 2 we learned that the Egyptians were handicapped by their inability to divide one number by another in an efficient manner. The Babylonians

FIGURE 3.1

A table of reciprocals: CBS 29 13 21 (obverse). (University Museum, Univ. of Pennsylvania).

FIGURE 3.1 (*Continued*)

(reverse)

were able to write

$$\tfrac{1}{25} = ;2,24$$

and therefore, they could write

$$\tfrac{23}{25} = (23) \times (;2,24)$$
$$= ;55,12$$

where we have carried out the multiplication in sexagesimal form. Consequently, most fractions posed no problem at all for the Babylonians. Here is a table reproduced from Neugebauer-Sachs [51], p. 11.

Table of Reciprocals of Standard Type

2	30	16	3,45	45	1,20
3	20	18	3,20	48	1,15
4	15	20	3	50	1,12
5	12	24	2,30	54	1,6,40
6	10	25	2,24	1	1
8	7,30	27	2,13,20	1,4	56,15
9	6,40	30	2	1,12	50
10	6	32	1,52,30	1,15	48
12	5	36	1,40	1,20	45
15	4	40	1,30	1,21	44,26,40

This table appeared on many tablets. We see clearly that it shows $\tfrac{1}{2} = ;30$, $\tfrac{1}{3} = ;20$, etc., omitting the notation for the sexagesimal point which is irrelevant. Figure 3.1 (CBS 29 13 21) displays an actual tablet that contains reciprocals. The top portion is composed of two columns and the bottom is divided into three. Each column contains a list of pairs of numbers that represent n and $\bar{n}$.

Note that the table omits the reciprocals of certain numbers, such as 7 and 11. Why is this? The reason is simple. For the most part (there was one exception) the tables contain only those reciprocals of numbers that could be easily written as a *finite* sexagesimal expression. But for 7

$$\bar{7} = 0;8,34,17,8,34,17 \ldots (8,34,17, \text{repeated}) \text{ (sexagesimal)}$$
$$= 0.142857142857 \ldots (142857 \text{ repeated}) \text{ (decimal)}$$

which repeats indefinitely and therefore has no terminating, or finite, sexagesimal representation. The Babylonians considered only reciprocals of *standard type,* that is, those that have no repeating sexagesimal expansion but can be expressed in a finite form. It is easy to see that a number n is of standard type if and only if it is of the form

$$n = 2^{\alpha}3^{\beta}5^{\gamma}$$

where α, β, γ, are positive integers or zero.

§3.3. The Babylonians were able to learn much more about the problem of measuring *geometric space* than the ancient Egyptians. They appear to have known the "Pythagorean theorem" from the earliest times. For instance, the old-Babylonian tablet BM 85 196 (*circa* −1700?) contains this problem (*cp.* van der Waerden [71], p. 76):

> A patu (=*beam?*) of length 0;30 (*stands against a wall*). The upper end has slipped down a distance 0;6. How far did the lower end move?

The situation is illustrated in Figure 3.2, where $d = 0;30$ and $d - h = 0;6$. The length b is determined by calculating $\sqrt{d^2 - h^2}$. This problem, with almost no variation, has been transmitted to the present day; you will probably recognize it from your own schooling. It also appeared during the Seleucid period in a cuneiform inscription (BM 34 568) which reads:

> A reed stands against a wall. If I go down 3 yards (*at the top*), the (*lower*) end slides away 9 yards. How long is the reed, how high the wall?

FIGURE 3.2

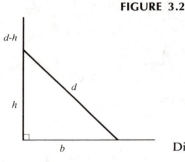

Diagram for BM 85 196.

Both problems have as solutions right triangles whose sides are of the form a/N, b/N, c/N, where N is some positive integer and $a^2 + b^2 = c^2$. Triples (a, b, c) of positive integers that satisfy $a^2 + b^2 = c^2$ are known today as *Pythagorean triples*. The argument given in §1.4 to prove that $\sqrt{3}$ is not rational also applies to $\sqrt{2}$ (see §1.4); this shows that we cannot have a Pythagorean triple with $a = 1$, $b = 1$, and c an integer, since c must then be $\sqrt{2}$, which is irrational—a contradiction. This example is studied in more detail below, but first we want to mention a controversial Babylonian tablet that contains a list of Pythagorean triples and provides some insight into the complexity of Babylonian mathematical thought. We are familiar with some examples of Pythagorean triples; for instance

$$(3,4,5): \qquad 3^2 + 4^2 = 5^2$$

and

$$(5,12,13): \qquad 5^2 + 12^2 = 13^2$$

Consider the old-Babylonian tablet Plimpton 322 shown in Figure 3.3, a translation of which is given in Figure 3.4; the headings b and d represent terms meaning "width" and "diagonal." It is easy to verify the remarkable fact that the numbers in the third, fourth, and sixth columns are all Pythagorean triples (h, b, d). On the

FIGURE 3.3

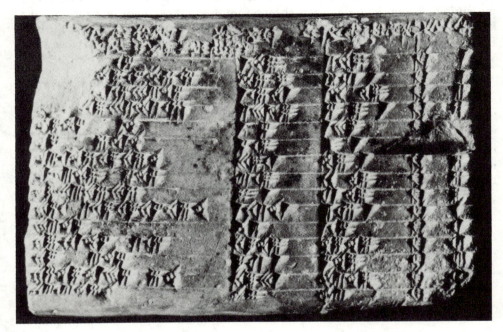

Pythagorean numbers: Plimpton 322. (Columbia University Libraries)

FIGURE 3.4

Row	$\dfrac{b^2}{h^2}$	b	d	No.	h
3	[59,0,]15	1,59	2,49	1	2,0
4	[56,56,]58,14,50,6,15	56,7	1,20,25	2	57,36
5	[55,7,]41,15,33,45	1,16,41	1,50,49	3	1,20,0
6	53,10,29,32,52,16	3,31,49	5,9,1	4	3,45,0
7	48,54,1,40	1,5	1,37	5	1,12
8	47,6,41,40	5,19	8,1	6	6,0
9	43,11,56,28,26,40	38,11	59,1	7	45,0
10	41,33,45,14,3,45	13,19	20,49	8	16,0
11	38,33,36,36	8,1	12,49	9	10,0
12	35,10,2,28,27,24,26,40	1,22,41	2,16,1	10	1,48,0
13	33,45	45	1,15	11	1,0
14	29,21,54,2,15	27,59	48,49	12	40,0
15	27,0,3,45	2,41	4,49	13	4,0
16	25,48,51,35,6,40	29,31	53,49	14	45,0
17	23,13,46,4[0]	28	53	15	45

Edited translation of Plimpton 322.

tablet itself (Figure 3.3) only b and d are given explicitly, but h is given in terms of the ratio b^2/h^2. We have listed the corresponding values of $h = \sqrt{d^2 - b^2}$ (which are integers!) in the rightmost column. For instance

$$(2, 0)^2 + (1, 59)^2 = (2, 49)^2$$

or, in base 10 representation,

$$(120)^2 + (119)^2 = (169)^2$$

The second row contains

$$(57, 36)^2 + (56, 7)^2 = (1, 20, 25)^2$$

or

$$(3456)^2 + (3367)^2 = (4825)^2$$

The numbers occurring in these examples are too large to be the result of casual observation, in contrast to the elementary examples of triples (3,4,5) and (5,12,13). One interesting point concerning the triples on Plimpton 322 is that the right triangles corresponding to the sides (b, d, h) have nearly uniformly decreasing values of the ratio b^2/h^2 (the first column), which means that the angle θ subtended by the sides d and h is decreasing at a nearly uniform rate as one proceeds down the rows of the tablet (Figure 3.5).

FIGURE 3.5

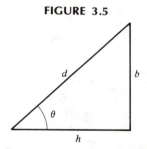

It is natural to wonder how the Babylonians found the Pythagorean numbers needed to construct the table. This is not a trivial problem, as some reflection will show. Some sort of mathematical analysis of the properties of integers must be performed in order to arrive at a scheme or algorithm that will generate these number triples in a systematic way. The Greeks and Arabs had such a formula available, but Plimpton 322 provides convincing evidence that much earlier the Babylonians were familiar with it (see Problem 3.11).

Since Plimpton 322 was written between -1900 and -1600, it is the oldest known document concerning *number theory* and therefore plays an important role in the study of the relative knowledge of the various ancient civilizations and especially of their understanding of and interest in abstract and sophisticated mathematical questions.

Mathematics does not look arcane or abstruse to everyone; a 1943 Columbia University catalog which contained a description of this tablet characterized it as a "commercial account." The real story of this exciting tablet was unfolded for the first time in the Neugebauer-Sachs 1945 publication [51] but the experts are still not agreed on the nature of the procedure used to derive the numbers in Plimpton 322. Neugebauer-Sachs [51] have a complicated explanation of the thought processes and mathematical techniques that could underlay the construction of the table of triples. The question is far from settled; their explanation has been criticized and modified by Vogel ([72], Vol. II, p. 37), and Bruins [76]. We think that Bruin's argument is the most convincing. He comes to the startling conclusion that the uniform decrease in b^2/h^2 is *accidental* and a consequence of the selection procedure, whereas Neugebauer-Sachs and Vogel assume that the Babylonian mathematician who worked out this table purposefully desired to construct both the regularly decreasing ratio and the Pythagorean triples, perhaps in order to construct a table of values of the tangent function; this demands a considerably more complicated mathematical derivation.

FIGURE 3.6

$\sqrt{2}$: YBC 7289. (Yale Babylonian Collection)

Whatever the truth about Plimpton 322 may be, it is certain that the Babylonians had a good command of the Pythagorean theorem when the three sides of the right triangle are rationally related (that is, are Pythagorean triples). But they knew more. Problem texts show that the Babylonians could also work with right triangles whose sides were not rationally related. For instance, YBC 7289 contains the diagram and numbers given in Figure 3.7. The number 1,24,51,10 is, if we assume that the "sexagesimal point" occurs between 1 and 24 (that is, 1;24, 51,10)

$$1 + \frac{24}{60} + \frac{51}{60^2} + \frac{10}{60^3} = 1 + \frac{2}{5} + \frac{51}{3600} + \frac{1}{21,600}$$

$$= 1 + 0.4000000 + 0.0141666 + 0.0000462^+$$

$$= 1.4142129^+$$

This is familiar; $\sqrt{2} = 1.414213562$ to 10 decimal figures. The Babylonian estimate

FIGURE 3.7

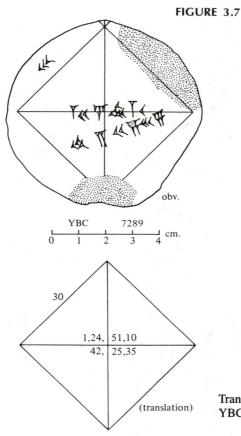

YBC 7289
0 1 2 3 4 cm.

obv.

1,24, 51,10
42, 25,35

(translation)

Transcription and transliteration of
YBC 7289. (Neugebauer-Sachs [51])

1;24,51,10 is about 0.0000007 too small. It is easy to check that our interpretation is correct, for if the side of a square is 30 its diagonal will be $30\sqrt{2}$ or in Babylonian notation, and using the presumed approximation,

$$(30) \times (1;24,51,10)$$

The product is computed like this:

$$30 \times 1 = 30$$

$$30 \times {;}24 = \frac{(30)\,(24)}{60} = 12$$

$$30 \times \frac{51}{60^2} = \frac{51}{2\,(60)} = \frac{25}{60} + \frac{1/2}{60}$$

$$= \frac{25}{60} + \frac{30}{(60)^2} = {;}25{,}30$$

and

$$30 \times \frac{10}{60^3} = \frac{1}{2}\,\frac{10}{60^2} = \frac{5}{60^2} = {;}0{,}5$$

Hence the product is $30; + 12; + ;25,30 + ;0,5 = 42;25,35$ which is written under the diagonal in Figure 3.7.

The Babylonians also used the less accurate estimate $1;25$ for $\sqrt{2}$; this is too large and not nearly so accurate.

How were these remarkable estimates discovered and what is their relationship? Nowadays digital computers are often programmed to calculate square roots by a repetitive procedure that was known to the Greeks. It works like this: suppose that $\sqrt{x}$ is wanted and let the positive number a_1 be a guess that is too small, that is, $x > a_1^2$. Then x/a_1 will be larger than $\sqrt{x}$, since this simply means $(x/a_1)^2 > x$ which is the same thing as $x > a_1^2$. It turns out that the *mean* (average) of these estimates, $a_2 = \frac{1}{2}(a_1 + x/a_1)$, will be closer to $\sqrt{x}$ than either a_1 or x/a_1. This requires proof, but the following diagram shows that it is at least intuitively reasonable.

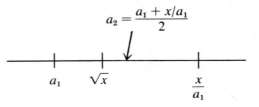

Therefore a_2 can be taken as a new estimate of $\sqrt{x}$. Repetition of this process provides increasingly more accurate estimates.

For instance, if we take $a_1 = 1$ as an approximation to $\sqrt{2}$, then

$$a_2 = \frac{a_1 + 2/a_1}{2} = \frac{1 + 2/1}{2} = \frac{3}{2}$$

is indeed better; continuing, we find

$$a_3 = \frac{1}{2}\left(\frac{3}{2} + \frac{2}{3/2}\right) = \frac{17}{12}$$
$$= 1.416 \ldots$$

and

$$a_4 = \frac{1}{2}\left(\frac{17}{12} + \frac{2}{17/12}\right) = 1\frac{169}{408} = 1.414215 \cdots$$

which is $0.000002 \ldots$ too large, and so on.

If this procedure is repeated in sexagesimal notation, we find

$$a_1 = 1$$

$$a_2 = \frac{1 + 2}{2} = 1;30$$

$$a_3 = \frac{1}{2}\left(1;30 + \frac{2}{1;30}\right) = 1;25$$

and

$$a_4 = \frac{1}{2}\left(1;25 + \frac{2}{1;25}\right)$$

Now $2/1;25 = 1;24,42,21, \ldots$, whence

$$a_4 = \frac{2;49,42,21, \cdots}{2} = 1;24,51,10, + \cdots$$

which is the number given in Figure 3.7. We have therefore discovered exactly the estimates of $\sqrt{2}$ found in the Babylonian mathematical texts!

§3.4.   Because the progress of geometry is intertwined with the state of development of algebra, it is appropriate to our theme to describe briefly the nature and state of sophistication of Babylonian algebra. The illustrative examples given below exhibit a particularly simple but successful technique that is representative of Babylonian capabilities throughout the entire period for which mathematical texts have been discovered.

Here is Neugebauer's translation (from his *Mathematische Keilschrifttexts*, rendered into English in van der Waerden [71]) of tablet AO 8862:

> *Length, width. I have multiplied length and width, thus obtaining the area. Then I added to the area, the excess of the length over the width: 3,3 (i.e., 183 was the result). Moreover, I have added length and width: 27. Required length, width and area.*

(given:)	*27 and 3,3*	*the sums*
(result:)	*15 length*	*3,0 area*	*12 width*

One follows this method:

$$27 + 3,3 = 3,30$$
$$2 + 27 = 29.$$

Take one-half of 29 (this gives 14;30).

$$14;30 \times 14;30 = 3,30;15$$
$$3,30;15 - 3,30 = 0;15$$

The square root of 0;15 is 0;30.

$$14;30 + 0;30 = 15 \; length$$
$$14;30 - 0;30 = 14 \; width$$

Subtract 2, which has been added to 27, from 14, the width. 12 is the actual width. I have multiplied 15 length by 12 width.

$$15 \times 12 = 3,0 \; area.$$
$$15 - 12 = 3$$
$$3,0 + 3 = 3,3.$$

The problem is stated in a form reminiscent of current high school mathematical "word problems"; indeed, the latter are lineal descendants of the Babylonian problem "books" transmitted to Western civilization by the Arabs. Notice the appearance of a square root.

The Babylonians had no symbols for the exclusive representation of unknowns, as we have x and y, but rather used the words *length, width,* and so on for this purpose. For instance, they had no qualms about adding a length to an area, which the Greeks would never have done, since the geometrical or physical interpretations would have had no meaning.

This tablet, as do the others, states the problem and provides a recipe for its solution. In our notation the problem in AO 8862 is

$$(3.1) \qquad \begin{aligned} xy + x - y &= 183 \\ x + y &= 27 \end{aligned}$$

The solution amounts to the introduction of a new variable, say y', defined by

$$y' = y + 2$$

(this "width" differs from the "actual width" by 2) which will simplify the original equations. This is a novel and ingenious notion: do not solve the original problem — simplify it. The new problem, expressed in terms of x and y', is indeed simpler because only the combination of unknowns xy' appears in the first equation:

$$(3.2) \qquad \begin{aligned} xy' &= 183 + 27 = 210 \\ x + y' &= 27 + 2 = 29 \end{aligned}$$

This expression of the problem was considered to be the standard form, and there was a standard prescription for solving it that can be described as follows. Suppose the equations

$$(3.3) \qquad \begin{aligned} xy' &= P \\ x + y' &= S \end{aligned}$$

are given. Then introduce a new unknown E, an "error," by putting

$$(3.4) \qquad \begin{aligned} x &= \frac{S}{2} + E \\ y' &= \frac{S}{2} - E \end{aligned}$$

The solution of the problem is obtained by taking

$$(3.5) \qquad E = \sqrt{(S/2)^2 - P}$$

We can easily verify it. Since $x + y' = S$, the second of the equations (3.3) is satisfied; substitute for x and y' in the first equation to find

$$(3.6) \qquad \left(\frac{S}{2} + E\right)\left(\frac{S}{2} - E\right) = P$$

This is the quadratic equation $E^2 = (S/2)^2 - P$, which agrees with (3.5).

For the problem given on the tablet $P = 210$ and $S = 29$; in the text E is explicitly computed as $E = 0;30 = \frac{1}{2}$; then

$$x = \frac{29}{2} + 0;30 = 14;30 + 0;30 = 15$$
$$y' = \frac{29}{2} - 0;30 = 14;30 - 0;30 = 14$$

and so $y = y' - 2 = 12$. In the last four lines of the tablet the solution $x = 15$, $y = 12$ is checked.

This procedure appears repeatedly in Babylonian problem texts and is actually more powerful than it appears at first sight. It was used as we today would use the "quadratic formula"

$$x = \frac{-b \pm \sqrt{b^2 - 4ac}}{2a}$$

to solve the equation

$$ax^2 + bx + c = 0$$

The quadratic formula was also known to the Babylonians and could have been used to solve (3.1) by substituting for y in the first of the two equations; this procedure is common today and was used by early Arab mathematicians but not by the Babylonians. Their procedure corresponds to a more intuitive and less formal technique. They might have reasoned as follows: one solution of the equation $x + y' = S$ would be $x = y' = S/2$, the *average value,* but the average value might not satisfy the second equation, $xy' = P$. Consider the difference between the correct value of x and the average $S/2$; this can be thought of as the error E, so that

$$x = \frac{S}{2} + E$$

Since $x + y' = S$, y' must therefore be given by

$$y' = \frac{S}{2} - E$$

So the problem has been reduced to finding the error E. As we have seen, this amounts to solving the special quadratic equation $(S/2 + E)(S/2 - E) = P$. Now it was known that $(a + b)(a - b) = a^2 - b^2$, so the equation can be "unfactored" to the simple quadratic equation $E^2 = (S/2)^2 - P$. We shall see later how the identity for the factorization of $a^2 - b^2$ could have been discovered in a geometric way.

Other algebraic problems were solved similarly. For instance, YBC 4697 contains a problem that may be stated as

$$\tfrac{1}{3}(x + y) - 0;1\,(x - y)^2 = 15$$
$$xy = 10,0$$

We would solve the second equation, say for x, and substitute the result in the first to find $x = 10,0/y$ and

$$\tfrac{1}{3}\left(x + \frac{10,0}{x}\right) - 0;1\left(x - \frac{10,0}{x}\right)^2 = 15$$

which simplifies to the *quartic* equation

$$\tfrac{1}{3}(x^3 + 10,0x) - (0;1)(x^4 - 20,0x^2 + (10,0)^2) = 15x^2$$

This looks as if it would be difficult for us to solve, but the Babylonians did not proceed by substitution from the second equation; rather they used their general technique by writing

$$x = A + E \qquad y = A - E$$

where A is the average value of the two solutions x and y. We can easily see that

$$x + y = 2A \quad \text{and} \quad x - y = 2E$$

The two equations of the problem then become

$$\tfrac{1}{3}A - 0;1(E)^2 = 15$$
$$A^2 - E^2 = 10,0$$

and the second equation shows that $E^2 = A^2 - 10,0$; substitution of this value in the first equation produces

$$\tfrac{1}{3}A - 0;1(A^2 - 10,0) = 15$$

which is only a *quadratic* equation and can be solved by using the quadratic formula. It is interesting to note that Neugebauer first thought that the problem led to a cubic equation; as van der Waerden pointed out, in this case the Babylonian method is preferable to our own usual technique.

Throughout our discussion we have used the algebraic identity $(a + b)(a - b) = a^2 - b^2$. Nowadays the usual application of this and the other important identities

$$(a + b)(a + b) = a^2 + 2ab + b^2$$
$$(a - b)(a - b) = a^2 - 2ab + b^2$$

replaces the right-hand side by the *factored* form given on the left. The Babylonians did the opposite: they replaced products of algebraic expressions by an expanded *un*factored form. As shown above, this is necessary to substitute and simplify expressions, as in (3.6) for instance. All three of these algebraic identities correspond to simple geometrical decompositions which may account for their early discovery. The identity $(a + b)(a - b) = a^2 - b^2$ can be read off in Figure 3.8, where the upper right-hand rectangle, of area $b(a - b)$, has been moved to the left and attached to the square as shown, to produce the hatched rectangle of area $(a - b)(a + b)$. The other two identities follow similarly (see Figure 3.8).

§**3.5.** The Babylonians applied their mathematical skills to the problems of astronomy. To set the stage we quote the following passage from Neugebauer's *The Exact Sciences in Antiquity* [50] p. 98:

> *We may now enumerate the tools which were available at the end of the "prehistory" of Babylonian astronomy which extends from about 1800 B.C. to about 500 B.C. The zodiac of 12 times 30 degrees as a reference system for solar and planetary motion.*

FIGURE 3.8

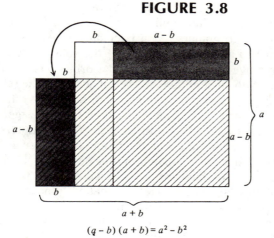

$$(q - b)(a + b) = a^2 - b^2$$

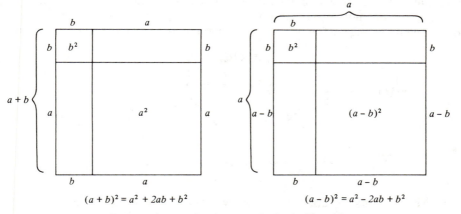

$$(a + b)^2 = a^2 + 2ab + b^2 \qquad (a - b)^2 = a^2 - 2ab + b^2$$

Geometric proofs of some algebraic identities.

A fixed luni-solar calendar and probably some of the basic period relations for the moon and the planets. An empirical insight into the main sequences of planetary and lunar phenomena and the variation of the length of daylight and night. The use of arithmetic progressions to describe periodically variable quantities. And, above all, a complete mastery of numerical methods which could immediately be applied to astronomical problems. The utilization of these possibilities marks indeed the crucial step.

From a mathematical viewpoint, Babylonian cosmology in effect placed the earth at the center of a sphere to which the stars were fixed because the motions of the sun, moon, and planets were studied in relation to the background of stars which served then, as it does now, as a frame of reference. The basic periodic motions of the sun, moon, and planets were measured, and an *arithmetic model* capable of predicting their conjunctions and other important features was established by the Babylonian astronomers. At a later date Claudius Ptolemy of Alexandria ($\sim +150$) described a geometric model of the solar system in which the planets, sun, and moon moved in compound circular orbits. Using this model, Ptolemy was able to predict with an admirable degree of accuracy the relative motion that is observed from the earth. Our present model of the solar system, except for certain minor refinements due to Einstein, is based on Newton's general laws of motion, which were created some 15 centuries later (see Chapter 12). The Babylonians, as far as we know, had no geometric model but rather used certain arithmetical functions whose values at an instant in time described the position of a celestial body with respect to the background of fixed stars as seen from the earth. These periodic functions are now sometimes called "linear zigzag functions," and they look something like Figure 3.9. The time interval P corresponds to one "period" of the zigzag function by which its future and past values are fixed. They are gross approximations of periodic functions like the sine and cosine. In the context of Babylonian astronomy Figure 3.9 represents a certain displacement (usually angular) of a celestial body, such as the moon, as it varies with time and from which its future displacement can be predicted, but the Babylonians did not display this information in graphical or other geometrical form. The linear zigzag functions were represented by a tabulation of their values.

FIGURE 3.9

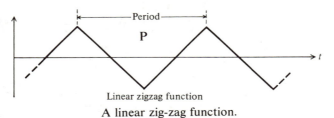

A linear zig-zag function.

These functions, or tables of values, were intellectual abstractions and not just the results of observations; they really had nothing intrinsic to do with observed phenomena. Linear zigzag functions were used, for instance, to determine how much of a celestial body would actually be obscured during a partial eclipse. At times other than the actual eclipse date these functions had no physical (or observable) meaning. We want to emphasize this point, for it appears to be a component of the astronomy of the Babylonians that has much in common with contemporary viewpoints.[1] If a physical system is unknown to us except for some small aspect that we can observe and measure and if a function of some sort can be

---

[1] *It is analogous to the present use of* complex-valued functions *in quantum mechanics to represent the "state of a system," whereas only the* absolute value *of the function, a real quantity, is a physical observable.*

found that accurately predicts the observations we make (as a function of time, for instance), then we have a "mathematical model" (*cp.* Chapter 12). In modern physics we measure the scattering angle of a beam of electrons which hits an "unknown substance" and try to find a "function" to represent this angle; this "function," if found, will, in a certain manner, be a "picture" of the "unknown substance."

The linear zigzag functions, hence the mathematical model of the cosmos used by the Babylonians, were based on arithmetical progressions (whence the term "arithmetic model") which capably predicted eclipses and the other astronomical events presumed by them to exert an influence on their everyday lives. Their achievements are a measure of their arithmetical capabilities and, above all, of the power and effectiveness of the sexagesimal positional system of notation, which permitted the laborious calculations that were and remain an integral part of astronomical theory.

§**3.6.** Astronomy has always fascinated but it has also always been important. Before its use for navigation, which we examine in Chapters 5 and 12, it was the sole means of regulating the calendar and thereby determining the proper time for annual agricultural activities such as crop planting and land irrigation. In still more remote times, perhaps before the invention of writing, the stars and other celestial bodies were worshiped as gods, and the study of their perfect motion formed part of the religious duties owed them. In this sense astronomy was initally a technical branch of religion that served as one of several channels of communication between the priestly hierarchy and the gods they served. In these early times no information about the *nature* or *structure* of celestial objects could be extracted from the simple visual observations made without instruments then possible. Therefore the astronomical study of these deified objects was virtually identical with the study and prediction of their paths of motion through the heavens. Thus freed from any necessity to examine questions related to the material composition or sources of energy of the stars and other celestial inhabitants, astronomy was exclusively mathematical, and mathematics thereby became the handmaiden of religion which in return fostered its development and ensured its prestige for thousands of years.

The practical component of religion which applies the achievements of mathematical astronomy to the welfare of mankind is *astrology,* the ancient art of divining the fate of human beings from the configurations and motions of the planets and stars. Astrology was already well developed in Babylonia in −3000. It was founded on the Babylonians' identification of personal deities with the various heavenly bodies: *Mercury* with *Nebo, Venus* with *Ishtar, Mars* with *Nergal, Jupiter* with *Marduk, Saturn* with *Ninib* (the reader cannot help but note that the current planetary names stem from a similar Roman identification), the *moon* with *Sin,* and the *sun* with *Shamash.* The movements of these five planets visible to the unaided eye, the moon, and the sun, were regarded as representing the activity of the corresponding gods, and therefore it required but a small leap of the imagination to conclude that if one could correctly "read" the heavenly motion of these divinities one would know what they were aiming to bring about on earth.[2]

The motion of the bodies in the solar system is viewed against the background of the fixed stars. From the earliest times the various star clusters were identified with familiar creatures and objects whose forms appeared to be similar to the patterns traced by the stars. Of principal importance were those constellations that lie, when viewed from earth, behind the paths of motion of the planets, moon, and sun. Since all these bodies move in nearly the same plane — the *plane of the ecliptic* — their motion against the background of stars appears to take place in a relatively narrow band. This imaginary zone of the heavens (illustrated on Figure 3.10), bounded by two circles equidistant from the ecliptic plane and separated by about 18°, is the *zodiac*. The zodiac is partitioned into 12 equal *signs*, each comprising 30° in the ecliptic plane; each sign is associated with a constellation that lies in the zodiacal band from which the sign draws its name, but because the constellations vary in size, as viewed from the earth, the signs do not coincide in angular aperture with the constellation to which each corresponds. The 12 signs are listed in Fig. 3.10 in an order opposite to that usual in astrology.

FIGURE 3.10

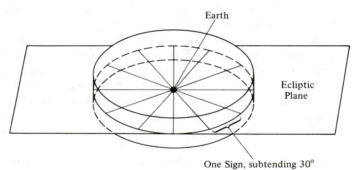

Earth

Ecliptic Plane

One Sign, subtending 30°

♑	Capricorn, the Goat	♋	Cancer, the Crab
♐	Saggitarius, the Archer	♊	Gemini, the Twins (Pollux and Castor)
♏	Scorpio, the Scorpion	♉	Taurus, the Bull
♎	Libra, the Balance	♈	Aries, the Ram
♍	Virgo, the Virgin	♓	Pisces, the Fish
♌	Leo, the Lion	♒	Aquarius, the Water Bearer

The Zodiac.

[2] *If this premise of astrology is valid, one could equally well predict the paths of motion of the heavenly bodies from a careful analysis of the actions and history of mankind. As mankind is more accessible to observation than the faraway stars, this consequently would seem to be the proper way to pursue the study of astronomy.*

FIGURE 3.11

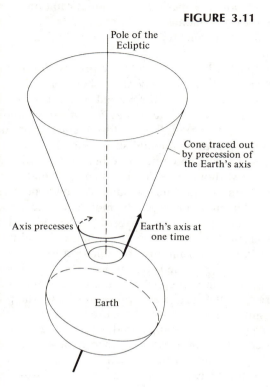

Pole of the
Ecliptic

Cone traced out
by precession of
the Earth's axis

Axis precesses

Earth's axis at
one time

Earth

Precession of the earth's axis.

The earth is a spinning "top," rotating on its axis once each day. If, after the annual circuit of its solar orbit, the direction of the axis of the spinning earth is compared with its direction when it was at the same location in space one year before,[3] a difference will be detected, for the earth, like a top, wobbles as it spins. The wobbling motion of its axis traces a circular cone in space as shown in Figure 3.11. Nearly 26,000 years pass before the earth's axis returns to its initial direction. This means that the axis of the earth has not always pointed to Polaris, the present polestar; in −2700 it pointed to α-Draconis, in Greek times it was about halfway between γ-Draconis and β-Ursae Minoris and there was no true polestar, whereas in +13,600 it will be near the brightest star of the northern sky, Vega in the constellation of the Lyre.

Another consequence of this wobbling *precession* of the earth's axis is a continual slow change in the annual appearance of the heavens. The navigator's map of the night sky evidently requires periodic revision to account for the changing location of the stars against which he frames his position; see Chapter 5. Man was apparently aware of this change in the most ancient times, although the reason for it

---

[3] *We neglect the rectilinear motion of the solar system through space which is irrelevant for this discussion because its effect is small compared with that due to other causes.*

and an estimate of the nearly 26,000-year period of this great rotation of the heavens was first given by the Greek astronomer Hipparchus of Nicaea (c. −180 to c. −125), whose work was closely connected with the Babylonian astronomical achievements of this period.

The apparent slow rotation of the heavens due to the precession of the earth's axis as the earth rotates gives rise to an equally slow change in the position at which the sun appears to rise each year on the vernal equinox relative to the zodiacal constellations. This phenomenon, the *precession of the equinoxes*, was viewed as responsible for the catastrophic fall and subsequent rise of successive "Ages" of the world. When the vernal equinoctal sun passed from one zodiacal constellation to the next, an Age ended and a new one began: the ruling gods laid down their scepters — often not without considerable struggle — and their places were assumed by newcomers. History has known many who have prophesied the "end of the world"; often their lamentations would be more accurately and significantly interpreted as the "end of an Age of the world," when violent and cataclysmic transitional events are to be expected as the transcendental power illuminating and guiding our world passes from one constellation to the next.

At this point the reader should refer to the list of zodiacal signs in Figure 3.10; the order of signs from top to bottom is the order of the constellations in which the (vernal equinoctal) sun rises as the earth's axis precesses about the pole of the ecliptic. Before −4000, the sun rose in Gemini; thereafter it moved into Taurus. The zodiacal constellations were probably named in this period because the most ancient religious myths associated the productive change wrought by the sun on the seasons and the produce of the earth with the Bull, but there is no trace of a corresponding connection with the Twins. The sun remained in Taurus until about −1800, when it began to rise in Aries, the Ram.[4] It passed into Pisces in −60, where it will remain until the Age of Aquarius dawns that fine spring day in +2740.

Our Age — Pisces — is marked by the advent of Christ the Fish, whereas Ram-crowned "two-horned" Moses (recall Michaelangelo's striking statue) descended from Mount Sinai with the new law in biblical −1491 to find his disobedient flock worshiping Taurus as the "Golden Calf," fallen to the ways of the preceding Age at whose beginning in −4004, according to the bible, the "world" had been created. These images are repeated in ancient myth and religion, from the bull worship of Apis-Osiris in Egypt and Zeus the Bull carrying off Europa to Hercules' defeat of the Cretan bull and Jason's triumphal capture of the Golden (ram's) Fleece which heralded the rise of another Age [23].

If the motion of heavenly bodies holds the secret of man's fate, it is no wonder that the ancients spared no effort to unlock this secret by systematic study of the heavens. The physical monuments to their expenditure of time and treasure rival

---

[4] *The "coordinate system" of zodiacal signs is continually readjusted so that the vernal equinoctal sun rises at the "first point of Aries," as it did for the early Babylonian astronomers; of course this has nothing to do with its position relative to the constellations.*

man's current works: great Egyptian pyramids built when α-Draconis was the polestar have interior passages aligned in the direction of that star as it passed the meridian of the pyramid. But there is today perhaps no more striking *physical* evidence of man's single-minded devotion to unraveling the enigma of the gods' intent than the colossal ruins of the precise astronomical *analog computer* that is Stonehenge in England. As Gerald Hawkins [39] has recently shown so convincingly, Stonehenge, built between -1900 and -1600 from stones weighing as much as 35 tons, was designed to keep track of the progress of such spectacular recurring celestial events as lunar eclipses.

But compare the power of a truly mathematical formulation: the Babylonian arithmetic model of celestial motion was an even more precise and flexible *digital computer* built from fragile clay tablet cuneiform mathematical tables and records of observations.

EXERCISES

3.1. Show how to construct a multiplication table for the sexagesimal number system. Indicate how large the table must be and give some sample entries. How did the Babylonians construct such a table? Discuss the use and usefulness (in your own opinion) of your multiplication table in conjunction with the table of reciprocals (Figure 3.1) for everyday arithmetic.

3.2. Using the table of reciprocals, compute

(a) $\dfrac{1,5}{4}$ (b) $\dfrac{20,1}{1,4}$ (c) $\dfrac{2}{5}$

*3.3. Prove that a number $1/n$ has a finite decimal expansion if and only if it is of the form

$$n = 2^\alpha 5^\beta$$

where α and β are nonnegative integers.

*3.4. Prove that $1/n$ has a finite sexagesimal expansion if and only if it is of the form

$$n = 2^\alpha 3^\beta 5^\gamma$$

where α, β, and γ are nonnegative integers.

3.5. Which numbers $1/n$ have finite expansions in base 2?

3.6. Did the Babylonians know the Pythagorean theorem? Why? Did they know how to prove it? Explain.

3.7. Using the Babylonian procedure, calculate the next three approximations to $\sqrt{3}$, following the (useless) approximation $\sqrt{3} \approx 1$. Express your results and your calculations in modern notation only.

3.8. Use the Babylonian procedure to estimate the following numbers, correct to two decimal places:

(a) $\sqrt{5}$ (b) $\sqrt{7}$ (c) $\sqrt{10}$ (d) $\sqrt{1.036}$

Why did you choose the starting value you did?

3.9. (a) Using the Babylonian procedure, devise a way to express $\sqrt{1-x^2}$ approximately (as a polynomial in x).
(b) Compare your result in (a) with that obtained by applying the binomial theorem.

3.10. Where in Figure 3.7 should there be "sexagesimal points" (which we represent as semicolons) or do all of the entries represent positive integers? Why?

3.11. Show that if p, q are positive integers, then $h = p^2 - q^2$, $b = 2pq$, and $d = p^2 + q^2$ are Pythagorean triples. Find the values of p and q which give the Pythagorean triple (120, 119, 169). (Note: *all* Pythagorean triples arise in this way; this is the "formula" referred to in the text on p. 75.)

3.12. Solve the following simultaneous equations by the Babylonian method:
(a) $xy = 720$, $x + y = 72$
(b) $xy = 1$, $x + y = a$
in which a is some given positive number.

3.13. State the conditions on P and S in the equations $xy = P$, $x + y = S$ that are necessary and sufficient such that the solution pair (x, y) is a pair of
(a) *positive* numbers
(b) *real* numbers

3.14. On page 81 it was proved that $(a + b)(a - b) = a^2 - b^2$ by using a *geometrical analog* of this *algebraic identity*. Use similar methods to prove
(a) $x(y + z) = xy + xz$ (the "distributive law")
(b) $x(yz) = (xy)z$ (the "associative law")
(c) $x^3 - y^3 = (x - y)(x^2 + xy + y^2)$
Hint: Consider three-dimensional objects.

CHAPTER 4

Greek Trigonometry: The Introduction of Inequalities and the Measurement of Area

This chapter is concerned with several important Greek contributions to mathematics. It begins by observing that a few qualitative mathematical statements can often replace a vast number of quantitative statements without any loss of precision or information. This fundamental and fruitful notion first appeared in the work of Aristarchus in which he estimated the distance from the earth to the moon and sun. In this same work Aristarchus introduced the use of inequalities to keep track of the range of error in his estimates. This second idea has also led a remarkable life and, together with other developments, forms a fundamental pillar upon which the theory of the calculus is supported (Chapter 9). Because of the importance of these two ideas, and also because this application of mathematics to astronomy is elegant and interesting, his paper is studied in detail. The reader who does not appreciate long calculations is nevertheless urged to try to find his way through them in order to understand the basic role that qualitative statements and inequalities play in mathematics and in the mathematical description of nature.

The use of inequalities to bound errors has become commonplace for users of computers, and an example is given to illustrate what can go wrong if this is not done. Next we study Archimedes' method of calculating an approximate value of π; the calculations are lengthy and not of real interest, but the method is another simple but important example of the application of inequalities; and it will occur again in Chapter 9.

Because everyone knows something about π but few have any idea how it can be calculated in a practical manner, we present examples of calculations that require much less imagination and ability than Archimedes' method; they are simple consequences taken from the calculus which help to suggest the power of that tool by contrast with Archimedes' effort.

An important theoretical advance, which suggests the principal technique of the integral calculus, is exhibited by Archimedes' method of computing the area bounded by two straight lines and an arc of a parabola. Here one of the basic ideas of calculus is unveiled in a particularly clear form. However, little further progress was made in that direction for more than 1500 years. We think that the inadequate notation employed by Archimedes and the other Greeks to represent variable quantities was responsible for the delay. Just as advances in numerical notation permitted progress, so too did advances in the means of representing variables and functional relationships, which took on their present form in Newton's time.

It is not generally realized by those who are not experts in mathematics that the concept of area is extremely difficult and beset with pitfalls for intuition derived from experience. The last section of Chapter 4 defines area in a rigorous way and exhibits some of the pathological aspects of this concept.

Although it is never explicitly mentioned, all the material in this chapter stresses implicitly that one fundamental purpose of mathematics is to convert problems whose solution requires an effort of imagination and creativity into other forms such that the solution can be obtained by rote "calculation". As mathematics continues to succeed in this process of conversion, the notion of what constitutes a "calculation" becomes increasingly sophisticated but its rote aspect remains. Compare, for instance, the modern techniques for calculating π, which can easily be prepared for computer calculation, with Archimedes' ingenious and creative method, which is in fact not automatic at all because of the unsystematic simplifications he makes at each stage in order to obtain numbers that could be easily expressed in the inadequate notation available to him.

§**4.1.** We have seen that Pythagoras' theorem was known in Old Babylonian times and that methods of approximating the values of certain irrational numbers, such as $\sqrt{2}$, were in use. With this as a foundation, later Babylonian mathematicians constructed tables of the lengths of chords of a circle of a given radius, which correspond to modern tables of sines and were particularly useful for astronomical purposes. Thus it appears that the essence of Euclidean trigonometry, that is, the trigonometry of the Euclidean right triangle, was known to the Babylonians. They did not draw any really profound cosmological conclusions from this knowledge because their mathematical models of the cosmos were solely arithmetical so they had to content themselves with making reliable determinations of the positions of heavenly bodies and predictions of their future positions and eclipses.

On the other hand, the Greek civilization developed a remarkably accurate geometric model of the universe. They assumed that the earth was a sphere, that the sun and moon were spherical physical bodies each at a fixed distance from the earth, all moving relative to one another, and held varying views on whether the sun or the earth was a fixed center of rotational motion.

Two scientists of Alexandria, the mathematician Eratosthenes (-276 to -198) and the astronomer Aristarchus of Samos (*circa* -310 to -230), made major contributions to man's understanding of the world beyond his immediate horizon. Eratosthenes was the first to estimate the circumference (hence the radius) of the earth, whereas Aristarchus, in a treatise entitled "The sizes and distance of the sun and moon," was able to derive a reasonable approximation of the distance between the earth and the moon (as well as other similar magnitudes), a rather ingenious effort, as we shall see. Eratosthenes' measurement showed Greek society how large the earth was (and how small the part they occupied and understood), and Aristarchus' measurements gave an entirely new perspective of the size of the earth in the larger universe of the heavens.

If the plate on which the shadow falls in a sundial is hollowed out to form a hemisphere, then it can easily be arranged to indicate the angular position of the sun; Aristarchus invented such a sundial, called a *scaphe* (see Figure 4.1). Era-

FIGURE 4.1

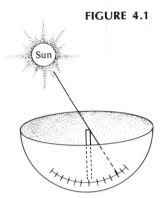

The scaphe.

tosthenes used the scaphe to make his remarkably accurate estimate of the earth's radius. He observed that at noon during the summer solstice the sun cast no shadow in the Egyptian town anciently called Syene (Sī.ē.nē), but today Aswan. At the same hour in Alexandria, which he took to lie on the same meridian as Syene, the shadow of the sun cast in the scaphe showed an angle of $2\pi/50$. Assuming the sun's rays to be parallel at Alexandria and Syene, as shown in Figure 4.2, the earth's circumference $2\pi R_e$ is given by 50 times the distance from Alexandria to Syene. The latter was estimated to be 5000 *stades,* 40 stades were supposed equal to the Egyptian σχοινος, and the σχοινος was 12,000 royal cubits of 0.525 meters each, which, all in all, yielded 516.73 feet for the length of a stade and about 25,000 miles for the circumference of the earth; Eratosthenes apparently improved his estimate to 252,000 stades, perhaps to obtain a number divisible by 360 and so more conveniently expressed in the sexagesimal system that Greek mathematicians invariably used for astronomical calculations, but also to obtain the easily remembered estimate of 700 stades per degree. This led to an estimate of 3925 miles for R_e, which is only about 25 miles shorter than the true polar radius. Eratosthenes' marvelous estimate of the circumference of the earth was by far the best made in all antiquity. It is true that it gained in accuracy from a fortunate cancellation of errors, but even if they had not canceled it would still have been better than other estimates of the times, and Eratosthenes' method was certainly objectionless.

FIGURE 4.2

$(2\pi/50)$

Alexandria

5000 stades

Syene

Sun

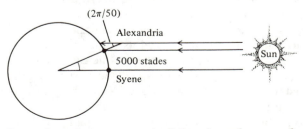

Eratosthenes' measurement of the circumference of the earth.

It is a remarkable twist of history, and one that had consequences even in the fifteenth century, that this highly accurate estimate of the size of the world was ignored and one that was grossly in error passed into general scientific use. It happened this way. The Stoic philosopher Poseidonius ($c. -130$ to -51) employed a method similar to Eratosthenes' to measure the size of the world, but his was based on observations of the star Canopus and the fact that at the city of Rhodes Canopus rose to be just barely visible on the horizon, whereas at Alexandria it rose to about $\frac{360}{48}$ degrees. The rest of his argument is the same as Eratosthenes', but by combining several inaccurate estimates he arrived at the good estimate of about 24,000 miles for the earth's circumference. Poseidonius' method is defective in one important way: the refractive properties of the atmosphere permit us to see objects that really lie below the horizon (see Figure 4.3) so that accurate angular measurements of a star's position cannot be made when it is close to the horizon. In this respect Eratosthenes' method, which used the position of a midday sun, was better. A compensating error in estimating the distance between Rhodes and Alexandria resulted in Poseidonius' 24,000-mile circumference. Had matters been left as they stood, future mathematicians and astronomers would have used one of these two accurate measurements. However, Strabo (-63 to $> +21$), the most influential compiler of geographic information of his time, wrote

> . . . *of Poseidonius, who estimates its [i.e., the Earth's] circumference at about one hundred and eighty thousand stadia . . . ,*

about 18,000 miles, and thus approximately 25 percent less than its true value. Strabo's misquotation of Poseidonius stuck, and was used by Columbus in his arguments about the possibility of reaching the Indies from Iberia.[1]

FIGURE 4.3

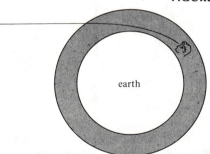

earth

Displacement of apparent position due to atmospheric refraction of light.

Ingenious arguments have been proposed to explain why Strabo misquoted Poseidonius. Heath [40] has suggested that the unnaturally conservative Strabo probably took the smallest estimate of the distance from Rhodes to Alexandria (ironically due to Eratosthenes!)—about 375 miles instead of Poseidonius', which was about 500 miles. Now $(500 - 375)/500 = 0.25$, which accounts for the 25 percent discrepancy. The reason that there were such varied estimates of the distance between Rhodes and Alexandria was that these cities were separated by water,

[1] *See Chapter 6.*

and until navigational astronomy was perfected more than 1500 years later distance measurements over water were but gross guesses.

We turn now from considering the size of the earth to the "size of the solar system," a problem of much greater complexity.

As mentioned earlier, Aristarchus of Samos appears to have been the first to combine trigonometrical theory and a mathematical model of the heavens with some simple physical measurements to investigate the metrical relationships of the earth, moon, and sun. Based on his observations, some of which were inaccurate, he proved, for instance, that if D_{em} denotes the distance from the center of the earth to the center of the moon and R_m the radius of the moon, then

(4.1)
$$\frac{45}{2} < \frac{D_{em}}{2R_m} < 30$$

In fact, $2R_m = 2160$ miles and $D_{em} = 238,857$ miles; therefore $D_{em}/2R_m \simeq 111$; Aristarchus' estimate is wrong by approximately a factor of 4. His error turns out to be the result of inaccurate estimates of certain angles, but the abstract formal argument is perfect, and it is on that argument which we wish to focus attention.

One of the basic tools used by Aristarchus is *trigonometry*, the study of the relations between angles and lengths of sides of triangles. We shall introduce a few trigonometric functions. Consider Figure 4.4, a diagram of a circle whose radius has length 1.

FIGURE 4.4

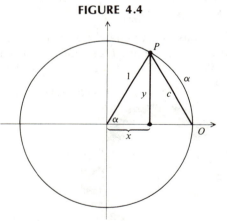

The *angle* α (measured in radians) is by definition the length of the arc of the circle running from O to P in a counterclockwise orientation. The segment c is the *chord* subtended by the angle, and the lengths x and y are vertical and horizontal distances from point P to the axes, as indicated. For each angle α the lengths x, y, and c are determined, and as the angle α varies so do these lengths, in terms of which we make the following definitions:

$$\sin \alpha = y$$

$$\tan \alpha = \frac{y}{x}$$

(4.2)
$$\text{chord } \alpha = c$$

If we imagine point P moving along the circle in a clockwise fashion toward point O, we can see that sin α (height y) goes to zero, chord α (length c) goes to zero, and tan α (the ratio y/x) goes to zero.

Throughout his analysis Aristarchus relies on two propositions that he assumes without proof (he proves the others that he uses); historians of mathematics think this means that the unproved propositions must have been generally known in his time. Both propositions and all his calculations were presented in terms of chords of circles and their corresponding arcs. Expressed in current terminology, the first proposition states that

$$\left(\frac{\sin \alpha}{\alpha}\right) \ decreases \quad and \quad \left(\frac{\tan \alpha}{\alpha}\right) \ increases$$

as α increases from 0 to $\pi/2$; see Figure 4.5 for the relevant part of the graph of sin α/α. These statements represent a basic advance in mathematics because they refer to the *entire behavior* of the sine and tangent *functions* rather than to particular values; this is the first time in this book that such a concept has appeared and you will soon see its power.

FIGURE 4.5

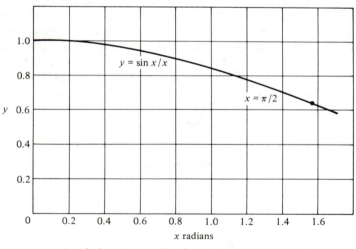

$y = \sin x/x$ is a decreasing function for $0 \leqslant x \leqslant \pi/2$.

The importance of considering the function as an entity rather than as a collection of values is intimately related to another basic advance in mathematics, which can be described as the replacement of numerous *quantitative* statements by a few *qualitative* statements. It is often thought that mathematics is exclusively devoted to quantitative analysis but rarely realized that the most powerful methods of attaining quantitative results are based on general qualitative properties. For instance, to assert that sin α/α is decreasing is a qualitative statement; it is not said how rapidly or to what extent this decrease occurs. Nevertheless, quantitative consequences flow from this qualitative observation to provide a degree of precision never before attainable. In order to explain this important idea further, we present Aristarchus' second unproved assumption, derive it from his first, and apply it to the determination of accurate estimates for the values of sines and tangents of small angles.

The second proposition states that if $0 < \beta < \alpha < \pi/2$ then

(4.3)
$$\frac{\sin \alpha}{\sin \beta} < \frac{\alpha}{\beta} < \frac{\tan \alpha}{\tan \beta}$$

This statement is partly quantitative, since it asserts something about the relative values of certain functions. It follows easily from the first proposition if the rules for operating with inequalities are known, for $\beta < \alpha$ implies $\sin \alpha/\alpha < \sin \beta/\beta$, since $\sin \alpha/\alpha$ is decreasing; therefore $\sin \alpha/\sin \beta < \alpha/\beta$ and similarly for the other inequality in (4.3).

In his estimation of the sizes and distances of the moon and sun Aristarchus has need of the values of the sines of certain special angles, which he obtained by using the second proposition; for instance, he states that

(4.4)
$$\tfrac{1}{45} > \sin 1° > \tfrac{1}{60}$$

that is, $0.022 \ldots > \sin 1° > 0.0166. \ldots$ This *quantitative* statement follows easily from (4.3) and knowledge of the natural right triangle with sides 1, 2, and $\sqrt{3}$. The idea is this: from an equilateral triangle, each of whose angles is $\pi/3$ and whose construction is simple, construct a right triangle with angles $\pi/2$, $\pi/3$, and $\pi/6$ by dropping a perpendicular from one vertex to the opposite side. If the side of the equilateral triangle is 2, the sides of the right triangle are, according to Pythagoras, 1, $\sqrt{3}$, and 2 (see Figure 4.6). Also, $1° = \pi/180$ in radian measure, and, since $\pi/6 > \pi/180$ and $\sin \alpha/\alpha$ is a decreasing function,

(4.5)
$$\frac{\sin \pi/6}{\pi/6} < \frac{\sin \pi/180}{\pi/180}$$

FIGURE 4.6

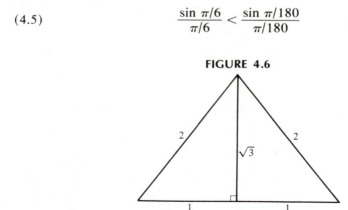

which reduces to $(\sin \pi/6)/30 < \sin 1°$. From Figure 4.6 $\sin \pi/6 = \tfrac{1}{2}$ whence finally $\tfrac{1}{60} < \sin 1°$, which is the right side of Aristarchus' estimate (4.4). To find the other part apply the same argument to the increasing function $\tan \alpha/\alpha$ to find

$$\frac{\tan \pi/6}{\pi/6} > \frac{\tan \pi/180}{\pi/180}$$

which reduces to $30 \sin 1° < \tan \pi/6 \cos 1°$. From Figure 4.6, $\tan \pi/6 = 1/\sqrt{3}$ and certainly $\cos 1° < 1$; therefore $\sin 1° < 1/30\sqrt{3}$. Finally $\sqrt{3} = 1.732 \ldots > 1.5$, whence

$$\sin 1° < \frac{1}{30(1.5)} = \frac{1}{45}$$

This is Aristarchus' other inequality.

The purpose of deriving these inequalities in detail is to show that they were quite within the grasp of elementary reckoning techniques, *provided that* the fundamental *qualitative* facts that $\sin \alpha/\alpha$ decreases and $\tan \alpha/\alpha$ increases (as α increases) are understood. We cannot stress too strongly the importance of the qualitative in mathematics; all the great concepts as well as the bulwark of precision and exactness in application are hidden here.

§**4.2.** We now turn to Aristarchus' application of these trigonometrical inequalities to the determination of the size of the sun and moon and their distance from the earth. The basic physical assumptions are that each of the bodies concerned is a sphere and that an eclipse of the moon occurs when the moon enters the shadow of the earth. An unstated assumption is that the sun is larger than the earth; this leads to Figure 4.7. The centers of the earth, moon, and sun are denoted, respectively, by O_e, O_m, and O_s. In Figure 4.7 the center of the moon lies on the line joining the centers of the earth and sun: D denotes the distance from O_e to the vertex O of the cone formed by the shadow of the earth and $2d$ is the diameter ($=$ width) of the shadow at the moon's center O_m. In Figure 4.7 the moon is shown as lying entirely within the shadow; this is correct because we know from observation that total eclipses of the moon are possible. The radii of the earth, moon, and sun are denoted by R_e, R_m, and R_s, respectively. One physical measurement associated with Figure 4.7 is necessary: the ratio $d/R_m \simeq 2$. One way to estimate

FIGURE 4.7

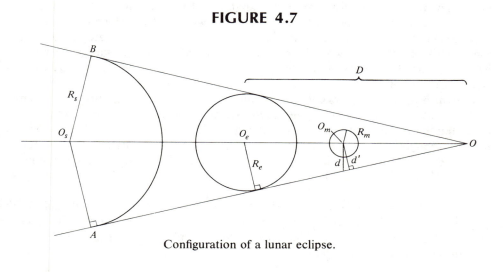

Configuration of a lunar eclipse.

this ratio is to time how long it takes for the moon to pass through the shadow region (more precisely, the time that elapses between the first obscuration of the moon on entering the shadow and its first reappearance as it leaves the shadow) and compare with the time it takes for the moon to become completely obscured as it enters the shadow. The ratio 2 is approximately correct; you can check it for yourself the next time you view a total eclipse of the moon.

There is a mathematical assumption that Aristarchus is careful to justify in great detail: the shadow width d and the perpendicular distance d' from O_m to the shadow boundary OA are nearly the same. This is true because of the great distance of the sun from the earth in relation to its size, so that its rays AO and BO are in reality nearly parallel; the errors introduced by this assumption are much smaller than the errors in physical measurements.

The second necessary physical measurement states that $D_{em}/R_m = D_{es}/R_s$; referring to Figure 4.8, note that this means that the sun and the moon appear to be the same size in the sky or, in more mathematical terminology, the angles that they subtend at the center of the earth are approximately equal. This can be verified by holding a coin at a distance from the eye where it just blocks the light of the sun; do the same for the moon. If the distances of the coin from the eye are nearly equal, then the subtended angles must also be nearly equal.

FIGURE 4.8

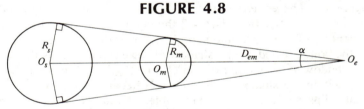

Sun and moon subtend equal angles when viewed from the earth.

In addition to this qualitative observation of equality of subtended angles, it is necessary to estimate how large the angle is. From Figure 4.8, R_m/D_{em} is $\sin(\alpha/2) = $ sine of half the subtended angle. Observation (with the coin if you are careful) shows that $\alpha \simeq (\frac{1}{2})°$. Aristarchus used the grossly inaccurate value 2° for the angle α, although it was claimed (by Archimedes) that he knew the better value. Perhaps he improved his observation only after his paper on distances and sizes was written.

The third necessary physical measurement is the angle θ in Figure 4.9, essentially the situation that occurs when there is a half moon. One measures the angle θ' subtended by the setting sun and the rising half moon and $\theta = 90° - \theta'$. Aristarchus thought that $\theta = 3°$, but its true value is more nearly 10', that is, $(\frac{1}{6})°$; therefore his estimate was in error by a factor of 18. This measurement implies that $D_{em}/D_{es} = \sin \theta = \sin 3°$.

§4.3.
Aristarchus gave a complete argument (see Heath [40]) that included careful estimates of d and d' (cp. Figure 4.7). Instead of reproducing them here we present a simplified version which leads to a result that is easier to follow, although not quite so good as his.

FIGURE 4.9

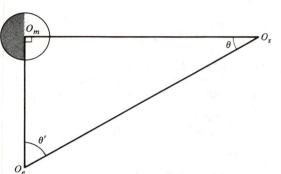

Configuration of half moon.

Under the assumption that $d \simeq d'$, we first summarize the relations derived above between the various distances and angles described in Figures 4.7, 4.8, and 4.9, using Aristarchus' measurements of the angles α and θ:

(4.6) $$\frac{d}{R_m} \simeq 2$$ (Figure 4.7)

(4.7) $$\frac{D_{em}}{R_m} = \frac{D_{es}}{R_s}$$ (Figure 4.8)

(4.8) $$\frac{R_m}{D_{em}} = \sin 1° \quad (= \sin \alpha/2)$$ (Figure 4.8)

(4.9) $$\frac{D_{em}}{D_{es}} = \sin 3° \quad (= \sin \theta)$$ (Figure 4.7)

We derive two more equations from Figure 4.7 by using similar triangles.

(4.10) $$\frac{D}{R_e} = \frac{D_{es}}{R_s - R_e}$$ (Figure 4.7)

(4.11) $$\frac{d}{R_e} = \frac{D - D_{em}}{D} \quad (d \simeq d')$$ (Figure 4.7)

We now have six equations for the six unknown quantities d, R_m, R_s, D_{em}, D_{es}, and D; note that a specific value for R_e is already known to us from Eratosthenes' work described earlier in the chapter. In principle, then, we can expect to solve these equations for each of the variables, including the distance from the earth to the moon, which is precisely what Aristarchus carried out.

Let us indicate how these equations are solved. To simplify the notation define $n = d/R_m$. Then by solving (4.10) for D, substituting the result in (4.11), and applying the second physical measurement (4.7) ($D_{em}/R_m = D_{es}/R_s$; moon and sun appear to be the same size) to eliminate the ratio D_{em}/D_{es} we find

(4.12) $$\frac{R_e}{R_m} + \frac{R_e}{R_s} = 1 + n$$

The next step eliminates R_s from (4.12) so that R_m is expressed in terms of the earth's radius and n. From (4.7) and (4.9) we easily derive $R_s = R_m/\sin 3°$. Substituting this in (4.12), we find

$$(4.13) \qquad \frac{R_m}{R_e} = \frac{1 + \sin \theta}{1 + n} = \frac{1 + \sin 3°}{1 + 2}$$

where we have used the second physical measurement, to the effect that $n = 2$.

All that is left to do is to estimate $\sin 3°$. We can use the same method that we used to estimate $\sin 1°$ at the beginning of this section: $\sin 3° = \sin(\pi/60)$, and $\pi/6 > \pi/60$; therefore

$$\frac{\sin \pi/60}{\pi/60} > \frac{\sin \pi/6}{\pi/6}$$

which reduces to

$$\sin \frac{\pi}{60} > \frac{\sin(\pi/6)}{10} = \frac{1}{20}$$

since $\sin(\pi/6) = \frac{1}{2}$; on the other hand

$$\frac{\tan(\pi/6)}{\pi/6} > \frac{\tan(\pi/60)}{\pi/60}$$

and $\tan(\pi/6)/10 > \tan(\pi/60)$. Now $\tan(\pi/6) = 1/\sqrt{3} < 1/1.7$; we have found $\frac{10}{17} \cdot \frac{1}{10} = \frac{1}{17} > \sin 3° > \frac{1}{20}$. Using a similar but more complicated method, Aristarchus was able to improve these estimates to find the inequalities

$$\tfrac{1}{18} > \sin 3° > \tfrac{1}{20}$$

Substituting these inequalities into (4.7) leads to

$$(4.14) \qquad \frac{54}{19} < \frac{R_e}{R_m} < \frac{20}{7}$$

which estimates the radius of the moon in terms of the radius of the earth. Aristarchus' more careful trigonometrical estimates led to a slightly better result, but what we have done conveys the flavor of his work.

Now to determine the distance from earth to moon! From Figure 4.8 and the second physical measurement (4.8) we see that if we can estimate $\sin 1°$ then we can combine (4.8) with the estimate of the radius of the moon (4.14) to get the desired distance. At the beginning of this section, in (4.3), bounds for $\sin 1°$ were indeed obtained. They imply that

$$(4.15) \qquad \frac{1}{60} < \frac{R_m}{D_{em}} < \frac{1}{45}$$

Multiply this chain of inequalities by that given in (4.14) and get

$$(4.16) \qquad \frac{54}{(19) \cdot (60)} < \frac{R_e}{D_{em}} < \frac{20}{(7) \cdot (45)}$$

The polar radius of the earth is, as has been already remarked, about 3950 miles, the radius of the moon is 1080 miles, and the distance between them is approximately 238,857 miles. Aristarchus' inequality, derived in (4.14),

$$\frac{54}{10} < \frac{R_e}{R_m} < \frac{20}{7}$$

implies that, using the value for R_e,

$$1390 > R_m > 1382$$

If his observations had been accurate, his result would have been correct within less than 10 miles! Similarly, from the *correct* value of R_m and Aristarchus' inequalities (4.15),

$$\frac{1}{30} < \frac{2R_m}{D_{em}} < \frac{2}{45}$$

it follows that $64,800 > D_{em} > 48,600$ — wrong by about a factor of 4. Even had his observations been accurate, the bounds imposed by the inequalities are not nearly so fine as those for the moon's radius.

These defects do not in any way detract from the beauty of the method or from its significance for the future; for the first time man had a method of estimating the size of the solar system he inhabits free from subjective opinion or mysticism and permitting endless refinement of accuracy. Aristarchus established a tradition that led Einstein and his followers in our own age to still grander conceptions of the universe and to the mathematical methods for estimating its size and shape (see Chapter 12).

§**4.4.**   You have noticed by now that the ancients always calculated the *circumference* of the earth but not its radius. The reason is this: the arc C of a circle of radius R, which is subtended by a central angle θ, has length $C = R\theta$; to estimate the total circumference it is necessary only to calculate the length of some given arc and the *fraction* of the total circle to which it corresponds. But given the circumference, the radius is obtained from the relation $C = 2\pi R$; therefore π must be known.

The Egyptians customarily used the estimate $\pi \simeq 3$, which would lead to an overestimate of the earth's radius by about 182 miles. They had the more accurate value $4(8/9)^2 = 3.160496 \ldots$, which, since $\pi = 3.14159 \ldots$, is 0.01890 too large. The Babylonians also generally took $\pi \simeq 3$, but the better estimate $\pi \simeq 3\frac{1}{8} = 3.125$ appears on an Old Babylonian tablet discovered at Susa, about 200 miles west of Babylon, in 1936. This estimate is 0.01659 too small, slightly closer to the true value of π than the best known Egyptian value. Its importance has to do not with its precision but rather with its method of derivation, which the Susa tablets imply.

One tablet contains a list of coefficients relating the area and circumference of the equilateral triangle, square, and the regular pentagon, hexagon, heptagon, and circle. Denoting the side and area of the regular n-gon by S_n and A_n, respectively, the relations

$$A_5 = 1;40 \, S_5^2$$
$$A_6 = 2;37,30 \, S_6^2$$
$$A_7 = 3;41 \, S_7^2$$

are given on the tablet. The equality for the hexagon implies the estimate $\sqrt{3} = 1;45 = 1.75$, since $A_6 = (3\sqrt{3})/2S_6^2$ is the true relation between side and area. It is worthwhile to note that $1;45$ is the third approximation to $\sqrt{3}$ in the standard Babylonian sequence $a_1 = 1$, $a_2 = \frac{1}{2}[1 + (3/1)] = 2$, $a_3 = \frac{1}{2}(2 + 3/2) = 1.75 = 1;45$ for estimating square roots. The famous astronomer Claudius Ptolemy (*fl.* about $+150$)[2] used the approximation $1;43,55,23 = 1.7320509$ for $\sqrt{3}$; the true value is 1.7320508 correct to eight significant figures. Ptolemy's estimate occurs in the standard sequence: we continue from $a_3 = 1;45 = \frac{7}{4}$ to find

$$a_4 = \frac{1}{2}\left(\frac{7}{4} + \frac{3 \cdot 4}{7}\right) = \frac{97}{56} = 1;43,55,42\,\frac{6}{7}$$

and

$$a_5 = \frac{1}{2}\left(\frac{97}{56} + \frac{3 \cdot 56}{97}\right) = \frac{18817}{10864} = 1\,\frac{7953}{10864}$$

Now

$$1\,\frac{7953}{10864} = 1;43,55,22\,\frac{662}{679}$$

The remainder $662/679$ is much closer to 1 than previous remainders (which were $\frac{6}{7}$, $\frac{5}{7}$, etc.), thus justifying rounding the last expression upward to 23, which is Ptolemy's expression.

The Susa tablet continues by giving an equation relating the circumference C_6 of the regular hexagon to the circumference C of the circumscribed circle, which is

$$C_6 = 0;57,36C$$

If the side of the hexagon is taken to be 1, then $C_6 = 6$; substitution of $C = 2\pi$ in the Susa relation produces the estimate

$$\pi \simeq \frac{3}{0;57,36} \simeq 3;7,30 = 3\frac{1}{8}$$

We do not know in detail how this excellent approximation was obtained. The significance of the result is that it probably is a special instance of a general and uniform procedure of comparing regular polygons with their circumscribing circles. The greater the number of sides of the polygon, the better it approximates its circumscribed circle in perimeter and area and the better the estimates of π that it determines. No general and progressively more accurate Egyptian techniques for estimating important constants are known; on the other hand, these Babylonian

---

[2] *See Chapters 5 and 6 for a further discussion of Ptolemy's work.*

methods are modern in spirit, with but one exception (bounding the errors), to which we now turn.

§**4.5.** Before the time of the Greeks there was little concern whether an estimate of some important number was too large or too small. For instance although the Babylonians knew how to obtain increasingly better approximations to square roots, they did not attempt to investigate their inherent errors and as a consequence they could not estimate the cumulative effect the residual errors of estimation might have in the course of a complicated calculation.

For instance, by eliminating $\sqrt{2}$ from the denominator we find

$$\frac{1}{\frac{17}{12} - \sqrt{2}} = 144\left(\frac{17}{12} + \sqrt{2}\right)$$

The standard Babylonian estimate is $\sqrt{2} \simeq 1;25 = \frac{17}{12}$; its substitution in the left-hand side produces a zero denominator, whereas substitution in the equal expression on the right yields the integer 408, which differs from the true value of $144\left(\frac{17}{12} + \sqrt{2}\right)$ by less than 1. It is obvious that the *order* of calculation is important when approximations are used. This is a critically important problem today because computers can store only approximations of most numbers, and they perform so many calculations in a second that it is often impossible to foresee the order in which each operation will occur in the calculation.

The technique that we use today to provide some assurance of the degree of accuracy of the result of a computer calculation is a more sophisticated version of Aristarchus' method in his astronomical calculations. The estimated quantities, like sin 1° or $\sqrt{2}$, are given *with bounds;* thus

$$\tfrac{1}{45} > \sin 1° > \tfrac{1}{60}$$

and these bounds are carried through the calculation to provide bounds for the accuracy of the final result. If, for example, we had noted that

$$\tfrac{14}{10} < \sqrt{2} < \tfrac{17}{12} = 1.41666 \ldots$$

then we could have calculated that

$$\tfrac{1}{60} = \tfrac{17}{12} - \tfrac{14}{10} > \tfrac{17}{12} - \sqrt{2} > 0$$

Therefore

$$60 = \frac{1}{1/60} < \frac{1}{(17/12) - \sqrt{2}} < \frac{1}{0} = \infty$$

This shows quite clearly the amount of error the approximation $\sqrt{2} \simeq 17/12$ permits in this calculation; it is too much. Applying the same technique to the equivalent expression $144\left(\frac{17}{12} + \sqrt{2}\right)$, we find

$$405\tfrac{3}{5} = 144\left(\tfrac{17}{12} + \tfrac{14}{10}\right) < 144\left(\tfrac{17}{12} + \sqrt{2}\right) < 144\left(\tfrac{17}{12} + \tfrac{17}{12}\right) = 408$$

which shows how far from the true value our calculation can be: not very far, this time.

The introduction of bounding inequalities for estimating the accuracy of calculations is an important Greek idea that is of even more significance today than it had ever been for them. It is not known who first used such techniques, but in the work of Aristarchus it is already well developed, and in that of his great younger contemporary Archimedes (−287 to −212) its use reached an artistic peak. Not only was Archimedes able to obtain the fine estimate

$$(3.1408 \ldots <)3\tfrac{10}{71} < \pi < 3\tfrac{1}{7}(<3.1429)$$

but he also anticipated the fundamental technique of the integral calculus, which was finally tamed by Newton and Leibniz to become the usually docile but still most powerful tool of modern mathematics.

§4.6.   Archimedes estimated π by generalizing the method of the Susa tablets to the regular polygon of 96 sides. It is worthwhile to reproduce his argument; recall that all of his calculations were carried out in the cumbersome Greek numerical notation. Let OA be a radius of a circle, O its center, and AC the tangent at A. Construct AOC so that it is one-third of a right angle ($\pi/6$). This is easily accomplished by constructing an equilateral triangle with side AO and bisecting the angle at O. Then, according to Pythagoras (Figure 4.10),

*(4.17) $$\frac{OA}{AC} = \sqrt{3}\left(=\frac{1}{\tan \pi/6}\right) > \frac{265}{153}$$

FIGURE 4.10

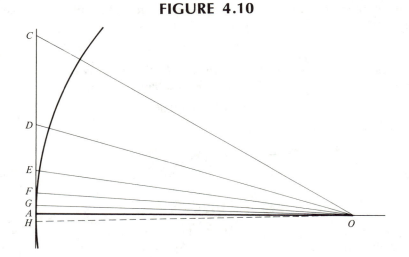

Since AOC is $\pi/6$, AC must be half the side of a regular hexagon circumscribed about the circle. The circumference of this hexagon is $12(AC)$ which is greater than the circumference $2\pi(OA)$ of the circle; that is, $12(AC) > 2\pi(OA)$. Therefore

$$\pi < 6\left(\frac{AC}{OA}\right)$$

Using Archimedes' inequality (4.17), this means that

$$\pi < 6\left(\tfrac{153}{265}\right) = 3\tfrac{123}{265} = 3.464 \ldots$$

Thus the simple geometric inequality (4.17) immediately gives an upper bound for π. (Each of the inequalities that provides information is indicated by an asterisk in the discussion that follows.)

Archimedes does not say how he found the inequality $\sqrt{3} > 265/153$, but we can easily check that he is correct because $3 = (\sqrt{3})^2 > (\tfrac{265}{153})^2$ if and only if $3(153)^2 > (265)^2$; in fact, this inequality simplifies to $70227 > 70225$, providing quite an accurate lower bound for $\sqrt{3}$.

We also see that

(4.18)
$$\frac{OC}{AC} = 2\left(=\frac{1}{\sin \pi/6}\right) = \frac{306}{153}$$

Now draw OD to bisect angle AOC and consider Archimedes' next step in his approximation of π. From Figure 4.11, it is clear that triangles AOD and COD' are similar; therefore $CO/OA = CD'/AD$, but triangle $D'CD$ is isoceles and $CD' = CD$. Then $CO/OA = CD/AD$, $CO/OA + 1 = (CD + DA)/AD$, and $(CO + DA)/OA = CA/AD$; consequently

$$\frac{CO + OA}{CA} = \frac{OA}{AD}$$

Using the inequalities (4.17) and (4.18),

*(4.19)
$$\frac{571}{153} = \frac{265}{153} + \frac{306}{153} < \frac{OA}{AC} + \frac{OC}{AC} = \frac{OA}{AD}$$

FIGURE 4.11

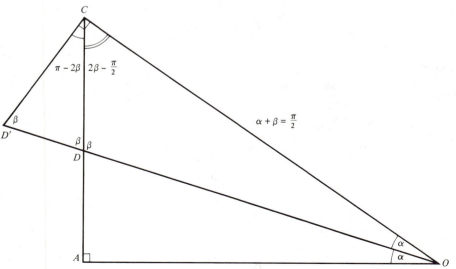

which implies an upper bound for π by comparing the circumference of the regular circumscribed 12-sided polygon with $2\pi(OA)$. For we know (as before) that the length of the circle of radius OA is less than the length of the circumscribing 12-sided regular polygon, each side of which has length $2 \cdot AD$; that is

$$2\pi \cdot AD < 24 \cdot AD$$

or

$$\pi < 12 \cdot \frac{AD}{OA}$$

which, by (4.19), gives

$$\pi < 12 \cdot \tfrac{153}{571} = 3\tfrac{123}{571} = 3.215 \ldots$$

Note that this is a rigorous upper bound for π which is an improvement on the upper bound $3\tfrac{123}{265}$ obtained before.

Let us proceed directly to a better approximation. Arguing as before, according to Pythagoras,

$$\frac{(OD)^2}{(AD)^2} = \frac{(OA)^2 + (AD)^2}{(AD)^2} > \left(\frac{571}{153}\right)^2 + 1 > \frac{349450}{23409}$$

Extracting the square root gives

(4.20)
$$\frac{OD}{AD} > \frac{591\tfrac{1}{8}}{153}.$$

Next draw OE to bisect angle AOD. Using the same argument as before and Figure 4.10,

$$\frac{OD}{OA} = \frac{DE}{AD}$$

whence

$$\frac{OD + OA}{AD} = \frac{OA}{AE}$$

Therefore

*(4.21)
$$\frac{OA}{AE} = \frac{OD}{AD} + \frac{OA}{AD} > \frac{591\tfrac{1}{8}}{153} + \frac{571}{153} > \frac{1162\tfrac{1}{8}}{153}$$

by (4.20) and (4.19). Arguing as before we have

$$\frac{(OE)^2}{(EA)^2} > \left(\frac{1162\tfrac{1}{8}}{153}\right)^2 + 1$$

$$> \frac{1373943\tfrac{33}{64}}{23409}$$

and extracting the square root produces

$$\frac{OE}{EA} > \frac{1172\tfrac{1}{8}}{153}$$

Bisect angle AOE with OF and reason in the same way to find

*(4.22)
$$\frac{OA}{AF} > \frac{1162\frac{1}{8}}{153} + \frac{1172\frac{1}{8}}{153} = \frac{2334\frac{1}{4}}{153}$$

which implies a still better approximation. Continuing,

$$\frac{(OF)^2}{(AF)^2} > \left(\frac{2334\frac{1}{4}}{153}\right)^2 + 1$$

$$> \frac{5472132\frac{1}{16}}{23409}$$

and therefore

$$\frac{OF}{AF} > \frac{2339\frac{1}{4}}{153}$$

Finally, let OG bisect angle AOF and in the same way find

*(4.23)
$$\frac{OA}{AG} > \frac{2334\frac{1}{4}}{153} + \frac{2339\frac{1}{4}}{153} = \frac{4673\frac{1}{2}}{153}$$

The original angle OAC was one-third of a right angle; therefore after four bisections angle AOG must be $\pi/96$. Now construct angle AOH equal to AOG but falling on the side of OA opposite AOG. Then angle GOH is $\pi/48$ and GH is one side of a regular polygon of 96 sides, circumscribed to the circle of radius OA. Then, if P denotes the perimeter of this polygon,

(4.24)
$$\frac{2(OA)}{P} = \frac{2(OA)}{96(GH)} = \frac{A}{96(AG)} > \frac{1}{96}\left(\frac{4673\frac{1}{2}}{153}\right) = \frac{4673\frac{1}{2}}{14688}$$

by (4.23). However,

$$\frac{14688}{4673\frac{1}{2}} = 3 + \frac{667\frac{1}{2}}{4673\frac{1}{2}} < 3 + \frac{667\frac{1}{2}}{4672\frac{1}{2}} = 3\frac{1}{7}$$

hence

(4.25)
$$\frac{2(OA)}{P} > 3\frac{1}{7}$$

P is greater than the circumference of the circle of radius OA; that is, $P > 2\pi(OA)$. Therefore $1/\pi > 2(OA)/P$, and (4.25) is the same as

*(4.26)
$$\pi < 3\frac{1}{7}$$

Archimedes makes a similar argument for a regular polygon of 96 sides *inscribed* in the circle of radius OA and concludes that its perimeter, say Q, satisfies the inequality

$$\frac{Q}{2(OA)} > \frac{6336}{2017\frac{1}{4}} > 3\frac{10}{71}$$

so that $3\frac{10}{71} < \pi$. Combining this with (4.26) gives his famous result

*(4.27)
$$3\frac{10}{71} < \pi < 3\frac{1}{7}$$

Although Archimedes' method is capable of extension to provide estimates of π as accurate as one pleases, it is apparent that the calculations are awkward and extensive. In fact, the utility of Archimedes' bounds derives from their simplicity, but he provides no hint about how he found the convenient bound $\frac{265}{153} > \sqrt{3}$, nor does he say (although historians have reconstructed his probable procedure) how he replaces the left side by the right in the inequality

$$\frac{6336}{2017\frac{1}{4}} > 3\frac{10}{71}$$

and so on. These arithmetical simplifications do not follow a regular procedure and therefore limit the *practical* generality of his procedure for estimating π.

§4.7.   There have been four distinct historical stages in ascertaining the value of π.

The early Egyptians and Babylonians obtained the value $\pi \simeq 3$ by guesswork, by measurement, or perhaps by comparing the perimeter or area of the circle with that of the inscribed regular hexagon.

The Susa tablets represent, and are the only known example of, the second stage. They indicate familiarity with a general procedure which in principle permits estimates as accurate as required.

Archimedes initiates the third stage, which lasted for more than 1500 years. It is characterized by complicated but general geometrical procedures that produce upper and lower *bounds* for π, thus providing automatic estimates of error as well as the required value.

The fourth, and considerably more recent, stage of development is based on the discoveries of the nongeometrical role played by π. This led to the discovery of *infinite series* representations of π, such as Gregory's series (James Gregory, 1638–1675)

$$\frac{\pi}{4} = 1 - \tfrac{1}{3} + \tfrac{1}{5} - \tfrac{1}{7} + \tfrac{1}{9} - \tfrac{1}{11} + \cdots ;$$

the finite partial sums of the series approach $\pi/4$ with increasing accuracy as more terms of the series are taken in the sum. Figure 10.4 shows how this happens. Here no new constructions or special abilities are required to improve a given estimate — just an enormous amount of arithmetic labor, suited more to mechanical than human calculation.

Gregory's series *converges* slowly; the first few sums give the estimates:

$$
\begin{aligned}
\pi \simeq\ &4 \\
&2.66666\ \ldots \\
&3.46666\ \ldots \\
&2.89524\ \ldots \\
&3.33968\ \ldots \\
&2.97608\ \ldots
\end{aligned}
$$

Notice that not even the first six terms suffice for the ancient approximation

$\pi \simeq 3$! This may place Archimedes' accomplishment in perspective for you. But think of making *these* calculations in the Greek or Egyptian systems of notation. Compared with them, Archimedes' method is perhaps not so complex after all.

Other series express π, many of which converge more rapidly; for instance,

$$\frac{\pi^2}{6} = 1 + \frac{1}{2^2} + \frac{1}{3^2} + \frac{1}{4^2} + \cdots$$

the first six terms of which lead to $\pi > 2.989 \ldots$. We are sure about the inequality because all of the summands of the series are positive. This series is not rapidly convergent.

The series

$$\frac{\pi^4}{90} = 1 + \frac{1}{2^4} + \frac{1}{3^4} + \frac{1}{4^4} + \cdots$$

converges more rapidly. In fact, just taking the first term implies $\pi^4 \sim 90$; therefore $\pi^2 \sim 3\sqrt{10}$ and $\pi \sim \sqrt{3}\sqrt{\sqrt{10}}$. Taking $\sqrt{3} \sim 1.732$ and $\sqrt{\sqrt{10}} \simeq \sqrt{3.1622} \simeq 1.778$ implies $\pi \simeq 3.079 \ldots$, not bad for one term.[3] A more accurate value of $\sqrt{3}$ leads to $\pi \simeq 3.080$, whereas two terms of the series yield the approximation $\pi \simeq (90 \cdot 17/16)^{1/4} = (9.625)^{1/4} \simeq 3.127$ and three terms yield $\pi \simeq 3.136$.

§**4.8.** Many problems in science and engineering and even in economics and the social sciences can ultimately be reduced to finding the area enclosed by a curve in the plane. This is the basic problem of the part of calculus called *integration theory*.

The first steps toward the development of a successful theory of integration were taken by Archimedes and his followers. They based their *techniques* on the methods of bounding estimates used by Aristarchus and later by Archimedes himself in his estimation of π. These techniques ultimately led, in the seventeenth and eighteenth centuries, to the enormously powerful and successful tool that the calculus is today.

The first paper to display the germ of integration theory is Archimedes' work on finding the area bounded by straight lines and a parabola. The typical case (expressed in current notation) is this: let $y = 1 - x^2$ be the equation of a parabola and consider the area bounded by it and that portion of the x-axis between -1 and $+1$, which is shaded in Figure 4.12. Partition the x-axis into equal segments marked by the points $0, \pm 1/N, \pm 2/N, \ldots, \pm k/N, \ldots, \pm 1$ and erect a rectangle R_k on the segment from k/N to $(k+1)/N$, of height $y_k = 1 - (k/N)^2 = y(k/N)$, for $k \geqslant 0$ (Figure 4.13). Then the area of R_k is

$$A(R_k) = \frac{1}{N}\left(1 - \frac{k^2}{N^2}\right).$$

---

[3] *These series are derived from the knowledge of a special function called the* Riemann zeta function; *see for example, Whittaker and Watson* [79].

FIGURE 4.12

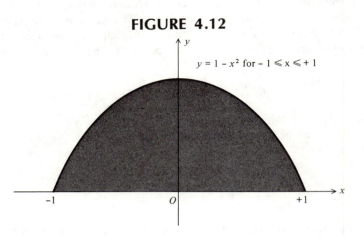

$y = 1 - x^2$ for $-1 \leqslant x \leqslant +1$

It is extremely useful at this point to introduce a new kind of notation to make the representation of complicated sums simpler. Suppose $x_1, x_2, \ldots, x_n$ are numbers. Abbreviate the sum of these numbers by

$$\sum_{i=1}^{n} x_i$$

that is, by definition,

$$\sum_{i=1}^{n} x_i = x_1 + x_2 + \cdots + x_n$$

For instance, if $x_k = k$, then $\displaystyle\sum_{k=1}^{n} x_k = \sum_{k=1}^{n} k = 1 + 2 + \cdots + n$. With this no-

FIGURE. 4.13

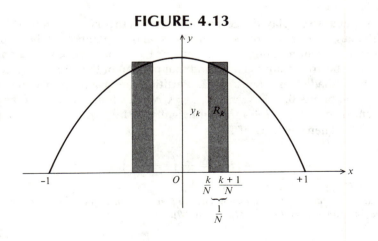

tation available we can write in an analogous way

$$A(R_0) + A(R_1) + \cdots + A(R_N) = \sum_{k=0}^{N} A(R_k)$$

Now $2[A(R_0) + A(R_1) + \cdots + A(R_N)] = 2\sum_{k=0}^{N} \frac{1}{N}\left(1 - \frac{k^2}{N^2}\right)$ is greater than the hatched area A in Figure 4.12; that is,

$$2\sum_{k=0}^{N} \frac{1}{N} - \frac{2}{N^3}\sum_{k=0}^{N} k^2 > A$$

which is just the same as

$$\frac{2(N+1)}{N} - \frac{2}{N^3}\sum_{k=1}^{N} k^2 > A$$

Now[4]

$$1^2 + 2^2 + 3^2 + \cdots + N^2 = \frac{N(N+1)(2N+1)}{6}$$

Then

$$\frac{2(N+1)}{N} - \frac{2}{N^3}\left[\frac{N(N+1)(2N+1)}{6}\right] > A$$

that is, after simplifying, for any positive integer N,

(4.28) $$\frac{4}{3} + \frac{1}{N} - \frac{1}{3N^2} > A$$

Let us hold this result in the background for a moment.

In a similar way the rectangle S_k with base k/N to $(k+1)/N$ and height $y[(k+1)/N] = 1 - [(k+1)/N]^2$, as in Figure 4.14, lies inside the area A and has

FIGURE 4.14

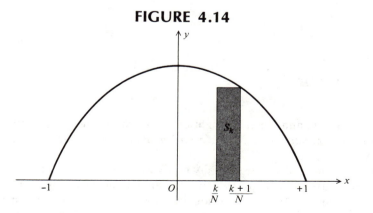

---

[4] *This can be proved by* mathematical induction.

area $A(S_k) = \dfrac{1}{N} (1 - [(k+1)/N]^2)$; therefore

$$A > 2[A(S_0) + A(S_1) + \cdots + A(S_{N-1})]$$

that is

$$A > 2 \sum_{k=0}^{N-1} \frac{1}{N} - 2 \sum_{k=0}^{N-1} \frac{(k+1)^2}{N^3}$$

This simplifies to

$$A > 2 - \frac{2}{N^3} (1^2 + 2^2 + \cdots + N^2)$$

So

$$A > 2 - \frac{2}{N^3} \left[\frac{N(N+1)(2N+1)}{6} \right]$$

Then

(4.29) $$A > \frac{4}{3} - \frac{1}{N} - \frac{1}{3N^2}$$

Comparing (4.28) and (4.29), we have the bounds

(4.30) $$\frac{4}{3} + \frac{1}{N} - \frac{1}{3N^2} > A > \frac{4}{3} - \frac{1}{N} - \frac{1}{3N^2}$$

N is an arbitrary positive integer. If it is very large, then $1/N - 1/3N^2$ and $1/N + 1/3N^2$ are very small and the approximation will look like

$$\tfrac{4}{3} + \text{small} > A > \tfrac{4}{3} - \text{small}$$

If we choose any small number $\epsilon > 0$, we can find an N so large that

$$\frac{1}{N} - \frac{1}{3N^2} < \epsilon$$

and

$$\frac{1}{N} + \frac{1}{3N^2} < \epsilon$$

Then we could write

(4.31) $$\tfrac{4}{3} + \epsilon > A > \tfrac{4}{3} - \epsilon$$

for *any* small $\epsilon > 0$, and these inequalities mean that A must be exactly $\tfrac{4}{3}$. Thus we have found the *exact* value of the area by using inequalities in a delicate and careful way.

Archimedes' method was essentially equivalent to ours but, because it was not known how to use coordinate systems for analytical purposes until the time of Descartes (1596–1650), he could not describe the rectangles drawn in Figures 4.13 and 4.14 in a simple way. Therefore, he worked with triangles (see Figure 4.15) and, using properties of the parabola, showed that each of the triangles ADC

FIGURE 4.15

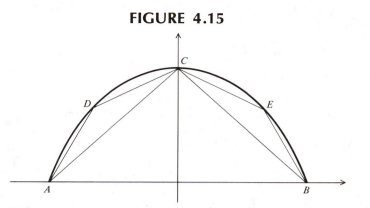

and BCE has an area equal to one-eighth of the area of ABC. He then constructed four triangles with vertices on the parabola and bases AD, DC, CE, and EB and showed that each has an area equal to one-eighth of the area of ADC. Continuing in this way, he found that if the area of ABC is $\triangle$ then the area A of the parabola satisfies

$$A > \triangle + 2\triangle \left(\frac{1}{8}\right) + 4\triangle \left(\frac{1}{8}\right)^2 + 8\triangle \left(\frac{1}{8}\right)^3 + \cdots + 2^N\triangle \left(\frac{1}{8}\right)^N \text{ for any } N$$

This series is the same as the geometric series

$$\triangle \left(1 + \frac{1}{4} + \frac{1}{4^2} + \frac{1}{4^3} + \cdots + \frac{1}{4^N}\right)$$

whose sum is

$$= \triangle \left(\tfrac{4}{3} - \tfrac{1}{3}\left(\tfrac{1}{4}\right)^N\right)$$

Now $\triangle = 1$ so $A > \tfrac{4}{3} - \tfrac{1}{3}(\tfrac{1}{4})^N$. As N is chosen larger and larger, $\tfrac{1}{3}(\tfrac{1}{4})^N$ becomes smaller and smaller; therefore we conclude that

$$A \geqslant \tfrac{4}{3}$$

In a way that would not be satisfactory today Archimedes proves that $\tfrac{4}{3} \geqslant A$ and, necessarily, $\tfrac{4}{3} = A$, the correct result.

Note that his method requires the summation of only the geometric series

$$\sum_{k=0}^{N} \left(\frac{1}{4}\right)^k$$

whereas the modern technique, which uses rectangles to approximate the parabola, requires sums of the form

$$\sum_{k=0}^{N} 1, \ \sum_{k=0}^{N} k, \ \sum_{k=0}^{N} k^2$$

the first two of which are easy to sum; the third is more difficult, but the results were essentially known in Archimedes' time. For instance, in his paper *On Spirals* he has the inequalities

$$(n-1)n + \frac{(n-1)^3}{n} + 1 < 1^2 + 2^2 + \cdots + n^2$$

and

$$1^2 + 2^2 + \cdots + (n-1)^2 < n(n-1) + \frac{(n-1)^3}{3}$$

(Equivalently, $1^2 + 2^2 + \cdots + n^2 < n(n+1) + n^3/3$.) Now

$$\frac{n(n+1)(2n+1)}{6} = \frac{n^3}{3} + \frac{n^2}{2} + \frac{n}{6}$$

therefore his inequalities could have been used if the use of rectangular coordinates had been understood.

§4.9. Filling out the area under the curve in Figure 4.12 with an increasing family of rectangles (or triangles, as done by Archimedes) is an example of the *method of exhaustion*. This principle was first employed by Eudoxus (*ca.* -408 to -355) and later developed into a precise tool by Archimedes in his studies of the area under a parabola and the volumes of various geometric objects such as the solid ball. Archimedes considered his greatest achievement to have been his discovery by the method of exhaustion that the volume of a ball is two-thirds of the volume of the circumscribed cylinder, and a representation of this geometric relation is said to have been inscribed on his gravestone at his request.

Figure 4.16 is a page of a handwritten Greek manuscript,[5] and Figure 4.17 is a printed version of the same page. Figure 4.18 is the English translation (the translation and printed text begin with the *second* line of the handwritten manuscript). This manuscript was copied sometime in the twelfth century and represents a facsimile of older no longer extant manuscripts. Almost all our knowledge of Greek mathematics comes from similar copies. The dry air of the Nile valley and the durable nature of the Babylonian tablets left original "manuscripts" for the scholar to study, but for information about Greek mathematics we are dependent in large degree on sources that are copies of documents that are often not originals themselves. Moreover, each copyist introduces manuscript changes which reflect his "style" in addition to errors of commission and omission.

A few things are worth noting on this particular page of Greek mathematical text "On the cylinder". First, the author was aware of Archimedes' work; therefore it was written later than *ca.* -257. At the end of the fourth paragraph we see that the post-Archimedian Greeks still used Egyptian unit fractions.[6] The

---

[5] *Folio number 26r from the* Codex Constantinopolitanus, *edited by Bruins* [77].

[6] *It appears that the Greeks were introduced to geometry by the Egyptians. Thales (ca. −636 to −546) is said to have studied geometry in Egypt and was the first to have set up a deductive system of proving abstract propositions regarding geometric objects such as lines, angles, and triangles. This had a profound influence on later Greek mathematics and philosophy and indeed all of mathematical history; that is, the concept of proof and logical deduction, so important in mathematics and science today, goes back to Thales.*

FIGURE 4.16

Greek manuscript describing work of Archimedes: in *Codex Constantinopolitanus* [77].

FIGURE 4.17

Δέδειχεν δέ Ἀρχιμήδης ὅτι κύλινδρος ὁ περιλαμβάνων τὴν σφαῖραν ἡμιόλιός ἐστι τῆς σφαίρας. εἰ οὖν ἥμιόν πρόσθενα τρίτον ἀφαίρενα, ἀφαιρῶ οὖν τοῦ κυλίνδρου, ὅ ἐστιν ἐπιφάνεια τῆς σφαίρας, τῶν $\bar{ν}$ καὶ $\bar{β}$ ἑβδόμων, τὸ γ'. καταλείπεται $\overline{λγ}$ γ' ζ' κα'. τοσούτων τὸ στερεὸν τῆς σφαίρας.

Ἐὰν δὲ τὸ β' λάβωμεν τῶν $\bar{ν}$ καὶ $\bar{β}$ ἑβδόμων ὁμοίως γίνονται $\overline{λγ}$ γ' ζ' κα'. καὶ ἔσται ἄρα ἡ μὲν ἐπιφάνεια τῆς σφαίρας ποδῶν $\bar{ν}$ καὶ $\bar{β}$ ἑβδόμων, τὸ δὲ στερεὸν ποδῶν $\overline{λγ}$ < γ' ζ' κα'. >

Καὶ ἔστω σφαίρας ἡ περίμετρος ποδῶν $\overline{ιη}$. εὑρεῖν αὐτῆς τὸ στερεόν. ποιῶ οὕτως· ἐπὶ τῶν κύκλων· τῶν $\overline{ιη}$ ἐπὶ τὰ $\bar{ζ}$· γίνονται $\overline{ρκς}$. τούτων τὸ κβ'· γίνονται $\bar{ε}$ καὶ ἑνδέκατα $\bar{η}$. ταῦτα $\overline{ια}$· γίνονται $\overline{ξγ}$. ταῦτα κύβισον· γίνονται $\underline{κε}$ καὶ $\overline{μζ}$. < . . . . > ταῦτα μέριζε παρὰ τὰ, β $\overline{φμα}$· γίνονται $\overline{?η}$ δ'ια' λγ' μα' ρκα' τξγ'.

Ἕτερον σφαῖραν εἰς μέρη $\bar{δ}$. καὶ εὑρέθη τὸ ἓν τμῆμα ἐξ ἀμφοτέρων τῶν μέρων ἀνὰ ποδῶν $\bar{ζ}$. εὑρεῖν τὸ στερεόν. ποιῶ οὕτως· κυβίζω τὰ $\bar{ζ}$· γίνονται $\overline{τμγ}$. ταῦτα δίς· γίνονται $\overline{χπς}$. ταῦτα $\overline{ια}$· γίνονται $_{\prime}$ζ $\overline{φμς}$. τούτων τὸ κα'· γίνονται $\overline{τνθ}$ γ'. τοσούτων ποδῶν τὸ στερεὸν τοῦ τμήματος. ‖

Transliteration of Figure 4.16.   117

FIGURE 4.18.

Archimed has shown that the cylinder which includes the sphere is one and a half times the sphere. If now half [of a quantity to form $1\frac{1}{2}$] is added a third [of the resulting number] is subtracted.

I take now away a third of the cylinder, which is [numerically] the surface of the sphere, of the 50 and two sevenths, one third. There is left $33 \frac{1}{3} \frac{1}{7} \frac{1}{21}$. So much is the volume of the sphere.

And if we take $\frac{2}{3}$ of the $50\frac{2}{7}$ likewise there results $33 \frac{1}{3} \frac{1}{7} \frac{1}{21}$. And the area of the sphere shall be $50\frac{2}{7}$ feet; and the volume $33 < \frac{1}{3} \frac{1}{7} \frac{1}{21} >$.

And let the perimeter of the sphere be 18-feet. To find its volume. I operate thus; for the circles of 18 [in-perimeter] into 7, result 126; a 22-nd of these, result $5\frac{8}{11}$[the diameter of the sphere]. These into 11, result 63. Cube these, result 250047. Divide these by 2541, result $98 \frac{1}{4} \frac{1}{14} \frac{1}{33} \frac{1}{44} \frac{1}{121} \frac{1}{363}$. [Should be $98 \frac{1029}{2541} = 98\frac{1}{3} \frac{1}{21} \frac{1}{77} \frac{1}{121} \frac{1}{363} = 98 \frac{49}{121}$.]

I did cut a sphere into four parts and one segment is found at both parts to be 7 feet.

To find its volume.

I operate thus: cube the 7, result 343; these two times, result 686; these 11 times, result 7546. The 21-st part of those becomes $359\frac{1}{3}$. So many feet shall be the volume of the segment.

Translation of Figure 4.16.

apostrophe after a number corresponded to the Egyptian to denote a unit fraction (note that the handwritten Greek has capital letters for numbers, but the numerical symbols in the printed text are in lower case). The English text in brackets are the editorial comments of the modern editor Bruins.

§**4.10.** The last two sections have shown how the technique of bounding errors by using inequalities can be adapted to find accurate estimates of the *length* of the circumference of a circle (hence of the value of π) and the *area* bounded by a parabola and a straight line. The same laborious techniques could be applied to find the "lengths" of other curves and the "areas" they enclose, but it turns out that the results are not always uniquely specified numbers and sometimes conflict with intuitive notions.

A more careful examination of this problem turns up some interesting results and startling examples and ultimately leads to a definition of area[7] that at once agrees with intuitive notions in all "reasonable" cases and shows, moreover, how

FIGURE 4.19

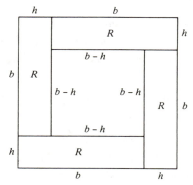

Area of rectangle expressed as difference of areas of squares.

[7] *And of the length of a curve; this subtle notion is more fully examined in §9.10.*

the value of the area of a given region can be estimated with as much precision as desired.

Let us first examine what we really know about area from elementary geometry. The area of a square of side s is, by *definition*, s^2.

If a rectangle R with base of length b and height h is given, then Figure 4.19 is a square composed of four copies of the rectangle R and a square of side $(b - h)$.[8] The area of the large square is $(b + h)^2$, whereas that of the small square is $(b - h)^2$; consequently four times the area of R equals

$$(b + h)^2 - (b - h)^2 = 4bh$$

Therefore the area of R is bh as we knew all along. Hence, knowledge of the area of squares entails knowledge of the area of rectangles.

If P is a parallelogram, the area of P is the same as the area of a rectangle constructed from P, as shown in Figure 4.20. The same idea shows that the area of a triangle is $bh/2$ because two copies of a triangle can be rearranged to form a parallelogram (See Figure 4.21).

FIGURE 4.20

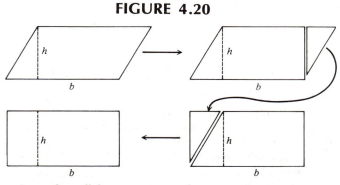

Area of parallelogram expressed as area of rectangle.

FIGURE 4.21

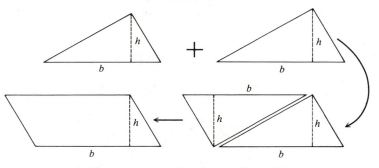

Area of triangle expressed as $\frac{1}{2}$ area of parallelogram.

[8] *If* h $>$ b, *interchange the letters* b *and* h *in the figure so that* b *becomes* h *and* h *becomes* b.

Any polygon can be subdivided into a collection of disjoint triangles, which means that knowledge of the formula for the area of a triangle is sufficient to determine the area of any polygon, although the actual calculation of the value of the area can be an onerous task. Figure 4.22 illustrates one possible decomposition of an 11-sided nondescript polygon into triangles.

FIGURE 4.22

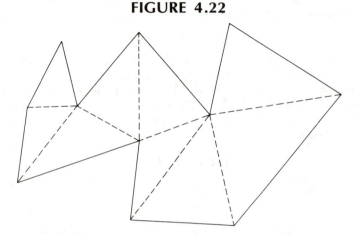

Next consider the "area" of a region enclosed by a curve such as a circle which does not consist entirely of segments of straight lines. It may turn out that the region can be ingeniously cut up and reassembled to form a rectangle, in which case the formulae derived above are applicable, the area of the region is well defined, and its value can be explicitly calculated, if only in principle. The majority of regions, however, including the interior of a circle, cannot be reassembled in such a way and therefore these formulae cannot be applied.

Archimedes' method contains a solution to this difficulty. Although it may not be possible to reassemble the region to form a rectangle, it may nevertheless be possible to put a rectangle entirely *inside* the region (see Figure 4.23), and in this case it seems sensible to assert that whatever the area A of the region bounded by the curve may be it is certainly at least as large as the area S_1 of the rectangle:

FIGURE 4.23

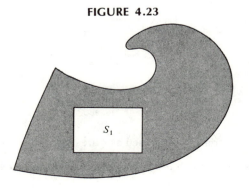

S_1

$$A \geqslant S_1$$

If S_1 does not exhaust all of the region A, it may be possible to fit other rectangles $S_2, S_3,...,S_n$ into A so that no two of the rectangles overlap (Figure 4.24); then certainly

(4.32) $$A \geqslant S_1 + S_2 + \cdots + S_n$$

FIGURE 4.24

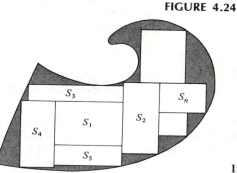

Inner approximation to a region.

As the size of the rectangles S_k is decreased, it becomes possible to approximate the irregular shape of the region A better by including more and more small rectangles (as we did in §4.8 to calculate the area bounded by the parabola and a particular line). One should therefore expect that the sum $S_1 + S_2 + \cdots + S_n$ can be made to approximate A as closely as desired, although (4.32) will, of course, always remain true.

In a similar way the region A can be approximated by a collection of disjoint rectangles which contains it; and for such an approximation (which corresponds to the "exterior" rectangles R_k in §4.8), the area A, whatever that means, should be no larger than the sum of the areas of the rectangles containing A (Figure 4.25); thus

(4.33) $$R_1 + R_2 + \cdots + R_n \geqslant A$$

FIGURE 4.25

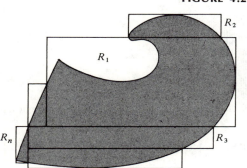

Outer approximation to a region.

Combination of (4.32) and (4.33) gives

$$R_1 + R_2 + \cdots + R_m \geqslant A \geqslant S_1 + S_2 + \cdots + S_n$$

Suppose that as sets of interior rectangles S_k which increasingly fill out the region A are chosen the number $S_1 + S_2 + \cdots + S_n$, which is the area of these rectangles, approaches evermore closely some real number S;[9] then certainly $A \geqslant S$. Also, it may turn out that as the rectangles R_k are shrunk so that they cover a smaller and smaller region but always completely cover A their total area $R_1 + R_2 + \cdots + R_m$ evermore closely approaches a number R; then certainly $R \geqslant A$. Combination of these observations leads to

(4.34) $$R \geqslant A \geqslant S$$

If $R = S$, then the area A of the region enclosed by the curve is trapped; it must equal R (and S). This statement can be used as the *definition* of area for regions enclosed by curved boundaries.

If the area of an expanding sequence of disjoint rectangles contained in *the region A approaches the area of a contracting sequence of disjoint rectangles* which *contain A, then A is said to possess an area, and the* value of its area *is the common value approached by the areas of the expanding interior and the contracting exterior sets of rectangles.*

The alert reader will observe that if A has an area, then its value is given by either the limiting value S of the area of the expanding interior rectangles or by the area R of the contracting exterior rectangles. Why bother with both? Will consideration of just one set of rectangles suffice? Unfortunately it will not; there are regions for which the "exterior area" R is actually larger than the "interior area" S; therefore the intuitive conception of area is contradicted and, according to the definition just set forth, such a region cannot be said to have an area in any meaningful sense.

FIGURE 4.26

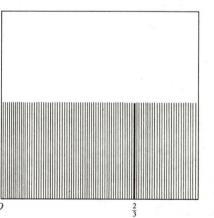

O $\frac{2}{3}$

The hairy square: a region for which area cannot be defined.

[9] *So that S is the* limit *of the sums* $S_1 + S_2 + \cdots + S_n$; *see Chapter 9.*

Consider, for example, a square of side 1. Draw lines of length $\frac{1}{2}$ within the square perpendicular to and touching the base at each point on the base whose distance from the vertex O (in Figure 4.26) is a rational number; the rational base point $\frac{2}{3}$ with the line segment erected above it is singled out in the figure as an example. Let us determine whether the interior H of this "hairy square" has an area (by *interior*, we mean all points inside the square but not lying on any of the "hair lines"). First calculate R, the limiting value of the area of decreasing rectangles which contain H. The square itself is a rectangle containing H, and obviously any rectangular chunk removed from the square will remove some points of H as well. Therefore the smallest value that can be taken by the contracting exterior rectangles is just the area of the original square of side 1; hence $R = 1$. In regard to the expanding interior rectangles, certainly a rectangle S_1 of base 1 and height $\frac{1}{2}$ will just fit into H above the hairs (Figure 4.27). This does not exhaust all of H because points between the hairs, lying above points on the base at an *irrational* distance from the vertex O (such as points between the hairs above $\sqrt{2}/2$), are not contained in that rectangle. But *no* rectangle lying entirely within H can contain such a point, for if such a point P were within the rectangle S_2 the base of S_2 must fall across a hair, since there are rational numbers as close as one pleases to P. Thus the largest possible area of interior expanding rectangles is the area of S_1, which is $\frac{1}{2}$. Consequently

$$1 = R \neq S = \tfrac{1}{2}$$

and H does *not* have an area.

FIGURE 4.27

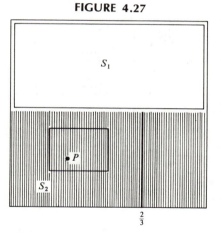

$$\tfrac{2}{3}$$

Other pathological examples are easily created. Consider a square U of side 1 successively subdivided and with a curve drawn as shown in the sequence of diagrams in Figure 4.28. Continue this process indefinitely and let C be the curve that results. After n repetitions of the subdivision process the small squares have side $(1/2)^n$. Since curve C runs through each small square, no point in U is more than $\sqrt{2}/2^n$ (the diagonal of the small squares at the nth stage) distant from the curve. As n gets large, that is, as the process of subdivision is continued, the distance of any point in the square U from the curve C becomes smaller than any positive number; in fact, according to the definition of real numbers in Chapter 1, it must be zero. Not every point of U lies *on* the curve C, however; for instance, the center of the square U is not on C. Let A be the set of all points inside U that are *not* on C.

FIGURE 4.28

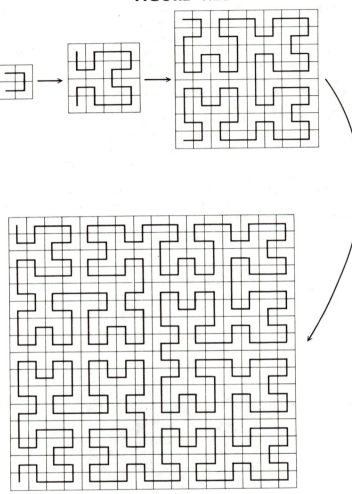

A curve which "fills" a square. The scale of successive squares has
been expanded by a factor of 2 to make the stages of construction
of the curve more apparent.

Just as we argued in regard to the "hairy square," we find that the contracting exterior
rectangles shrink down on U so that the limiting value of the external area approximation
must be the area of U, which is 1. On the other hand, since every point in U is at zero distance from C (although not every point need be on C!), there can be no rectangle larger
than a point inside U. Hence the value of the interior area approximation must be 0 and
$A = U - C$ does not have area.

Intuition is a poor guide in any consideration of the properties of all the real
numbers rather than just the rational numbers. This is perhaps nowhere better

illustrated than by the disconcerting examples of the area, or volume, of geometrical objects whose properties are not solely determined by finitistic "rational" descriptions. Archimedes' methods of calculating areas, refined and extended to provide an effective definition of this geometric concept, form an important bridge between the arithmetic of real numbers and the fundamental concepts of geometry. This bridge is completely open, and over it pass not only those properties of the real numbers that introduce precision and measure into geometry; the pathology and complexity of the real numbers pass over as well, thus demanding a sophistication of geometric intuition and enforcing an appreciation of the subtle web of contradiction that stands ever ready to envelop the unwary mathematical practitioner.

EXERCISES

4.1. What was the most important feature of Greek mathematics, beginning with Thales, not present in earlier cultures? Why is this feature important?

4.2. Aristarchus gave several estimates of physical and mathematical quantities. What is important about the *form* of his statements?

4.3. From Table 4 at the end of the book show that sin x/x is a decreasing function of x by calculating approximately its values for x corresponding to 1°, 10°, 20°, 30°, 40°, 50°, 60°, 70°, and 80°, where x is in *radian* measure.

4.4. Proceed as in Problem 4.3 to show that tan x/x is an *increasing* function.

4.5. Using the formula

$$\frac{\sin x}{\sin y} < \frac{x}{y} < \frac{\tan x}{\tan y} \quad \text{for} \quad 0 < y < x < \frac{\pi}{2}$$

show that sin $(\pi/3) < 0.94$. From a diagram find the exact value of sin $(\pi/3)$ and compare the two results.

4.6. Show that sin $(2y) < 2$ sin (y) and 2 tan $(y) <$ tan $(2y)$ for $0 < y < \pi/2$.

*4.7. Using a right triangle with one angle $= \pi/12$, find an "upper bound" for sin 1° that is better than Aristarchus' upper bound sin $1° < \frac{1}{45}$

4.8. Are Aristarchus' basic assumptions reasonable? Why? Give evidence for each.

4.9. Could the "size of the universe" have been reasonably estimated in early times if the earth had no moon?

4.10. Express the Egyptian estimates $\pi \approx 3$ and $\pi \approx 4(\frac{8}{9})^2$ in Egyptian notation, using unit fractions and Egyptian multiplication ($\pi \approx 3$, of course, is trivial).

4.11. Show that $(\sqrt{2} - 1)^{-1} = \sqrt{2} + 1$, and then compare the values of

$$(\tfrac{17}{12} - \sqrt{2})^{-1} \quad \text{and} \quad 144(\tfrac{17}{12} + \sqrt{2})$$

for the first three Babylonian approximations to $\sqrt{2}$, starting with $a_1 = 1$.

4.12. Compare the values of

$$(\sqrt{2} - 1)^{-1} \quad \text{and} \quad \sqrt{2} + 1$$

for the first three approximations, as in Problem 4.11.

4.13. Estimate the errors in (a) Problem 4.11 and (b) Problem 4.12 by providing *bounds* for the accuracy of $\sqrt{2}$.

4.14. Why is it important to provide bounds as in Problem 4.13?

4.15. Define the concept of *area* as carefully as you can.

4.16. Show how your definition applies to the area of (a) a square, (b) a triangle, and (c) a "hairy square."

*4.17. Is it just as easy to "define" the *length of a curve* (or "arc length") as it is to define area? (See Problem 4.15.)
    (a) If you answer "yes," define it and show that your definition applies to the perimeters of the geometric objects mentioned in (a) and (b) of Problem 4.16 and to the length of a circle.
    (b) If you answer "no," give examples illustrative of the serious problems you envisage.

*4.18. Using Babylonian approximations for square roots and Archimedes' method of comparison of circumferences, estimate π by considering a regular 12-sided polygon. (See pp. 106 through 108.) Find both an "upper" and a "lower" bound for π.

4.19. (a) Draw the graph of $y = \sin(1/x)$ for $0 < x < 2/\pi$ on graph paper.
    (b) Has this curve a length? Why?

*4.20. (a) Draw the graph of $y = 1/x^2$ for $x \geqslant 1$ on graph paper.
    (b) Denote by S the set of points (x, y) such that $1 \leqslant x$ and $0 \leqslant y \leqslant 1/x^2$. Has S an area? Why? Try to be as exact as possible.

4.21. Compute the first four approximations to π, using the "infinite series":

    (a) $\dfrac{\pi^2}{6} = 1 + \dfrac{1}{2^2} + \dfrac{1}{3^2} + \dfrac{1}{4^2} + \cdots$

and

    (b) $\dfrac{\pi^4}{90} = 1 + \dfrac{1}{2^4} + \dfrac{1}{3^4} + \dfrac{1}{4^4} + \cdots$

4.22. Prove that $1 + 2 + 3 \cdots + n = n(n+1)/2$.

4.23. Find the area A bounded by the x-axis, the line $x = 1$, and the curve $y = x^2$ (see Figure E4.23).

FIGURE E4.23

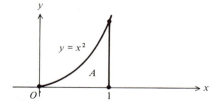

4.24. What would you have to know to find the area "under" part of a curve $y = x^m$, m being a positive integer?

*4.25. Show how to find an irrational number between the two rational numbers:

$$1.4141137777 \cdots$$
$$1.4141147777 \cdots$$

Do you think that between any two rational numbers there will be an irrational number? Why?

*4.26. Prove that between any two irrational numbers there is a rational number.

4.27. What is the rule that describes the progression from one approximation of the curve C to the next in Figure 4.28? Draw the next stage of the approximating sequence (use rectangular graph paper with $\frac{1}{4}$ inch squares).

4.28. Using only elementary geometry and the definition of the area of a square, as in §4.12, find the formula for the area of a trapezoid of height h whose bases have lengths b_1 and b_2.

4.29. Use a procedure similar to that given on pp. 118–119 to show that the volume of a rectangular solid of height h whose base is a square of side b is the product b^2h.

4.30. Use the result of Problem 4.29 to show that the volume of a rectangular solid with sides a,b,c is the product abc.

PART TWO

The Adolescence of Computation

CHAPTER 5

Navigation

This chapter is concerned with the navigational problems of seafarers of the fifteenth century, considered from the mathematical standpoint. Open-ocean voyages of long duration depended on the ability to determine position on the earth's surface, which in those times was possible only by observing the positions of stars. Thus the Age of Exploration required refinements of astronomical observations and an understanding of the geometry of spheres expressed in terms of spherical trigonometry. The chapter opens with the study of latitude, *discusses the relation of* longitude *to accurate measurement of time, and describes the* navigational triangle. *Implications of accurate latitude measurement for estimating the size of the universe are briefly noted, and the concept of* curvature *is introduced in an intuitive way to prepare the reader for a more detailed study of differential geometry and the problems of navigation in the universe which is undertaken in Chapters 11 and 12.*

§**5.1.** *Celestial navigation* is the art of transforming observations of celestial objects into useful information about the observer's position on earth. It is an ancient subject; the earliest recorded history indicates the use of stars to guide the traveler. Celestial navigation became increasingly scientific, complicated, and reliable because of the introduction of increasingly sophisticated mathematical techniques for transforming observations into directly usable information in a rote, if necessarily recondite, manner.

Isaac Newton once wrote[1]:

> *If, instead of sending the observations of seamen to able mathematicians at land, the land would send mathematicians to sea, it would signify much more to the improvement of navigation and safety of men's lives and estates upon that element.*

Newton's hopes in this direction were fulfilled a century and a half later when Nathaniel Bowditch was born in 1773 in Salem, Massachusetts. Bowditch was a capable mathematician who became interested in the problems of navigation. His book on the subject, called *The New American Practical Navigator,* was published in 1802, and is still periodically revised (as *U.S. Navy Hydrographic Office Publication No. 9*). It has passed through some 70 editions since its first appearance. Basically, to return to Newton's words, Bowditch was a mathematician who went to sea and in so doing revolutionized the practice of navigation in his time.

It is necessary to distinguish between astronomy and navigation. The science of astronomy is usually practiced by a small group of people in a fixed location, whose purpose is to learn as much as *possible* about the celestial universe. The "art," which was gradually transformed into the "science" and "technology" of navigation, is and has been practiced by a much larger (and therefore, on the average, less educated) group of seamen. Always on the move, they want to find out from their study of the skies only as much as is *necessary* in order to determine their own position. Consequently, many techniques and problems that are quite simple for the astronomer only gradually become part of the repertoire of the practical navigator on a ship in the middle of the ocean. This, of course, was more

---

[1] *Attributed to Isaac Newton by James Phinney Baxter 3rd in* Scientists Against Time *[4], p. 404.*

true in earlier times than it is today because of the recent remarkable improvements in communications and the advances in technology.

One of the first mathematical contributions to the theory of navigation was made by Ptolemy, whom we have mentioned before. Of his two outstanding books, the *Almagest* dealt with astronomy and remained the dominant influence on astronomical thought until the time of Copernicus. The other, the *Geographia,* was the first book to deal in a scientific manner with the problem of formulating a description of the world and the location of the parts known to man at that time. One of its basic contributions was the concept of angular coordinates to describe position on the earth's surface. Ptolemy's cosmological picture put the earth at the center of everything, which we now know to be false, but he knew that the earth itself is spherical in shape, a notion generally abandoned by enlightened Europeans in the Dark Ages and rediscovered, accompanied by considerable fanfare, during the Renaissance. He introduced the system of coordinates for points on the earth known today as latitude and longitude and assigned them to each of the places known to him.

Because Ptolemy used Strabo's estimate of the size (circumference) of the earth, he estimated the length of a degree (that is, $\frac{1}{360}$ of the circumference) on the earth's surface to be 50 miles instead of the true value of 69.2 miles, an error of 27.7 percent, which was to be later reflected in Columbus' gross underestimation of the distance to Cipangu (Japan) measured westward from Europe. Knowledge of the actual distance to be covered during such a long journey was crucial for Columbus, since the total distance that he could travel was limited by the storage capacity (and the technology) of his ships; this is more fully discussed at the end of Chapter 6.

For many purposes it was less important to know the actual distance to be covered than it was to answer the question: how do we know when we have reached a particular predesignated location (point) on the globe? Angular coordinates are independent of the size of the earth, and the means for computing them in open sea were always far superior in accuracy to those for computing the actual distance traversed in a given direction.

There are three fundamental and interconnected problems of navigation:

1. The determination of *latitude.*
2. The determination of *longitude.*
3. The determination of *time.*

Each had its own extensive and interesting scientific development and historical consequences. It is impossible for us to describe any of them in detail, but we have selected one aspect of each to illustrate the role mathematics played in their development.

§5.2. Although Ptolemy's notion that the earth is at the center of the universe is false in an absolute sense, nevertheless, it still *appears* that way to an earthbound observer. This is the point of view we have adopted in our discussion of navigation, and we use Ptolemy's concept of a *celestial sphere,* with the earth at the center and the stars attached to a large concentric transparent sphere, as in Figure 5.1.

FIGURE 5.1

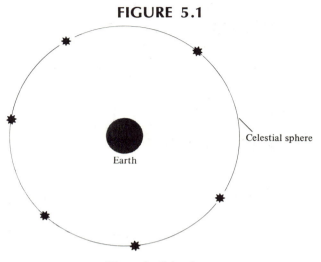

The celestial sphere.

The basic idea is to place independent coordinate systems on the earth and the celestial sphere; by using the relation between them at a given time, as determined by observation, we can determine position on the earth. First consider the latitude and longitude coordinates on the earth's surface. The earth is spinning about an axis, and the two points on its surface that are fixed with respect to this motion are called the *North Pole* and the *South Pole*. The *equator* is the great circle obtained as the intersection of a plane that passes through the center of the earth perpendicular to the axis of rotation. Letting G be a fixed point on the equator (in practice, the intersection of the equator with the meridian through Greenwich, England), define *latitude* angle α and *longitude* angle β of a point on the earth's surface by reference to Figure 5.2. The angle of latitude α varies from 0 to $\pm 90°$, corresponding to positions varying from the equator to the North or South Pole; for example we write 35°N or 22°S to denote the difference. Longitude varies through 360°, and we could let β vary from 0 to 360° in a fixed direction (say east on the diagram), or, as it is done in navigation, we can measure longitude varying from 0 to 180 degrees east and west of the meridian of G; for instance, longitude 150°E corresponds to the meridian reached by traveling from G around the equator 150° in the easterly direction, and 25°W refers to the meridian 25 degrees from the point G in the westerly direction. With these ranges the angles α and β uniquely determine a point on the surface of the earth. In fact, they uniquely determine a point on *any* sphere concentric with the earth and, in particular, they uniquely determine a point on the celestial sphere.

We shall consider, for this discussion, that the celestial sphere is fixed with the stars attached and that the earth is a concentric sphere rotating about its axis. The axis of the earth's rotation intersects the celestial sphere in two points, which we call the *celestial north pole* and the *celestial south pole,* respectively, as shown in Figure 5.3.

FIGURE 5.2

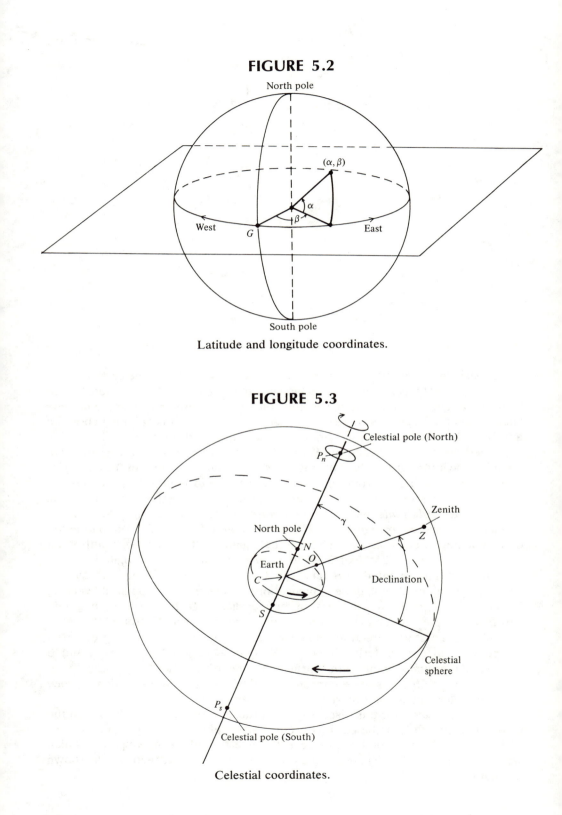

Latitude and longitude coordinates.

FIGURE 5.3

Celestial coordinates.

Note that the earth and celestial sphere appear to be rotating in opposite directions. For an observer on the fictitious celestial sphere the earth would be rotating inside it approximately once a day. On the earth an observer would see the celestial sphere rotating about him, with a particular star rising in the east and setting in the west as it rotates. If an observer were standing at point O in Figure 5.3, he could observe the celestial pole, and it would appear to be stationary as the earth rotated; stars in the neighborhood of that point would appear to rotate around it in small circles.[2] The problem of determining the latitude for an observer located at point O (Figure 5.3) is one of measuring the angle γ, since that angle determines the latitude on the earth sphere in a unique manner by means of the relation $\alpha = 90° - \gamma$. The point directly above the observer on the celestial sphere (that is, the point defined by the intersection of the line segment passing through the observer and joining the center of the earth to the celestial sphere) is called the *zenith*, whose location, of course, is a function of time, due to the rotation. The angle γ is not a function of time, however, or rather it is constant with time.

How can an observer at point O determine the angle γ? One simple way is to use a star located at (or very near) the celestial north pole, if one happens to be there, as described in Figure 5.4. It is clear that we can reduce the problem to a two-dimensional one by restricting our attention to the plane passing through the three points C, P_n, and Z; that plane is the plane of the figure in Figure 5.4. By using some sort of measuring device, measure the angle δ as accurately as possible; this is the *altitude* of the star (in this case, and for the current era, the polestar, *Polaris*). That is, sight the star and the horizon at the same time and find the angle between them. A modern device for performing this measurement accurately is the *sextant*, invented by Isaac Newton in 1700 (see Figure 5.5). Earlier but

FIGURE 5.4

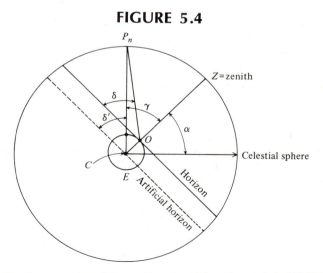

[2] *We are neglecting the* precession *of the earth's axis which, with a period of 26,000 years, is undetectable for navigational purposes but of central importance for astrological considerations; recall the discussion in §3.7.*

FIGURE 5.5

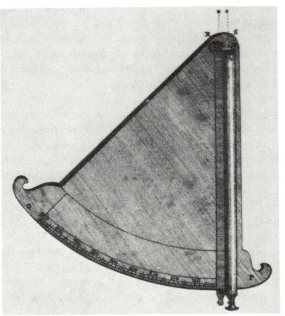

Isaac Newton's sextant, discovered, with a description in Newton's handwriting, among the papers of the astronomer Halley and first published in 1742.

less accurate and convenient instruments worked on the protractor principle, say by using two sticks as in Figure 5.6, and measuring the angle by superimposing a scale on one of the sticks.

In Figure 5.4 observe that $\delta' + \gamma = 90° = \gamma + \alpha$; hence the latitude α is the same as the angle δ'. However, the altitude δ of the star Polaris, which is the angle between the direction of Polaris and the horizon, does not always coincide with the desired angle δ'. The relative sizes of the quantities involved, and therefore the structure of the physical world, come in here. The following assumptions are made:

A1. The polestar is precisely at point P_n. This, in fact, is not quite true. Polaris is very near point P_n, and for a first approximation to latitude the difference can be ignored (see Figure 5.3). There are procedures for taking care of this discrepancy by using trigonometry and a knowledge of the motion of Polaris in a small circle about P_n as a function of time. In Figure 5.4 assumption A1 is part of the diagram.

FIGURE 5.6

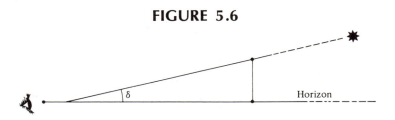

A2. The angle δ is about the same as δ'. This is a sensible assumption because the stars are so far away relative to the radius of the earth; that is, the *horizon* may be assumed to coincide with the *artificial horizon*, which passes through the center of the earth.

Using these two assumptions, we find that the latitude α is approximately the same as the altitude δ of Polaris above the horizon.

This procedure has been known for some time, and many a traveler has observed and controlled his progress north or south in a rough manner by this means. In early times a similar use of the sun's declination (altitude) at midday gave more accurate information if the proper declination for a given latitude at any day of the year were known. During the fifteenth century experienced seamen were determining their latitude at sea within one or two degrees. The use of Polaris, with suitable corrections requiring more mathematics than we have presented here, became the predominant method and is still used. Today, however, celestial navigation has been simplified by the precomputation and reproduction in tabular form of all possible positions of convenient stars. One just reads off the latitude after observing the apparent altitude of some convenient star. This avoids computation at the time the information is needed.

We should make one comment about assumption A1. Let us recall (*cp.* §3.7) the generally well-known fact that the earth's axis precesses and the solar system is moving with respect to the stars in our galaxy (the Milky Way). It follows that the relative positions of the stars in the sky are changing ever so slightly; that is, the relative positions of the stars on our idealized celestial sphere are slowly changing with respect to time. A reflection of this fact is that Polaris was much farther from its present position near the celestial north pole (an idealized point) some 5000 years ago than it is today and was not then considered an appropriate choice for the polestar. The history of scientific navigation has been brief in terms of this motion, whose influence is small over periods of a few hundred years. Modern astronomers can compute the precise position of the stars as a function of time, and the varying position of Polaris is often shown in planetariums. Such knowledge can be used to establish the relationship of ancient calendrical systems to our own. For instance, early Babylonian astronomers observed the position of certain stars on the celestial sphere and recorded this information and the date of observation according to their own calendar. By calculating the positions of stars on the celestial sphere as a function of time we can determine when an observation was made, thereby providing a way of dating historical events, such as the reign of a certain king, in an accurate manner. The ancients' method of dating can thus be linked to ours, and the study of history gains an important technique for verifying the origin and antiquity of some of its material.

§**5.3.** The determination of latitude by the method described above depends on the near equality of the angles δ and δ'. Their difference depends on the ratio of the radius of the celestial sphere to the radius of the earth, that is, on the ratio of the distance to the closest stars to the radius of the earth. If this ratio is large, then δ–δ' will be small. On the other hand, if the stars are close to the earth and the ratio is small, then δ–δ' will be sensibly large, and the determination of latitude by this method will be in substantial error. Furthermore, the amount of error in the difference $\delta-\delta'$ will depend on the location of the observer on earth. If, for instance, he is at the North Pole, both δ and δ' will be 90°, and there will be no error no matter how close the stars are to us, but if he is at a point O on the earth, where he can just see the polestar on the horizon, then $\delta = 0$, but $\delta' \neq 0$ and their difference will be as large as it ever gets as the observer travels about (Figure 5.7).

FIGURE 5.7

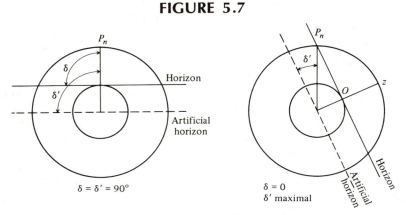

Limiting positions of artificial horizon.

Suppose that the observer knows the true latitude of a location on the earth but discovers by the measurement technique described above that there is some error; that is, $\delta-\delta'$ is not zero. If he attributes this error entirely to the fact that the stars are at a finite distance from the earth, then he can estimate how far away they must be. To simplify matters suppose that the polestar is just visible on the horizon from the observer's position; then $\delta = 0$ and δ' is maximal. Figure 5.8 shows how to calculate the distance to the polestar: $R_e/D = \sin \delta'$, where R_e is the earth's radius and D is the distance from the center of the earth to the polestar. Hence $D = R_e/\sin \delta'$. Now an error of 1° in the measurement of latitude or longitude can correspond to an error of almost 70 miles on the earth's surface. By the early eighteenth century the British government required accuracy within 1° in the measurement of longitude during long trips; the accuracy demanded of latitude measurements must have been much greater. If an average long sea voyage lasted 100 days and if errors of longitude measurement were cumulative, which is likely, since they arose from poor timekeeping properties of clocks as described later in this chapter, then accuracy of about $(\frac{1}{100})°$ per day must have been attainable for longitude measurements. If a similar level of accuracy is assumed for measurements of latitude, then

$$D \geqslant \frac{R_e}{\sin \frac{1}{100}°}$$

From a table of sines, or better, using the methods of Chapter 10, we find 0.01° = 0.0001745 . . . radians and

$$\sin 0.01° = 0.0001745 \ldots$$

FIGURE 5.8

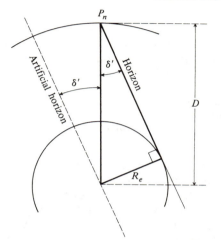

Distance to the polestar.

Therefore, recalling that R_e is approximately 4000 miles,

$$D > \frac{4000}{0.0001745} \text{ miles } \simeq 23 \text{ million miles}$$

These calculations mean that once navigators were able to control their travels across the seas with accuracy sufficient, say, to locate a small island after a journey of thousands of miles, they also must have known that the universe was vast compared with the size of the earth. The estimate found above is, of course, still much too small; the true distance from the earth to the sun is about 92,000,000 miles, and certainly the polestar is farther away. This technique simply gives *lower bounds* for the distance and in any event represents a result that must have been known to a large number of people, since it was available to seamen. Incidentally, Aristarchus estimated the distance from the earth to the sun, concluding that they were about 250,000 miles apart; in his day it may not have been clear that the other stars are much more distant.

In conclusion, it appears that the increased precision of navigation in the fifteenth and sixteenth centuries must have led to an expanded view of the size of the physical universe, which, coupled with the expanded views of cultures and societies that resulted from the discoveries of new peoples by the Europeans, must surely have greatly influenced that flowering of progress that we call the Renaissance.

§5.4. The problem of determining longitude can be reduced to the problem of accurately measuring time. As it turns out, the problem of measuring time was the most difficult of all overcome by scientists concerned with navigation. Before we examine the details of the problem of determining longitude, let us mention a few of the developments in the measurement of time. A succinct statement of the problem is found in Bowditch's *American Practical Navigator* [70], 1958 edition, p. 44:

> *A statement once quite common was, "The navigator always knows his latitude." A more accurate statement would have been, "The navigator never knows his longitude." In 1594 Davis wrote: "Now there be some that are very inquisitive to have a way to get the longitude, but that is too tedious for seamen, since it requireth the deep knowledge of astronomy, wherefore I would not have any man think that the longitude is to be found at sea by any instrument, so let no seaman trouble themselves with any such rule, but let them keep a perfect account and reckoning of the way of their ship." In speaking of conditions of his day, he was correct, for it was not until the 19th century that the average navigator was able to determine his longitude with accuracy.*

As a result of his inability to determine longitude the early mariner used the method of *parallel sailing;* this consisted of sailing to a particular latitude, then proceeding east or west, as the case might be. Such a method wasted a lot of time, but at least one had some idea of how to cross the Atlantic, for example. The early Portuguese explorers voyaging up and down the west African coast had no need for longitude, only latitude, and therefore had fewer problems than their New World counterparts. Once again we quote from Bowditch [70], p. 45:

> *During the Age of Discovery, Spain and Holland posted rewards for solution to the problem [of determining longitude], but in vain. When 2,000 men were lost as a squadron of British men-of-war ran aground on a foggy night in 1707, officers of the Royal Navy and Merchant Navy petitioned Parliament for action. As a result the Board of Longitude was established in 1714, empowered to reward the person who could solve the problem of "discovering" longitude at sea. A voyage to the West Indies and back was to be the test of proposed methods which were deemed worthy. The discoverer of a system which could determine the longitude within 1° by the end of the voyage was to receive £10,000; within 40', £15,000; and within 30', £20,000. These would be handsome sums today. In the 18th century they were fortunes.*

An interesting attempt to build a chronometer was made by the Dutch mathematician and scientist Christian Huygens (1629–1695), who used the solution to a nontrivial mathematical problem concerning a special type of pendulum (see Figure 5.9). This was inadequate for long-range sailing because of the variation of size of certain metallic components with temperature. The first effective naval chronometer was constructed by John Harrison, who finally received the above-mentioned reward on his eightieth birthday but only after the intervention of the English king. The chronometer that became the model for modern naval timepieces was invented in 1766 by Pierre LeRoy of France.

Having noted that many scientists made great efforts to learn to measure time accurately, let us determine how this knowledge can be utilized to measure longitude. We have to start with a brief mathematical digression. Recall the fundamental mathematical fact in the study of plane trigonometry: a triangle consists of three sides and three angles and the knowledge of any three of these sides or angles mathematically determines the remaining three with the single exception that three angles do not determine the three sides of the triangle. This is proved by

FIGURE 5.9

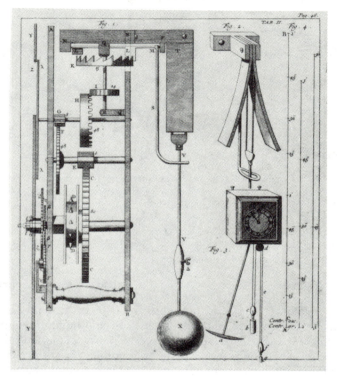

Huygens' pendulum clock, perfected in 1657, made it possible for the first time to determine longitude accurately. Notice the cycloidal jaws.

using the various relations amongst the trigonometric functions such as the law of cosines. To determine three unknown quantities (angles and sides) *practically* from three that are given requires the use of *trigonometric tables*. The result, of course, is only as accurate as the tables. As you also probably recall, the computations can sometimes be laborious if extreme accuracy is required. We shall return to this point when we study the invention of logarithms in Chapter 7. Now step over to *spherical trigonometry*, a subject that has been crowded out of the high school curriculum by calculus and other subjects of modern mathematics. Consider a sphere. A *spherical triangle* is determined by three distinct points on the sphere; the segments of great circles joining these points form the *sides* of the spherical triangle, as in Figure 5.10, which exhibits the spherical triangle PQR determined by the three points P, Q, and R and having sides a, b, and c. The first question is: what do we mean by an *angle* between two sides? Let α be the angle (to be defined) between sides a and b at point P. Then α is *defined* as equal to the (plane) angle (also denoted α) between the projections of segments a and b onto

FIGURE 5.10

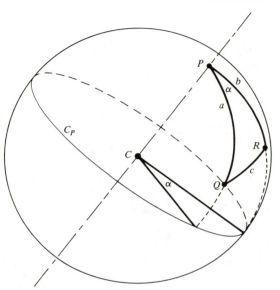

Definition of angle between two great circles on a sphere.

the plane perpendicular to the axis through P and C (C is the center of the sphere), as in Figure 5.10; thus α is equal to the angle between the meridian planes determined by segments a and b. Given these concepts of angle and side it can be proved that the fundamental result of plane trigonometry carries over to this case: knowledge of three of the sides or angles of a spherical triangle mathematically determines the three remaining quantities (and this time there are no exceptions, for three angles uniquely determine the three sides). Once again, using relations between angles and sides of spherical triangles, much as in the planar case, we can derive formulae for the solution of such a triangle problem and, by using the appropriate trigonometric tables, obtain solutions to any desired degree of accuracy.

The solution to the longitude problem can be described if these formulas are known for specific triangles on the celestial sphere. First, a reference point will be needed for measurement of longitude. That reference point has been standardized only recently because in earlier times no one could measure longitude; hence the lack of the need for a standard. In 1810 the Americans concerned with navigation wanted to establish the meridian of Washington, D.C. as a reference standard, but this was resisted by the British, who had been using Greenwich for some time because it was the first astronomical observatory to tabulate the positions of stars as a function of time and to publish this information in almanacs, the first of which appeared in 1767. As we shall see, this kind of information plays an important role in the measurement of longitude. At a conference in 1884 in Washington, D.C., 25 nations agreed to use the meridian of Greenwich, England as the *prime meridian* and designated it as *0° longitude*.

An *almanac* consists of listings of various data about certain celestial bodies. For our purposes it suffices to know that an almanac gives the position of celestial objects on the celestial sphere as a function of time, with the observer at a reference point, in this case Greenwich (see Figure 5.11 for a sample from an early almanac). The objects for which celestial positions are provided include certain easily located stars, as well as the sun, moon, and various planets. The data are obtained from observations, and then by the use of celestial mechanics calculated future positions are tabulated. Roughly, all the stars rotate about 15° per hour (that is, the celestial sphere rotates at this speed, completing one full revolution every 24 hours), and by observing a star at Greenwich at a particular time and knowing how fast it is moving and in what direction we can compute its position at later times. These data are committed to the tables the navigator on a ship has at his disposal. They are also equipped with error estimates so that bounds for the final error can be determined.

Suppose we are observing from a given point O on the surface of the earth whose latitude is known and we wish to determine its longitude from a measure-

FIGURE 5.11

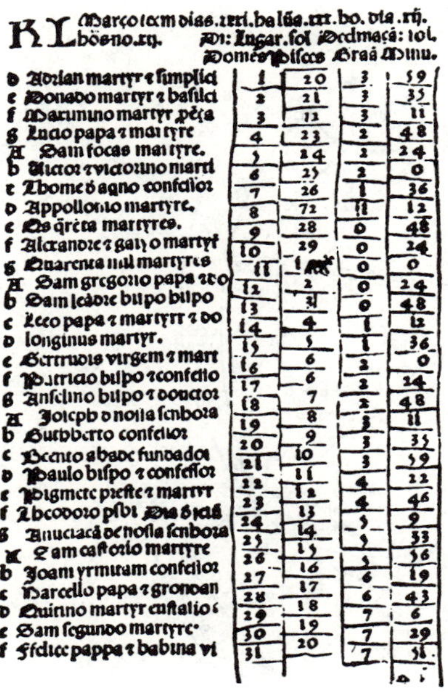

FIXED STARS, 1855.

MEAN PLACES OF 100 PRINCIPAL FIXED STARS, FOR JANUARY 1, 1855.					
Star's Name.	Mag.	Right Ascension.	An. Variation.	Declination.	An. Variation.
		h. m.		° ' "	
α ANDROMEDÆ	2	0 0 53.97	+ 3.087	+28 17 23.3	+19.93
γ PEGASI (*Algenib*)	3.2	0 5 46.37	3.065	+14 22 38.1	20.05
β Hydri	3	0 18 3.62	3.292	−78 4 23.1	20.23
α CASSIOPEÆ	var.	0 32 18.36	3.356	+55 44 29.2	19.83
β Ceti	2	0 36 18.45	3.016	−18 47 0.1	19.86
ε URS. MIN. (*Polaris*)	2	1 6 29.82	+18.117	+86 22 11.3	+19.23
δ' Ceti	3	1 16 46.57	3.000	− 8 55 58.6	18.74
α Eridani (*Achernar*)	1	1 32 18.42	2.238	−57 58 28.2	18.59
α ARIETIS	2	1 59 0.44	3.365	+22 46 28.4	17.29
γ Ceti	3.4	2 35 47.42	3.102	2 37 19.4	15.44
α CETI	2.3	2 54 42.21	+ 3.129	+ 3 31 4.7	+14.40
α PERSEI	2	3 13 59.52	4.243	49 20 26.8	13.25
η Tauri	3	3 38 52.31	3.553	+23 39 11.0	11.54
γ' Eridani	3	3 51 15.91	2.796	−13 55 26.7	10.59
α TAURI (*Aldebaran*)	1	4 27 36.26	3.436	+16 12 49.4	7.72
α AURIGÆ (*Capella*)	1	5 5 59.03	+ 4.423	+45 50 41.8	+ 4.27
β ORIONIS (*Rigel*)	1	5 7 34.23	2.884	− 8 22 22.5	4.54
β TAURI	2	5 17 7.72	3.791	+28 26 48.3	3.55
δ ORIONIS	2	5 24 36.06	3.066	− 0 24 37.8	3.05
α Leporis	3	5 26 20.19	2.648	−17 55 46.0	2.94
ε ORIONIS	2	5 28 51.43	+ 3.044	− 1 17 54.6	+ 2.71
α Columbæ	2	5 34 24.05	2.177	−34 9 13.3	2.23
α ORIONIS	var.	5 47 19.35	3.249	+ 7 22 32.6	+ 1.11
μ Geminorum	3	6 14 11.30	3.636	+22 34 59.9	− 1.37
α Argus (*Canopus*)	1	6 20 44.13	1.330	−52 37 4.7	− 1.81
51 (Hev.) Cephei	5	6 31 6.10	+30.650	+87 15 7.9	− 2.80
α CANIS MAJ. (*Sirius*)	1	6 38 45.60	2.646	−16 31 12.8	4.52
ε Canis Majoris	2.1	6 52 55.69	2.360	−28 46 40.3	4.58
δ Geminorum	3.4	7 11 27.65	3.597	+22 14 41.7	6.16
α' GEMINOR. (*Castor*)	2.1	7 25 20.49	3.841	32 12 6.2	7.37
α CAN.MIN. (*Procyon*)	1	7 31 42.52	+ 3.145	+ 5 35 35.7	− 8.79
β GEMINOR. (*Pollux*)	1.2	7 36 26.23	3.681	+28 22 19.9	8.26
15 Argus	3	8 1 22.22	2.557	−23 53 20.5	10.06
ε Hydræ	3.4	8 39 5.74	3.189	+ 6 56 52.2	12.86
ι Ursæ Majoris	3	8 49 15.44	4.123	+48 36 26.7	13.78
ι Argus	2	9 13 12.52	+ 1.602	−58 40 3.3	−14.91
α HYDRÆ	2	9 20 27.65	2.951	− 8 1 56.8	15.36
θ Ursæ Majoris	3	9 23 7.85	4.048	+52 20 6.3	16.13
ε Leonis	3	9 37 36.82	3.424	24 26 22.0	16.34
α LEONIS (*Regulus*)	1.2	10 0 38.72	3.205	+12 40 26.4	17.40
η Argus	2	10 39 26.75	+ 2.306	−56 55 21.5	−18.74

Sixteenth and nineteenth century nautical almanacs.

FIGURE 5.12

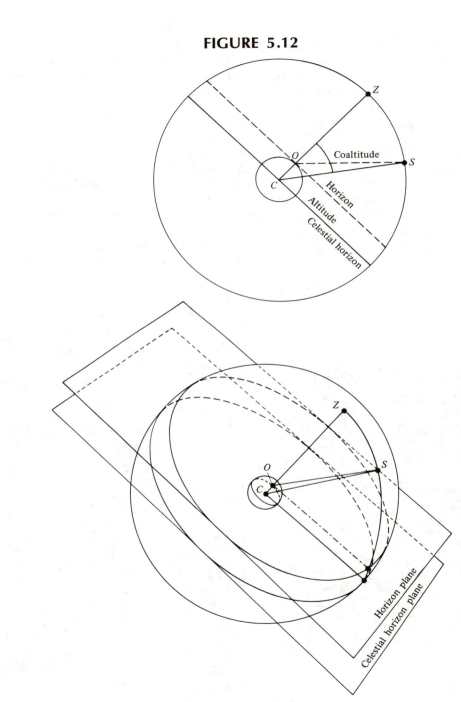

Horizon plane and celestial horizon plane.

ment of the altitude of a star (perhaps the sun) at a particular time. The altitude of the polestar was discussed earlier, and in general the notion is the same for any other celestial body, as shown in Figure 5.12: the altitude of any star is its angular deviation from the horizon and, once again, no distinction is made between the actual and artificial horizon. The complement of the altitude, that is, $90° - \alpha$, is called the *coaltitude,* the angle between the line of sight (the line OS or more precisely the line CS, since we assume that the angle SOZ is about equal to SCZ because of the distances involved) and the normal at point O, that is, the line OZ. The arc subtended by the coaltitude of the star is a segment of a great circle, joining S to Z.

Now we are prepared to define a *navigational triangle;* it is simply a triangle whose vertices are the celestial north pole, the zenith of an observer, and a particular celestial body S, not coincident with these first two points (see Figure 5.13).

If for this triangle we know our latitude and the altitude (or coaltitude) of the star S at a particular time t_0, then we know all three sides of the triangle. Here we say we "know" the sides of the triangle if we know the angles at the center of the celestial sphere that subtend the arcs which constitute the triangle's sides. Up to a constant of proportionality (which is the radius of the sphere) the three sides are determined uniquely by the central angles. These three sides in turn determine the three angles of the navigational triangle.

How do we know the three sides? First, the arc P_nZ is subtended by the latitude of the observer, which we assume we know. Second, the arc SZ is subtended by the coaltitude of the star S, as seen from the observer's position (see Figure 5.12). Third, the *codeclination* of the star (angle SCP_n) is known from the almanac. This is essentially the *latitude coordinate* of the star in the celestial sphere. Thus we

FIGURE 5.13

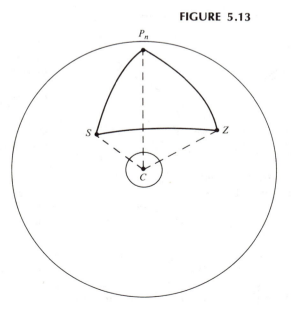

A navigational triangle.

know all three sides of the triangle. The only side that changes its length with time is the side SZ due to the relative motion of the star and the observer. If the observer is assumed to be at Greenwich, there is a P_nSZ_g, formed the same way; this situation is described in Figure 5.14. We make the basic observation that the angle $\beta_g(t) - \beta(t)$ is a constant [here $\beta_g(t)$, $\beta(t)$ indicates that the angles β_g, β are varying with time t] and is indeed the longitude difference between points G and O on the earth. In the almanacs the angle $\beta_g(t)$ is given as a function of time for a given star, in this case our S. On the other hand, by using spherical trigonometry we can measure the angle $\beta(t)$, since we know the three sides of the navigational triangle P_nZS from our observation of the altitude of the star S and the latitude of the observer at point O. Thus the longitude is found.

This gives some indication of the manner by which navigators can compute both latitude and longitude. It should be quite clear how important these calculations were to the development of the maritime nations and the growth of world commerce.

FIGURE 5.14

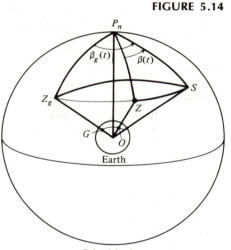

Celestial sphere   Calculation of longitude.

§5.5. The kind of geometry man has used to describe the world about him has changed with the scope of his knowledge. The Egyptians had an indifferent notion of a flat world; the Greeks developed an accurate and axiomatic *plane geometry* to describe reality and introduced *spherical geometry*. With the later development of navigation, spherical geometry was extended and essentially completed. It describes the geometry of a real object (the earth) which is *positively curved* like a spherical surface. What is the geometry that must be used to describe our *universe?*

Should we use a flat three-dimensional space, similar to our conception when we think of the solar system with planets circling the sun in some expanse of space that may be likened to a big room with many small objects floating around in it? This is adequate for some purposes, but suppose we consider traveling to other galaxies. Then it becomes necessary to know if the geometry of the "big room" is appropriate to reality on so large a scale.

It turns out to be necessary to consider *time* as a fourth dimension (it has already entered into our discussion of the measurement of longitude) and to study the properties of a certain *four dimensional* "space" and to ask what its natural geometry is. We examine this concept in some detail in Chapters 11 and 12. At the present time it is not known what the best geometry is; the situation is similar to that confronted by the earliest cosmologists who did not know whether the world was flat or round. There are various current theories about the shape of the universe, and we mention one of them briefly here. First we need a new geometric concept, that of a *negatively curved surface,* which can be intuitively understood by consideration of an example. Part of the surface defined by the equation $z = x^2 - y^2$ in ordinary three dimensional space is shown in Figure 5.15. It is something like a saddle near the point labeled O, which therefore is called a *saddle point* of the surface. We say that this surface is *negatively curved* at point O. In general, a negatively curved surface "goes down" in one direction and "up" in another at each of its points. *Positive curvature* of a surface means that the surface either "goes up" in every direction or "down" in every direction at each of its points. This is the case on the surface of a sphere, where "down" and "up" refer to the directions perpendicular to the surface at a particular point. Using this heuristic definition, we call a surface *flat* if it goes neither "up" nor "down" in any direction: a plane in three-space provides an obvious and typical example.

FIGURE 5.15

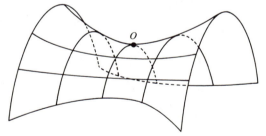

A surface of negative curvature.

It is possible to consider these concepts of positive and negative curvature for higher dimensional spaces, but it is impossible to illustrate them in this case. It *may* be true that the four dimensional space representing the universe moving through time is, in some natural way, a *negatively curved space.*[3] At some future time this concept may be as easy for the average educated person to comprehend as the concept of a globular earth is today, but that will no doubt come to pass only if it has practical consequences that will affect the lives of civilized people, as air travel, weather satellites, and global communication systems do today.

[3] *The conceptual framework in which these concepts are studied is called* differential geometry; *its essence is the application of* calculus *to geometry. Differential geometry was the fundamental mathematical tool used by Albert Einstein (1879–1955) in his general relativity which is a theory of the geometrical structure of the universe. See Chapters 11 and 12.*

To an observer standing on the earth its surface nearby looks flat, but if he looks farther off the surface appears to be (and indeed is) positively curved. Suppose an observer on the earth peers into the universe. Near the observer, in our solar system, for instance, the universe appears to be "flat," but if the observer looks "farther" (and considers time) it is possible that what he "sees" is negatively curved.

Just as it was necessary to understand the spherical geometry of the earth and of the celestial sphere in order to navigate on the earth, it will be equally necessary to understand the nature of the geometry of the universe if we want to navigate through it. Figure 6.2 shows contemporary navigators at work charting the course of a grand voyage of exploration and discovery. Their traditional and time-honored task remains unchanged, but the orderly increase of our mathematical knowledge and technological skill enables us to attain the vaulting ambition of our dreams.

EXERCISES

5.1.   What is the justification for using a model of the universe (the celestial sphere) which assumes that the earth is at its center and the stars, sun, and planets are attached to a sphere of fixed (large) radius that revolves about the earth, even though we know for a fact that these heavenly bodies vary greatly in distance from the earth?

5.2.   What modification in the "celestial sphere" model will be necessary to make it useful for analyzing interplanetary travel? Describe in general terms how such a model could be used.

5.3.   Describe how an astronomical laboratory (at Greenwich, England, for example) might construct a table of declinations (with errors) such as that given in Figure 5.11 for a single "fixed" star. Can all fixed stars have their positions in the celestial sphere determined by a single observatory? Why?

5.4.   Suppose you are standing at a point on the surface of the earth and you can accurately measure the altitude of Polaris to be 45° 2′. What is your latitude, assuming that Polaris is precisely at the celestial north pole?

5.5.   Approximate the error in your latitude in Problem 5.4 by using the data in the Table of Fixed Stars for 1855 (assuming that it is up-to-date) given in Figure 5.11.

5.6.   Suppose that you are in the southern hemisphere, and unable to see Polaris but that you can see Sirius, one of the brighter stars in the sky. Under what conditions will a measurement of its altitude determine your latitude? Assuming these conditions and assuming that you have measured the altitude of Sirius and have found it to be 59° (the angle of the star above the horizon), determine your latitude from the data in Figure 5.11. Would a sailor be able to tell that your conditions have been met, assuming that no other measurements have been made? Why?

5.7.   Using a reference other than this book, determine how a ship's chronometer differs from an ordinary clock. Were the problems associated with the "invention of an accurate chronometer" of a mathematical nature or were they problems of a different kind? Explain. What are the most accurate clocks, to your knowledge, available today? Would they be useful for modern navigation?

5.8.   Can a star located at the celestial north pole be used in determining longitude? Give a reason for your answer.

5.9.   Suppose you are located at a point P on the surface of the earth and you know that your latitude is approximately 45°N. Suppose you have the opportunity to measure the altitude of various stars which have latitudes that are (a) near the celestial north pole, (b) near 45°N in latitude, and (c) near 0°N in latitude (over the equator). Which of these choices would minimize your error in determining your latitude? Which would tend to contribute the largest amount of error? Why?

*5.10. (a) Find the formulas necessary for solving the navigational triangle in order to determine longitude, once the latitude is known. (Use a trigonometry text that includes a discussion of spherical trigonometry.)

(b) Give a hypothetical example of the solution to such a problem.

5.11. Suppose that the earth did not rotate on its axis. Invent a method for determining longitude. Could a sixteenth-century seaman have used this method? Why?

5.12. Consider the spherical triangle formed by the area between the meridians of 0° and 90°W and above the equator.

(a) Find the sum of the angles of this triangle.

(b) How does this compare with the sum of the angles of a planar triangle?

*5.13. Does the result in Problem 14b always hold for a spherical triangle (or for any triangle) on the surface of the earth? Give an intuitive reason for your answer.

CHAPTER 6

Cartography

Associated with problems of navigation are those of recording the course of a voyage and of determining those paths that are most efficient or simplest to navigate. This chapter is devoted to cartography, *studied from the mathematical viewpoint. The main problem is that of representing a portion of a curved surface on a plane, for example, on a sheet of paper.* Cylindrical *and* stereographic projections *are introduced, but the main burden of study is* Mercator's projection. *Its important connection with course plotting is investigated in detail. Cartography exhibited rapid advances during periods when mathematics was used in an integral manner, and declined during others when maps were simply drawn by travelers based on their experiences or devised by scholars to illustrate philosophical theories of world order rather than to exhibit a correspondence between places on the earth and points on a sheet of paper. A number of interesting early maps are appended to illustrate these matters. Finally, the reasons that Columbus sailed under the Spanish flag rather than the Portuguese are presented in terms of the astronomical and technological problems that were involved.*

§**6.1.** Cartography, the science of map making, is a highly mathematical subject because the region mapped is most often a portion or all of the surface of the earth, and the result is its depiction on a part of the Euclidean plane — that is, on a flat surface. The surface of the earth is nearly spherical, but a sphere cannot be truly mapped onto a part of a plane with complete preservation of all its geometrical properties. This is an informal way of stating a precise mathematical theorem: there is no one-to-one correspondence between a sphere and a portion of a plane that preserves two of the following three quantities: angles, areas, and distances. A map can represent angles *or* areas accurately but not angles *and* areas; and a map of a sphere or even of part of a sphere on a plane can *never* preserve distances. Different practical requirements will dictate which, if any, of the quantities that *can* be preserved *should* be preserved for a particular application. Some useful maps do not preserve any of them but are helpful because all of them are nearly preserved so that the errors are small, at least for small portions of the earth's surface.

Serious mathematical map making was given its start by Claudius Ptolemy's work, about +150, but it was only when the practical navigational requirements of the late fifteenth century had to be met that useful mathematical solutions to map-making problems were discovered. Foremost amongst these solutions was the famous *Mercator projection,* illustrated in Figure 6.1 and familiar to everyone. Its application to the problem of mapping the entire surface of the earth (considered to be spherical) was made by Mercator (Gerard Kremer or De Cremer, 1512–1594), who, after obtaining a degree at the University of Louvain, took further private lessons in advanced mathematics from the Dutch astronomer Reiner Gemma Frisius. It was no accident that Mercator made his useful discovery. His engraved *Great World Map* appeared in 1569; in 1969 it celebrated its four hundredth anniversary of continued utility to mankind in the Mission Operations Control Room of the Manned Spacecraft Center as man first stepped on the moon (Figure 6.2).

There are few products of human ingenuity that have enjoyed so long a lifespan; therefore it is worthwhile to consider what special properties the Mercator projec-

FIGURE 6.1

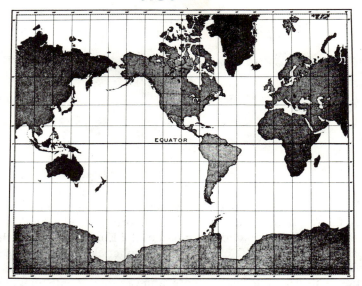

A Mercator map of the world.

FIGURE 6.2

Mission Operations Control Room, Manned Spacecraft Center.

tion must have which sustain its utility and life. It turns out that one special property, although of a purely mathematical nature, nevertheless has great practical significance.

Before turning to a detailed discussion of this remarkable property, it will be instructive to examine the general nature of the Mercator projection and other projections that are related to it. Suppose that a sphere is given and that a cylinder is constructed tangent to the sphere, as shown in Figure 6.3. The tangent curve lying on both the sphere and the cylinder will be a great circle and the axis of the cylinder will pass through the center of the sphere perpendicular to the plane of this circle. The axis meets the sphere in two points, which may be called the *axial north pole* and the *axial south pole;* their location obviously depends on how the cylinder is oriented with respect to the sphere, and indeed, given two diametrically opposite points on the sphere, there is just one cylinder tangent to the sphere such that these points are its axial poles.

Denote the center of the sphere by O and let P be an arbitrary point on the sphere. The ray from O extended through P will meet the cylinder in just one point, but when P is one of the axial poles the ray will not meet the cylinder at all

FIGURE 6.3

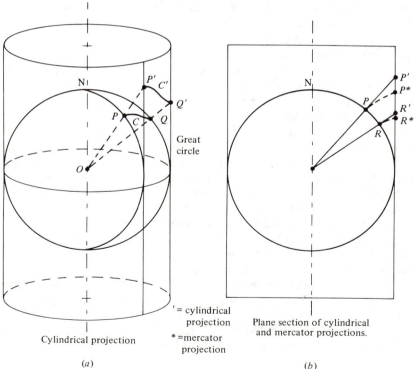

' = cylindrical projection
* = mercator projection

Cylindrical projection

(a)

Plane section of cylindrical and mercator projections.

(b)

(a) Cylindrical projection; (b) plane section of cylindrical and Mercator projections.

because it will lie on the axis. Except, then, for the axial poles, each point P on the sphere can be made to correspond in this way with a unique point on the cylinder. Denote this point on the cylinder by P'. Conversely, if P' denotes *any* point on the cylinder, the line segment joining P' to O will meet the sphere at one point, say P. This technique sets up a *one-to-one correspondence* between the points of the cylinder and the points of the sphere other than the two axial poles; such a correspondence is called a *map* of the sphere (except for the axial poles) onto the cylinder.[1] Now cut the cylinder along a line parallel to its axis and roll it out flat. This produces a map of the sphere, except for the two axial poles, on a strip in the Euclidean plane, which looks superficially like the Mercator map shown in Figure 6.1 (if the cylinder is tangent to the equator of the earth, considered to be the sphere under study), except that the map in the figure has been trimmed at top and bottom (it should extend to infinity in each of these directions) and the detailed transformation of distances, angles, and areas is not quite the same as that produced by a true cylindrical projection.

Examination of the technique of projection shown in Figure 6.3*a* shows that points near the axial poles will be mapped onto points that are far from the points of the tangent circle and that areas will be increasingly and unreasonably magnified the nearer they lie to the axial points. This is why Greenland in the northern hemisphere and Antarctica in the southern appear so large in this and in the related Mercator projection, although in reality they are relatively insignificant.

The earth's surface is shown oriented in the usual way in Figure 6.1, with the cylinder tangent to the equator. This orientation is not necessary, although we show later why it is particularly convenient for the Mercator projection. In any event, the projection obtained from this configuration exhibits areas and distances that are nearly accurate on and near the equator but increasingly inaccurate the farther from the equator one moves. If the cylinder is chosen tangent to some other great circle, then it will provide a map that accurately represents distances and areas near the circle of tangency but not near the equator. Figure 6.4 (actually a Mercator projection) shows what happens if a cylinder is chosen tangent to the great circle that passes through the North and South Poles and also through the equator at 90°W longitude. The corresponding map provides an accurate representation of the part of the earth's surface that lies near the 90°W longitude line. The map in this figure is incomplete; only slightly more than one quarter of the cylinder has been displayed; Europe, Asia, and other interesting places are left out. The map does show the relative size of North and South America, Greenland, and Antarctica in a more nearly honest manner than usual.

Figure 6.5 shows another projection in which the cylinder is tangent along the great circle connecting Washington D.C. to Moscow. The lower figure (*b*) shows part of this (Mercator) map; the upper (*a*) outlines the part shown in (*b*) as it normally appears on the standard Mercator map whose cylinder is tangent at the equator (Figure 6.1). Things look different. The straight line in (*a*), which connects

---

[1] *In Figure 6.3a two points* P *and* Q *joined by a curve* C *are projected to the points* P', Q' *joined by the projected curve* C'. *The curve* C' *might represent a river or boundary and becomes a part of the map of the sphere on the cylinder.*

FIGURE 6.4

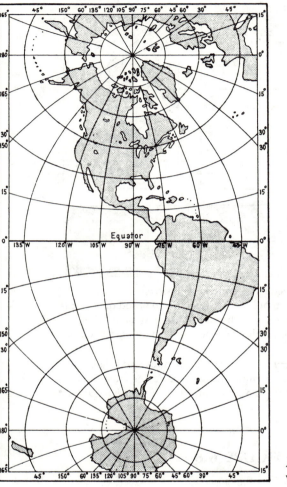

A transverse Mercator map of the western hemisphere.

Washington and Moscow, passes through France or perhaps Belgium; the great circle between those cities, which is the straight line in (*b*), passes through Iceland.

It is easy to describe the location of the point P' on a cylindrical projection map that corresponds to point P on the surface of the earth. Suppose the cylinder is tangent to the sphere along the equator. Every meridian on the earth's surface is mapped by a cylindrical projection onto a line on the cylinder parallel to the cylinder's axis. Cut the cylinder along the line corresponding to $\pm 180°$ longitude and spread it out to obtain a flat map. The equator becomes a straight line of length $2\pi R_e$ (R_e = radius of the earth) perpendicular to the edges of the map. The meridian corresponding to $0°$ longitude is mapped onto a straight line perpendicular to the equator and equidistant from the edges of the map (Figure 6.6).

FIGURE 6.5

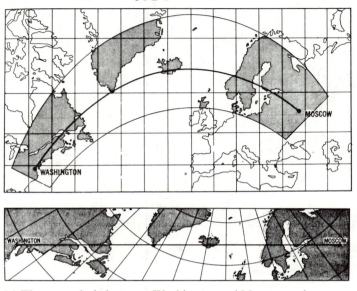

(*a*) The great circle between Washington and Moscow as it appears on a Mercator map; (*b*) an oblique Mercator map based on a cylinder tangent along the great circle through Washington and Moscow and includes an area 500 miles on each side of the great circle.

This map is too large: its width, $2\pi R_e$, is about 25,000 miles and it has infinite height. In order to obtain a practical map, the horizontal and vertical scales must be reduced. Let the scale factor in the horizontal direction be S_H and introduce a *map horizontal coordinate x* by defining

$$x = S_H R_e \beta$$

FIGURE 6.6

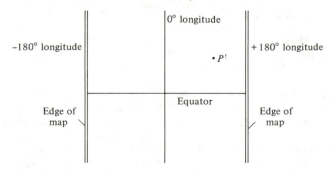

A cylindrical or Mercator projection spread flat to form a map.

where β is the longitude of P (measured in radians). We choose S_H to make the map whatever width we desire; for instance, to obtain a map 1 foot wide take $S_H = (2\pi R_e)^{-1}$ feet/mile.

The height of P' above the plane of the equator is determined by the latitude α of P. This is easy to see by redrawing part of Figure 6.3b as shown in Figure 6.7.

The side h of right triangle $OP'E$ is the height in question. Angle POE is the latitude of P and OE is a radius of the earth, hence of length R_e. Then $h = R_e \tan \alpha$. To make this length small enough to fit on a practical map introduce a vertical scale compression factor S_V and a *map vertical coordinate y* by defining

$$y = S_V R_e \tan \alpha$$

FIGURE 6.7

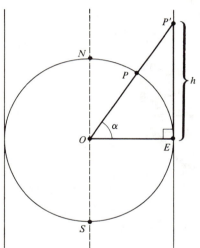

Map vertical coordinate for a cylindrical projection.

By appropriately fixing a value for S_V the distance from the equator to any prescribed circle of constant latitude can be adjusted to whatever size is convenient; for example, if it be desired that the distance on the map from the equator to the horizontal line corresponding to the circle of 45° latitude be 6 inches, then, recalling that $\tan 45° = 1$, choose $S_V = 6/R_e$ inches/mile.

The two equations combined,

$$x = S_H R_e \beta$$
$$y = S_V R_e \tan \alpha$$

completely prescribe the location on the (practical) map of the point P on the earth's surface having latitude and longitude coordinates α and β as long as P is not one of the poles. The elongation of regions near the poles is expressed in analytic terms by the statement that $\tan \alpha$ increases without bound as α approaches 90°.

§6.2.

It is absolutely clear that neither distances nor areas are preserved by the cylindrical or by Mercator's projection, but it turns out that Mercator's projection, which is obtained from the cylindrical projection by compressing that map in the vertical direction in a certain complicated way described later, *preserves the angles between curves*. For instance, the lines of latitude intersect the lines of longitude at right angles on the surface of the earth, and they do also on a Mercator map (*cp.* Figure 6.1) and, perhaps surprisingly, in Figures 6.4 and 6.5*b* as well. This is an important property of the Mercator projection, although some other maps have it too. For instance, if the sphere that is the earth is placed on a plane

FIGURE 6.8

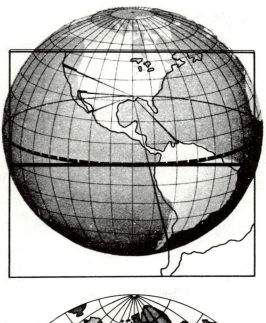

(*a*) An equatorial stereographic projection; (*b*) a stereographic map of the western hemisphere.

so that the South Pole is the point of tangency and rays are drawn from the North Pole through points of the sphere to where they meet the tangent plane, then another kind of map is obtained (called *stereographic projection*) which also preserves angles between curves. A map like this (but tangent at the North Pole and projected from the South Pole) appears on the flag of the United Nations. It is otherwise unlike the Mercator projection and has no use in navigation. Figure 6.8 illustrates a stereographic projection map.

We shall prove that stereographic projection preserves angles between curves. Let C and C' be two curves on the sphere that meet at point P. Denote the tangent to C at P by T and the tangent to C' at P by T'. The two tangent lines T and T' determine a plane and form an angle in that plane, say α. Define the angle between the curves C and C' at P to be α. With this definition, what is required is to prove that $\alpha = \alpha^*$ in Figure 6.9, where α^* is the angle between the stereographic image curves. To prove this first observe that T^* lies in the plane determined by T and PP^* and T'^* lies in the plane determined by T' and PP^*. In fact, T^* is the image of T under stereographic projection and therefore the plane determined by N and T (see Figure 6.10) must also contain T^*, and similarly for T'^*, which is the stereographic projection of T'. Because of these facts BB^* and AA^* can be drawn parallel to PP^*, as shown in Figure 6.10 and, as shown, planes π and π^* can be constructed perpendicular to the plane ABA^*B^*. It follows from the geometry of solid angles that $\alpha = \alpha^*$ if the angles between planes PAB and π, and between $P^*A^*B^*$ and π^* are equal. To determine whether these angles are equal, draw the plane through line PP^* perpendicular to plane ABB^*A^* and see where it intersects the original sphere; this intersection is shown in Figure 6.11. We must prove that angle $\gamma = \gamma^*$, which is easy. First note that $\gamma^* + \pi/2 = \delta + \pi/2$, where δ is the angle ONP because it is an exterior angle of triangle NSP^*; hence $\gamma^* = \delta$. On the other hand, $ON = OP$, since both are radii of the circle, and the base angles of triangle ONP are equal, as shown in the figure. Then

FIGURE 6.9

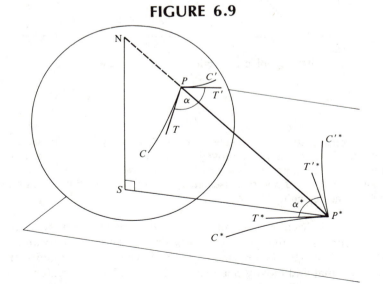

Stereographic projection preserves angles: part 1.

FIGURE 6.10

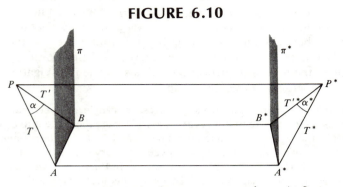

Stereographic projection preserves angles: part 2.

FIGURE 6.11

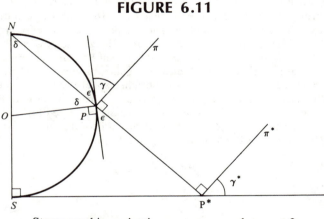

Stereographic projection preserves angles: part 3.

$$\delta + \epsilon + \frac{\pi}{2} = \gamma + \epsilon + \frac{\pi}{2}$$

follows from consideration of the angles at point P, so $\delta = \gamma$. Combining the results gives $\gamma = \gamma^*$, which is what was to be proved. This slightly involved geometrical proof that stereographic projection preserves angles between curves could be replaced by a simpler and more straightforward one that uses analytical techniques, but they are too "sophisticated" for us at this point.

§6.3. The cylindrical projection of the sphere discussed in §6.1 has a general appearance similar to that of Mercator's projection, but important differences make Mercator's projection enormously useful, whereas the cylindrical projection is only of marginal and mostly pedagogical interest. The Mercator projection has two remarkable properties:

1. It preserves the angles between curves.

2. Curves on the surface of the earth which make a fixed angle with all the lines of longitude appear on a Mercator map as straight lines.

The second property is significant because the easiest way to direct the motion of a ship is to require the helmsman to steer a course that makes a fixed angle with the direction shown by a compass needle. If we assume that a compass needle points to the North Pole, then a course that makes a fixed angle with the compass direction is the same thing as a course that makes a fixed angle with the longitude lines. Consequently, such *a compass course will appear on the Mercator map as a straight line,* and the task of plotting it from one point on the earth's surface to another is reduced to drawing the straight line connecting the images of the two points on the Mercator map and measuring (with a protractor, for instance) the angle this line makes with the lines of longitude (which are all parallel on the map). As already noted, Mercator's projection preserves angles, whence the angle measured on the map between the course line and the longitude lines will be the same as the angle that the compass needle will show between the direction north and the direction of the ship's motion.

Compass needles point (approximately) to the North Pole, which explains why Mercator maps are made by compressing cylinders that are tangent to the earth's surface *along the equator;* the axis of the cylinder coincides with the line joining the poles. If a compass needle pointed to some other point, say P, then the convenient Mercator projection would be obtained by choosing a cylinder whose axis went through P instead. It is the action of the compass needle that determines the natural and useful axis for the tangent cylinder.

It has been said several times that the Mercator projection is obtained by an appropriate compression of the cylindrical projection, as illustrated schematically in Figure 6.3b.

This can be described in a precise manner by writing out the equations that determine the coordinates of a point on the map in terms of the coordinates (latitude and longitude) of a point on the earth's surface. To do this suppose that a point P on the earth's surface has latitude α and longitude β. Define a point P^* in the Euclidean plane by giving its Cartesian coordinates (x, y) as follows (see Chapter 7 for the logarithm function and Chapter 8 regarding coordinate systems):

(6.1)
$$x = S_H R_e \beta$$
$$y = S_V R_e \log \tan \left(\frac{\alpha}{2} + \frac{\pi}{4} \right)$$

where the fixed numbers S_H and S_V are the horizontal and vertical scale factors that determine how large the map will be — that is, whether it will fit on an ordinary sheet of paper or will require a larger wall chart or whatever.[2] The obvious properties of the Mercator projection can be easily verified from these equations. For instance, the poles correspond to latitudes $\pm \pi/2$ in radian measure and therefore the corresponding points on the map must have y-coordinates

$$y = S_V R_e \log \tan \left(\frac{\pi}{4} + \frac{\pi}{4} \right) = S_V R_e \log \infty = \infty$$

[2] *To be sure, the Mercator (and cylindrical) projection is a strip that is infinite in the "vertical" directions, but it is usually cut to size at a convenient latitude, say $\pm 80°$.*

and

$$y = S_V R_e \log \tan 0 \qquad = S_V R_e \log 0 = -\infty$$

that is, the North and South Poles appear at infinite distances above and below the equatorial line no matter what the scale of the map. The longitude lines correspond to points on the sphere having a fixed β coordinate, and therefore their images on the map will have a fixed value of $x = S_H R_e$; that is, their images will be straight lines parallel to the y-axis. Similarly, lines of latitude on the sphere correspond to constant α, and their images on the map will have the equation $y = $ constant, which describes a horizontal line; the images of the latitude and longitude lines form a Cartesian coordinate network on the map.

The remarkable property of the Mercator projection is that compass courses, that is, curves on the earth's surface that meet all longitude lines at a fixed angle, are transformed into straight lines on the map. This raises the question: what kind of curve on the sphere is a compass course? Is it a *geodesic*[3] (that is, part of a great circle on the sphere) or some other easily described curve? Unfortunately it is not. A compass course is generally a complex curve of a spiral type that is not at all easy to describe without the aid of equations. Such a curve is illustrated in Figure 6.12 and further described in Chapter 8 in connection with Figure 8.29.

Compass courses have some interesting properties that can be readily derived from the equations, which we now set out to discover. A straight line on the Mercator map will have an equation of the form $ax + by = c$, in which a, b, c are certain constants. Equations 6.1 tell how to express x and y in terms of the latitude and longitude coordinates on the sphere; substitution of these expressions in the equation of the line produces the equation of the compass course on the sphere. To simplify matters choose $S_H R_e = 1$ and $S_V R_e = 1$; the equation of a compass course becomes

$$\beta = \frac{c}{a} - \frac{b}{a} \log \tan \left(\frac{\alpha}{2} + \frac{\pi}{4} \right)$$

and is called a *loxodrome*. For example, the straight line $y = 0$ corresponds to constants

FIGURE 6.12

Constant compass
course = straight
line on a Mercator
projection = "loxodrome"

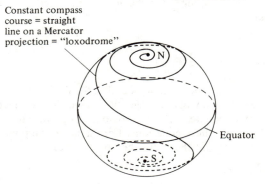

Equator

Constant compass course.

[3] *A geodesic curve is a curve that traces out the path of shortest distance between any two nearby points; for example, great circles on a sphere have this property. Latitude circles, except for the equator, have not. See Chapter 11.*

$a = 0$, $b = 1$, $c = 0$, *hence to the curve* $0 = \log \tan (\alpha/2 + \pi/4)$; the logarithm of a number is zero only if the number is 1 (*cp*. Chapter 7), and therefore $\tan (\alpha/2 + \pi/4)$ must be 1, which means that $\alpha = 0$. Therefore this particular straight line corresponds to the equator on the earth's surface. Similarly, the line $y = x$ corresponds to the loxodrome whose equation is

$$(6.2) \qquad \beta = \log \tan \left(\frac{\alpha}{2} + \frac{\pi}{4} \right)$$

What does this curve look like on the sphere? If $\alpha = 0$, then $\beta = \log 1 = 0$; therefore the point where the Greenwich meridian intersects the equator is certainly on the curve. As β increases from 0 to 2π the curve winds around the sphere in the northern hemisphere (compare Figure 6.12) to a point lying above the equator on the Greenwich (0°) longitude line; increasing β by another 2π corresponds to another turn of the curve around the sphere to a still higher latitude. As β increases indefinitely the expression on the right-hand side of (6.2) must also increase, which means that $\tan (\alpha/2 + \pi/4)$ increases indefinitely. Since $\tan \theta$ approaches infinity as θ approaches $\pi/2$ (*cp*. Figure 8.20 on p. 221), it follows that $(\alpha/2 + \pi/4)$ must approach $\pi/2$ as β approaches infinity; that is, α must approach $\pi/2$. There is only one point at latitude $\pi/2$: the North Pole. This shows that part of the loxodrome curve which corresponds to the line $y = x$ on the Mercator map is a spiral in the northern hemisphere that winds around the North Pole. The rest of the curve is obtained by seeing what happens as β becomes increasingly *negative;* this corresponds to a spiral in the southern hemisphere that winds about the South Pole. Figure 6.12 illustrates this loxodrome. All loxodromes other than the lines of latitude and longitude are spirals of the same sort, as we can easily verify.

Since compass courses on the Mercator map represent spiral curves on the earth's surface, it is clear that compass courses are *not* geodesic curves unless they are in the direction of the equator or a line of longitude. Although geodesic courses, that is, great circle routes, are shorter than other courses, they cannot be easily plotted on a map nor can they be easily maintained by the helmsman of a ship who would be required to make continuously changing course corrections in order to follow a great circle route between two points. The increased cost of a voyage due to the greater length of a compass course is apparently offset by the simplicity of loxodromic navigation.

Great circles on the sphere do not appear as straight lines in the Mercator projection, but their equations can, of course, be obtained by writing the equation of a great circle in terms of α and β, solving (6.1) for α and β in terms of x and y, and then substituting these results in the great circle equation. The final expression is complicated; it would be extremely awkward to plot geodesic courses on Mercator maps.

The theory of *functions of a complex variable,* which is not discussed in this book, shows how to relate the Mercator projection to stereographic projection. It turns out that Mercator's projection is the result of first projecting the sphere stereographically onto the plane and then transforming the points of the plane by means of the *complex logarithm function,* a natural generalization of the ordinary logarithm function (described in Chapter 7) to complex numbers. Looked at from the point of view of complex function theory, it turns out to be simple to prove that angles are preserved by the Mercator projection because both stereographic projection and the complex logarithm transformation have this property.

It is instructive to compare the cylindrical and Mercator projections. Recall the pairs of equations defining them:

$$\text{Cylindrical projection} \quad \begin{aligned} x &= S_H R_e \beta \\ y &= S_V R_e \tan \alpha \end{aligned}$$

$$\text{Mercator projection} \quad \begin{aligned} x &= S_H R_e \beta \\ y &= S_V R_e \log \tan \left(\frac{\alpha}{2} + \frac{\pi}{4} \right) \end{aligned}$$

The first important fact to be noted is that a knowledge of trigonometrical functions, including tables of their values, suffices for the practical construction of a cylindrical projection map. The location (Cartesian coordinates x, y) of a place on the map is determined from the latitude and longitude of that place, but the trigonometric functions are not sufficient for the construction of a Mercator map; it is *necessary* to introduce a new kind of function — the logarithm — and to tabulate its values for convenience of application. Introduction of this new and relatively complicated function cannot be avoided. The repertoire of functions has been extended many times in mathematical history, following this procedure as if it were a paradigm: in the search for a solution to an important problem a new function automatically appears. It is studied, its values are tabulated, and, with increasing use, it becomes more familiar and better understood. Often the study of these "new" functions suggests more sophisticated problems whose solution in turn leads to the introduction of other still more complicated functions, a process that apparently will continue forever.

Figure 6.3*b* illustrates the remark that the Mercator map point P^* corresponding to a point P on the earth's surface lies closer to the equatorial plane than the cylindrically projected image P' of P. It is instructive to compare the positions of P^* and P' and a convenient way to do so is by comparing their corresponding vertical map coordinates y. From the equations the ratio of y for cylindrical projection to the corresponding y for the Mercator projection is

$$\frac{\tan \alpha}{\log \tan (\alpha/2 + \pi/4)}$$

The vertical scale factor S_V and the radius of the earth R_e cancel in numerator and denominator; hence the ratio of the y's does not depend on either the choice of scale or on the size of the earth.

It can be shown, once something is known about the properties of the logarithm function, that this ratio is always greater than 1 when $0 < \alpha < \pi/2$. This means that, although the y-map coordinate of P approaches infinity as P approaches the North Pole for *both* types of map, nevertheless the map point P' (on the cylindrical projection) races out to infinity so much more rapidly than P^* (the map point on the Mercator projection) as P approaches the pole that the distance between P' and P^* also increases without bound.

§**6.4.** The history of cartography is fascinating. Little interpretation is necessary to form a reasonable opinion of the view of the world held by men hundreds of years ago; we have only to look at the maps they left, which express far more strikingly than words their outlook and state of knowledge. Regress is exhibited as well as progress in man's continuing efforts to construct an accurate represention of the nature and location of geographic places, and an entire cosmological viewpoint is contained in the description of the global shape of the earth.

Ptolemy's maps have a well-developed system of latitude and longitude coordinates, and he was obviously aware of the necessity of utilizing different mapping transformations for different purposes; Figure 6.13 shows two projections he used. Figure 6.14 reproduces a beautiful map attributed to him by its sixteenth-century Venetian publisher, but between Ptolemy's era (*ca.* + 150) and the sixteenth-century reproduction of his map there were many peculiar world views committed to posterity in this form. Ptolemy's model of the world was a ball, its spherical surface incapable of accurate representation on a portion of the Euclidean plane. Sixteenth-century geographers held the same view. In +535, however,

FIGURE 6.13

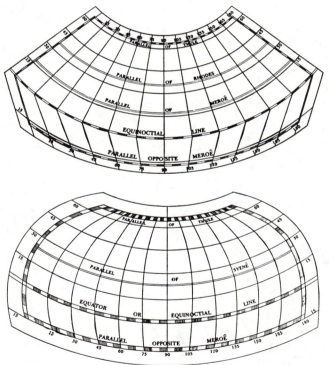

(*a*) A reconstruction of Ptolemy's conic projection, suggested for the construction of a map of the habitable world (after a sixteenth-century copy); (*b*) Ptolemy's modified spherical projection of the world, giving a superior likeness of the earth's surface on a sheet of paper. Though preferable to the conic projection, Ptolemy confessed that it was far more difficult to construct (courtesy Enoch Pratt Free Library).

FIGURE 6.14

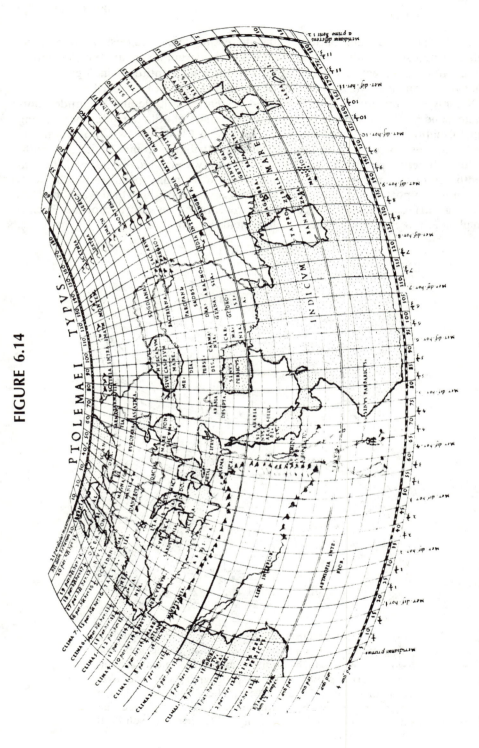

The world of Ptolemy according to a Venetian editor in 1561. Longitude is expressed in fractions of hours east of the Fortunate Islands; latitudes are designated by the number of hours in the longest day of the year (courtesy Enoch Pratt Free Library).

FIGURE 6.15

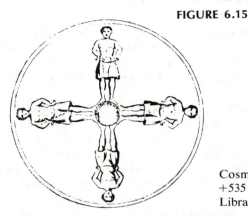

Cosmas' denial of the antipodes, +535 (courtesy Enoch Pratt Free Library).

the worldly merchant traveler turned monk, Cosmas of Alexandria, argued that

If two men on opposite sides [of the earth] placed the soles of their feet against [it] how could both be found standing upright?

(*cp.* Figure 6.15). If the earth were not spherical, the geometrical model of Aristarchus, used with good effect to compute the distances from the earth to the moon and sun, would not be applicable. Here is a clear illustration of the role of geometrical models in the interpretation of physical reality. Choice of the wrong model entails wrong numerical estimates of interesting physical quantities in almost all cases. Cosmas supplied his world view in a map made about +548 (Figure 6.16); the geometry that enabled Aristarchus to compute astronomical distances is entirely irrelevant here.

FIGURE 6.16

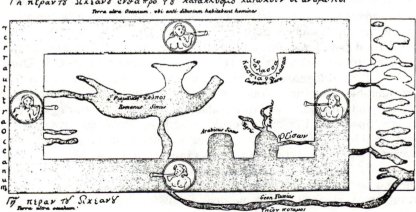

The world of Cosmas, about +548, was patterned after the Tabernacle. Four great rivers in Paradise supply the waters of the earth (from an original in the Library of Congress).

The symmetry inherent in Cosmas' map is typical of medieval European world views. It is carried to even further extremes by Isidore (Figure 6.17), the sixth-century Bishop of Seville, in a schematic map that is more suggestive of philosophical cosmology than a reduction of observations. The rectangular world of Cosmas' era apparently was unsatisfactory; it was replaced by a disk-shaped flat world, responsive to the same outlook as Cosmas' but frequently showing more details.

The map in Figure 6.18 includes Paradise and Eden; unfortunately, neither distances nor coordinates are supplied. Figure 6.19 is based on the same general geometric model, but introduces two disjoint massive islands, the one in the southern hemisphere free of life because the biblical ark was presumed to have landed on the other. Returning for a moment to Figure 6.15, we wonder if Cosmas realized that his argument would be nullified if similar assumptions were made about hypothesized antipodal islands on the surface of a spherical earth.

The views expressed by Cosmas and Isidore have always had adherents. Many in our own time, including the Flat Earth Society of London, have found the notion of a flat earth more consistent with naïve perception and congenial to superficial reasoning than the subtle concept of universal gravitation that is automatically entailed by a spherical earth. The proponents of naïve perception as the measure of truth were dealt a mortal blow by perception itself when photographs such as that shown in Figure 6.20 became commonplace. One wonders how the sophisticated Cosmas would have responded.

In general, it can be concluded that the theologians and philosophers of the Middle Ages, whose geometrical models were arrived at on the basis of symmetry and simplicity rather than observation and measurement, conformed to the stereotyped image of the abstract mathematician as one ever reasoning but never looking to a far greater extent than the mathematicians themselves.

§6.5. A map of a portion of the earth's surface made according to some mathematical principle contains within it a certain model of the earth, a notion of the true appearance of the globe. In addition, such a map implies a systematic means of determining the distances between points on the earth, measured at least in terms of latitude and longitude, and finally, utilizing a knowledge of the radius of the earth, in terms of meters or miles or some other conventional unit. During the early years of the Renaissance there was uncertainty about the size of the earth and considerable ignorance about the relative positions of famous geographical places. As a consequence it was often difficult for rulers of seafaring nations, such as Portugal, Spain, and England, to decide whether a proposed voyage of exploration was sensible in terms of the capabilities of ships to sustain, seamen to endure, and navigators to guide extended journeys into the unknown.

The story of Columbus' discovery of America in 1492 illustrates the array of opinion and partly understood facts current in his time as well as the risks of opportunities overlooked that sometimes follows acceptance of the counsel of learned men.

By 1492 Portugal was the recognized leader in exploring the seas. From 1434 on intrepid captains guided their small ships from Portugal along the west coast of

FIGURE 6.17

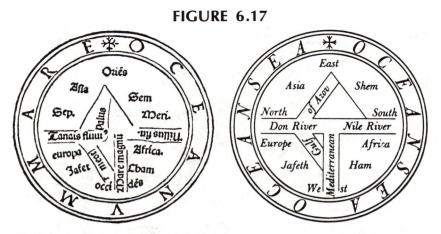

The world of Isidore, Bishop of Seville (+570–636), was extremely simple. The T–O map at the left, from his *Etymologies,* is explained by the diagram at the right (courtesy The Walters Art Gallery).

FIGURE 6.18

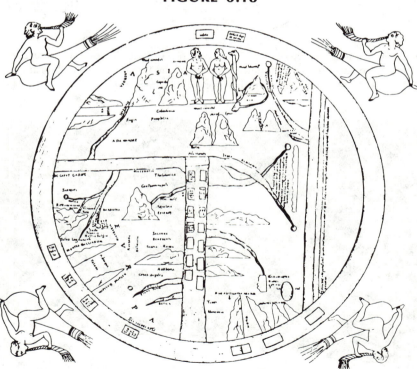

The world of +900. Paradise and Eden are in the Far East. Four wind blowers are releasing the winds of the earth from their Aeolus bags (from an original in the Library of Congress).

FIGURE 6.19

Life on Antichthon, below the equatorial ocean, was impossible, said the Church, because the ark with all survivors of the flood landed north of it on Mt. Ararat (from an original in the Library of Congress).

FIGURE 6.20

Earth as seen from the moon (Apollo 8 photograph).

Africa, frequently out of sight of land and in treacherous waters. They had no effective means of determining longitude, and their use of Polaris and other northern stars for latitude determinations became less and less reliable as they approached the equator. At about 9°N latitude Polaris sank out of sight below the horizon. As described in Chapter 5, other stars can be used for latitude determination *if* the time of observation is known, but because of the unavailability of accurate seaworthy timepieces this method was of little help to the fifteenth-century navigator; with the exception of the sun at noon, he could not determine the time of transit of a star through the zenith. The calculation of latitude by measurement of the sun proceeds along the same general path outlined in Chapter 5, but the technical problems of observation are greater and the necessary calculations much more complex because the path of the sun as seen from the earth varies with the season and indeed the *daily* differences are significant for navigational purposes. In addition, extensive tables of the sun's apparent position are necessary for the application of this *Rule of the Sun*, as it was then known, but these tables could not be prepared nor their use prescribed except by trained mathematicians and astronomers. We quote J. H. Parry's excellent book, *The Age of Reconnaissance* [53]:

> The exact processes whereby late medieval science was made available to seamen are, in general, little known. Chroniclers often stated that princes . . . invited astronomers to their courts in order to pick their brains; but they very rarely explained precisely what the learned men taught, to whom, and with what result. The Rule of the Sun is an exception: a clear and rare instance of a group of scientists, deliberately employed by the State, applying theoretical knowledge to the solution of a particular and urgent practical problem. John II of Portugal in 1484 convened a commission of mathematical experts to work out a method of finding latitude by solar observation.

> The work of John II's commission was summarized for the benefit of practical navigators in a manual compiled in Portugal under the title Regimento do astrolabio e do quadrante. This was the first European manual of navigation and nautical almanac. The earliest known copy was printed in 1509, but there were probably earlier editions, and the work seems to have been circulated in manuscript form from the fourteen-eighties. [See Figure 5.11.]

Columbus was not a professional seaman; Parry calls him a "self-taught and extremely persuasive geographical theorist, with some knowledge of hydrography and a grounding in navigation." In 1484, while John II's committee of mathematicians was working, Columbus approached the Portuguese crown with a proposal to voyage, at crown expense, westward from Europe over an unbroken expanse of water to Cipangu (Japan). He estimated the distance at about 3000 miles; it is actually more nearly 11,000. According to Parry, this tremendous discrepancy was the consequence of combining the Ptolemy-Poseidonius underestimate of the radius of the earth, overestimates of the size of Asia (due to Ptolemy and Marco Polo), and Marco Polo's estimate that Japan was 1500 miles distant from the mainland. When Columbus, sailing under the Spanish flag, found land after about 3000 miles of travel, he was not surprised. John II's experts, who had a relatively accurate knowledge of the size of the earth and a justified scepticism of reports of distances obtained from travelers, apparently cautioned the king to decline the proposal. At the more provincial Spanish court, free from the eminent mathematicians' burdensome knowledge and scepticism, Columbus' reception was, as we all know, warmer, and consequently all of South America, with the sole exception of Brazil, speaks Spanish today, not Portuguese.[4]

In Figure 6.21 we reproduce (from [14]) the map reputedly used by Columbus with the North American continent superimposed; Japan is shown extending from the actual location of Baja California almost to the equator.

John II's support of scientific research activity related to practical state problems should be familiar to us who live under an umbrella of federally supported research programs. State-supported research is not new now nor was it then. Archimedes did some Defense Department work from time to time (although he did not like it much), and several famous Elizabethan mathematicians were also cryptanalysts for the sixteenth-century equivalent of the CIA. That it may be undesirable does not make it necessarily a product of our era.

FIGURE 6.21

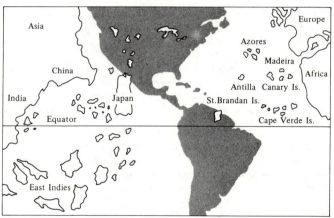

The shaded portions show what Columbus and his successors found superimposed on the Toscanelli map that Columbus used.

Of greater interest is the recognition that scientific *opinion* about scientific possibilities is essentially conservative, although the research activity itself is often revolutionary. Thus the opinion of John II's commission, that small ships could not sail to Japan over an unbroken sea, was scientifically valid but a discouragement to exploration of the western seas and ultimately a great limitation of Portugal's future.

---

[4] *The Guianas, with a total population of less than 1 million, are anomolies which we exclude from consideration.*

EXERCISES

6.1. Consider the cylindrical projection of a sphere which is then rolled out onto the plane, as shown in Figure 6.3a,b and described in the text. Suppose a region F on the sphere is mapped onto a *square* in the plane by such a projection. What can you say about the relations between the lengths of the "sides" of the region R on the sphere, assuming that two of the sides lie on meridians and one of the sides lies on the equator (the circle of tangency)? Draw a schematic diagram to illustrate the relationship.

6.2. Consider an equilateral triangle $\triangle$ on the plane as in Figure E6.2 where the segment AB lies on the equator. Describe the corresponding region in the sphere under the cylindrical and Mercator projections. What happens to the lengths of the sides under these projections? What happens to the angles α, β, and γ under these projections? Draw a schematic three-dimensional diagram of the image of the triangle.

FIGURE E6.2

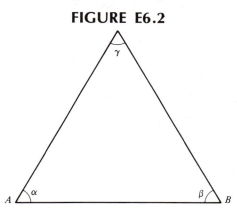

6.3. What can you say about the relationship between the areas of the regions considered in Problems 6.1 and 6.2. Give (intuitive) reasons for your answer.

6.4. Consider a sphere and suppose that the scale factors S_H and S_V are equal. What size map must we have in order that all of the sphere except between latitudes 88 and 90° (north and south) be represented if the image of the equator on the map is 1 foot long (a) under the cylindrical projection, (b) under the Mercator projection.

6.5. Consider a loxodrome on a sphere that spirals about the North Pole as described in the text. By using the equations that describe this curve plot the *projection* of the curve onto the plane passing through the center of the sphere and the equator by means of rays perpendicular to the plane. This will illustrate the spiral behavior of a straight line on a Mercator map.

6.6. Consider stereographic projection of the lower hemisphere onto a plane tangent to the South Pole. Can all of the hemisphere be mapped onto a finite region of the plane? Explain and illustrate your answer.

*6.7. Consider any map of the sphere onto a planar region. Give an argument to show that there cannot be any mapping that could *preserve all distances*. Recall that distance on the sphere is measured by the length of the great circle joining two points, and the length in the plane is measured by measuring the straight-line distance between two points. (*Note:* This is a hard question to answer rigorously; see Chapter 11. Give as complete an answer as you can by referring to examples, for instance.)

6.8. Do you agree with the assertion that the maps made by man at a given time in history reflect his view of cosmology and the universe? Explain.

6.9. Construct a rough scale model of a Mercator map of the outline of the United States by finding the latitude and longitude of about 15 to 20 representative cities and plotting them on an appropriately scaled sheet of graph paper, using the Mercator mapping equations (start with Seattle, San Francisco, San Diego, El Paso, Brownsville (Texas), and Miami).

6.10. By drawing a schematic diagram illustrate what the assertion "the Mercator map preserves angles" means. Indicate what kind of "computations" would have to be performed to prove this assertion.

6.11. How do you explain the fact that Ptolemy's map of the world (Figure 6.14) is much more accurate and complete than maps such as those in Figures 6.16–6.19 which appeared much later?

CHAPTER 7

The Invention of Logarithms

This chapter is concerned with logarithms, *a calculational device designed to ease the effort required to perform multiplication and division of large numbers. Calculations with logarithms utilizes tables in a manner similar to the way the Babylonian tables of reciprocals (described in Chapter 3) were used, and for much the same purpose. The demands of astronomy on calculational techniques were responsible for the invention of logarithms. In modern times their principal significance is their connection with* the calculus *and its application to* complex functions, *since extensive numerical calculations are now normally performed by computing machines.*

§7.1. In Chapter 2 we learned why the positional notation of the Babylonian mathematician gave him a great advantage over his contemporaries when it came to problems of computation, especially when fractions were involved. This was particularly true in astronomy. As time went on astronomy became a highly developed science whose predictions were increasingly precise, and consequently the computations grew more cumbersome. A need arose for methods that would simplify the ever more complicated computations such as the multiplication and division of 10- and 12-place decimal numbers. At the end of the sixteenth century a method for simplifying computations grew out of the study of trigonometric functions. Until the invention of logarithms by John Napier (1550–1617) in 1614 this method, called *prosthaphaeresis,* was the principal means of simplifying multiplication used in the major astronomical laboratories of Europe, of which that of Tycho Brahe (1546–1601) of Denmark was undoubtedly the most important for the development of modern science. It was the data accumulated in that laboratory, and the computations made there, that led Johann Kepler (1571–1630) in the next scientific generation to his formulation of the three laws of motion governing planetary movement. These laws were later incorporated into Isaac Newton's (1642–1727) theory of gravitational motion, which, for practical purposes (such as sending men to the moon!) is still the theory we use today to predict the motions of objects in space. Without adequate computational capability, Kepler's laws could never have been discovered.

It therefore seems appropriate to examine the means Tycho Brahe had for carrying out his complex and laborious computations. Until the advent of mechanical devices for computation, which later led to the development of electronic computers, the guiding principle behind computational methods was that it is easier to add and subtract than it is to multiply and divide; "easier" here means quicker and more convenient. Positional notation makes multiplication of large numbers feasible, but there is a limit to an individual's ability to compute the solution to a problem with large numbers in a given amount of time, although from time to time individuals possessing remarkable powers of mental calculation have appeared on the scene. Such a phenomenon was the German lightning calculator Zacharias Dase (1824–1861), whose extraordinary abilities included the (mental) calculation in 54 seconds of the product of two 8-digit numbers, two 20-digit numbers in 6 minutes, two 40-digit numbers in 40 minutes, and two 100-digit numbers in 8 hours and 45 minutes; Figure 7.1 displays these data on full logarithmic graph paper, from which it is easy to see that the time he required for these tasks varied approximately as $n^{2.7}$, where n denotes the number of digits in each factor. No one knows how Dase or the other rare calculating prodigies differ from ordinary persons nor whether the procedures they use introduce any essential simplifications that would, if they were known, significantly improve the calculating abilities of the average person. There are temptations to think of the calculating prodigy as a "biological computing machine" which by some happy quirk of the laws of probability and genetics has been "wired" in a remarkably more efficient way than the rest of us.

FIGURE 7.1

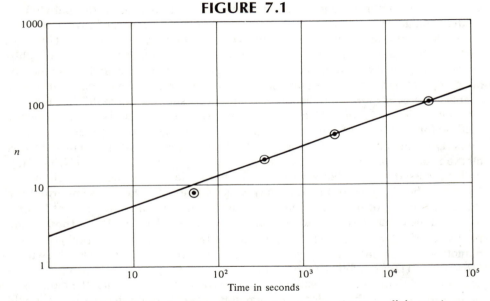

Time required for Zacharias Dase to calculate product of two n-digit numbers.

It is easy to make a theoretical estimate of how long it "should" take to calculate the product of two n-digit numbers. Let the numbers be $a = a_n \cdots a_2 a_1$ and $b = b_n \cdots b_2 b_1$. If multiplication is performed in the usual fashion, a diagram of the following kind is obtained:

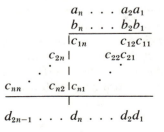

in which $d_{2n-1} \cdots d_2 d_1$ is the product; for example, 12×34 corresponds to the diagram

$$
\begin{array}{r}
12 \\
\underline{34} \\
48 \\
\underline{36} \\
408
\end{array}
$$

To calculate the product $a \times b$ each digit of a must be multiplied by each digit of b, a total of $n \times n = n^2$ elementary multiplications. This provides the digits c_{ij}. Next the c_{ij} in each column must be summed; we will ignore problems of "carrying" when the sum of two digits is greater than 9. There will be approximately n^2 elementary sums, as is easily seen by moving the triangle of numbers c_{ij} to the left of the dotted line to fill the space under c_{11}. A more careful analysis would account for the possibility that some of the products $a_j b_i$ may be greater than 9, but this is not important.

Let us assume that it takes p seconds to find the product of two 1-digit numbers and s seconds to find the sum of two 1-digit numbers in our memory. Then it must take *at least*

$$ T(n) = pn^2 + sn^2 = (p+s)n^2 $$

seconds to calculate the product of two n-digit numbers. Notice that the time Dase required increased more rapidly with increasing n than this theoretically minimal time n^2. If we assume that our memory does not know the difference between storing elementary products and elementary sums, then we can assume that $p = s$. Then

$$ T(n) = 2pn^2 $$

for the theoretically minimum time to compute the product of two n-digit numbers; therefore, if $n = 1$, $T(1) = 2p$, hence the memory time for recalling an elementary product can be estimated as

$$ p = \tfrac{1}{2} T(1) $$

The graph in Figure 7.1 does not show a time for the product of two 1-digit numbers for Dase: he was too fast to clock at this level and moreover the time needed to speak the answer was much greater than the time required to extract it from memory. But p can be estimated from Figure 7.1 in the following way. From the known values

$$ T(20) = 360 \text{ seconds} $$
$$ T(100) = 28{,}845 \text{ seconds} $$

we find the formula

$$ T(n) = (2p)n^{2.7237} $$

for Dase, from which p is determined by substituting, say, $n = 20$; therefore

$$ 360 = (2p)(20)^{2.7237} $$

Now $(20)^{2.7237} \simeq 3496$; consequently

$$p \simeq \frac{360}{(2)(3496)} = 0.051 \text{ second}$$

A similar argument can be made to estimate how many seconds are required to add two n-digit numbers. If s has the same meaning as before, then there will be n elementary additions of pairs of 1-digit numbers as well as about $n/2$ further elementary additions to account for "carrying" from one column to its left neighbor, since nearly half (exactly 25 out of 55) of the elementary sums exceed 9. This means that it must take at least

$$s(n + \tfrac{25}{55}n)$$

or about $\tfrac{3}{2}sn$ seconds.

§7.2. Since multiplication of two n-digit numbers requires about $(p + s)n^2$ seconds, whereas addition of two n-digit numbers requires only about $\tfrac{3}{2}sn$ seconds, including the time required to account for "carrying" from one column to the left adjacent column, it is clear that addition is easier and faster to perform than multiplication. Any procedure that can replace multiplication by addition (and/or subtraction) and a moderate use of tabulated values is likely to reduce the amount of labor required to calculate the products of multidigit numbers which are a natural feature of navigational and astronomical calculations.

There are numerous different ways of converting multiplication to addition accompanied by the use of tables. In this chapter we discuss three of them. The first has never been put into practice, as far as we know, but it is simple and illustrates the essential features of all methods. The second is the prosthaphaeretic technique which did play an important historical role as mentioned above. Logarithms, the third method, is the most efficient of all and also provides a solution to the arithmetical problem of calculating nonintegral powers and arbitrary roots of numbers as well as a means of computing locations on Mercator maps.

The idea underlying all methods of converting multiplication to addition is displayed in the following example. If x and y are any two numbers, then (*cp.* §4.10)

$$(7.1) \qquad (x + y)^2 - (x - y)^2 = 4xy$$

Therefore, if we define a function S by

$$S(x) = \tfrac{1}{4}x^2$$

and make a table of the values of $S(x)$, we will be able to calculate the product xy by performing only addition and subtraction and extracting certain values from the table of values of the function S. Indeed, when expressed in terms of S, the algebraic identity (7.1) shows that

$$xy = S(x + y) - S(x - y)$$

In order to calculate xy, we first calculate $x + y$ and $x - y$, then refer to the table of values of S to determine $S(x + y)$ and $S(x - y)$, and finally perform subtraction to obtain the result $S(x+y) - S(x-y) = xy$. It will be enough if the table lists the value of $x^2/4$ for x between 0 and 1; if x and/or y are larger than 1, then shift the decimal points to make both numbers lie in the range from 0 to 1 and, after using the table to perform the multiplication, shift the decimal back again in the answer.

An example will clarify this procedure. Consider the product 64.1×520. Here is an extract from a table of values of the function $S(x) = x^2/4$ in which x varies by 0.001 from entry to entry:

x	$S(x)$
0.103	0.00265225
0.104	0.00270400
.	.
.	.
.	.
0.585	0.08555625
0.586	0.08584900
0.587	0.08614225
0.588	0.08643600
0.589	0.08673025
.	.
.	.
.	.
0.690	0.11902500
0.691	0.11937025
0.692	0.11971600
0.693	0.12006225
0.694	0.12040900
.	.
.	.
.	.
0.999	0.24950025

To calculate 64.1×520, $64.1 = 0.641 \times 10^2$ and $520 = 0.052 \times 10^4$; therefore

$$64.1 \times 520 = (0.641 \times 0.052) \times 10^6$$

Take $x = 0.641$ and $y = 0.052$. Then

$$x + y = 0.693$$
$$x - y = 0.589$$

where we have been careful to write 520 as 0.052×10^4 rather than 0.52×10^3 to ensure that $x + y$ lies within the range of the table. From the table we find

$$S(x+y) = S(0.693) = 0.12006225$$
$$S(x-y) = S(0.589) = 0.08673025$$

186 THE ADOLESCENCE OF COMPUTATION

and
$$xy = S(x + y) - S(x - y) = 0.03333200$$
Consequently
$$64.1 \times 520 = (xy) \times 10^6 = 33{,}332$$
which is correct.

In Chapter 3 it was noted that the Babylonians had tables of squares of integers; since their notation for numbers was positional, they certainly could have used this method to simplify multiplication in their astronomical calculations, but there is no evidence that they actually used the tables for that purpose.

A table of values of the function S makes it easier to multiply pairs of numbers but it does not help the calculation of fractional powers, such as $2^{1/2}$. This is one reason why a table of logarithms is a much more useful invention than a table of squares.

§7.3. The term *prosthaphaeresis* originally referred to the correction necessary to find the "true," that is, actual apparent, position of a planet, from its average estimated position and later came to be used also for the mathematical methods employed in calculating the correction.

A *prosthaphaeretic rule* is a trigonometric identity that converts a product of two numbers into the sum of two other numbers. These identities are familiar to the student who knows some trigonometry, but their use for simplifying multiplication is an application of trigonometry that is not as well known as it should be. One prosthaphaeretic rule, for instance, is,

(7.2) $$2 \cos A \cos B = \cos (A + B) + \cos (A - B)$$

which is easily derived from the standard addition formula

$$\cos (A + B) = \cos A \cos B - \sin A \sin B$$

To use formula (7.2) for multiplication we need a table of cosines that will conform to the desired degree of accuracy of the computation. Suppose we are interested in the product 80765×97803. We use a table of cosines as follows: first write

(7.3) $$80765 \times 97803 = (0.80765 \times 0.97803) \times 10^{10}$$

so that the numbers under consideration are replaced by decimals between 0 and 1. Now compute the product in the parentheses this way: set

$$\cos A = \frac{(0.80765)}{2} = 0.40383$$

$$\cos B = 0.97803$$

and find angles A and B from the table of cosines with these values, namely (in radians),

$$A \simeq 1.152$$
$$B \simeq 0.208$$

Next compute

$$A + B = 1.360$$
$$A - B = 0.944$$

and again from the table compute

$$\cos (A + B) \simeq 0.208$$
$$\cos (A - B) \simeq 0.588$$

Now we obtain

$$\cos (A + B) + \cos (A - B) = 0.796$$

which is, by (7.2), the desired product in parentheses in (7.3). By relocating the decimal point we have the (approximate) result

$$80765 \times 97803 \simeq 7.96 \times 10^9$$

Actual multiplication shows that the product is exactly 7,899,059,295; use of more accurate five-place trigonometrical tables improves the prosthaphaeretic result to 7.9008×10^9.

Which method is simpler? Even in our example it is easier to refer to the tables and make three additions and one division by 2 than it is to carry out the full multiplication. To get a more accurate result it is necessary only to use better tables than those we applied here. On the other hand, a computation involving scientific data is only as accurate as the original measurements. The extra work involved in full multiplication to obtain the complete 10-digit product above is worthless if the initial data are only accurate to, say, five significant figures.

§7.4. Logarithms are another and more efficient means of converting a product to a sum and a quotient to a difference, but they have the additional important virtue of converting an exponential power to a product, including square and cube roots, etc. By using a table of logarithms we can reduce the computation of complicated multiplications to sums, and powers to products, and save an enormous amount of time.

Since it appears to be the nature of physical reality as well as society that we cannot get anything free, not even a simplified computation, it is natural to suspect that a large amount of work and energy must have gone into the preparation of tables of logarithms, if they are themselves such labor-saving devices. In fact, the first table of logarithms, *Mirifici logarithmorum canonis descriptio,* was published in 1614 by John Napier after he had worked 20 years to compile it. It has been translated (into English by W. R. McDonald, 1889) as *The Construction of the Wonderful Canon of Logarithms.*

The essential property of the logarithm is that it is a function which converts a product to a sum and a quotient to a difference according to the following rules:

(7.4) $$\log (x \cdot y) = \log x + \log y$$

(7.5) $$\log \left(\frac{x}{y}\right) = \log x - \log y$$

Although many alternative definitions are available, depending on the sophistication of the student and the application of interest, the logarithm is generally defined this way. Let b be a positive real number other than 1. Then define

$$\log_b x = y \quad \text{if and only if} \quad b^y = x$$

that is, $\log_b x$ is that number y such that $b^y = x$. For instance,

$$\log_{10} 1000 = 3 \text{ because } 1000 = 10^3$$
$$\log_2 16 = 4 \text{ because } 16 = 2^4$$

and

$$\log_2 \sqrt{2} = 0.5 \text{ because } \sqrt{2} = 2^{1/2}$$

This is a perfectly good definition, provided we have an idea what b^y means for an arbitrary y. The number b is called the *base* of the logarithm system. If $y = n$ is a positive integer, then

$$b^n = b \cdot b \cdots b \quad (n\text{-times})$$

So $\log_b (b^n) = n$, according to the definition. Similarly, if $y = p/q$, where p and q are positive integers, we can define

$$b^{p/q} = \sqrt[q]{b^p}$$

where we assume that we understand how to find the qth root of a given number. Note that

$$\log_b (b^n \cdot b^m) = \log_b (b^{n+m}) = n + m$$
$$\log_b (b^n) = n$$

and

$$\log_b (b^m) = m$$

Therefore

$$\log_b (b^n \cdot b^m) = \log_b (b^n) + \log_b (b^m)$$

hence the logarithm of the *product* $(b^n \cdot b^m)$ is the *sum* of the logarithms of the factors b^n and b^m. This is what the logarithm function does in general. The law of exponents still holds for fractional powers, since

$$b^{p/q} \cdot b^{m/n} = b^{(p/q)+(m/n)}$$

so

$$\log_b (b^{p/q} \cdot b^{m/n}) = \frac{p}{q} + \frac{m}{n} = \log_b (b^{p/q}) + \log_b (b^{m/n})$$

If we know something about limits and continuity[1], then we know that any real number y is the limit of rational numbers like p/q (indeed, cut off the decimal expansion of the real number[2] at some digit and obtain a rational approximation.) Let p_n/q_n be a sequence of

---

[1] *Discussed in Chapter 9.*

[2] *Recall the definition and properties of real numbers presented in Chapter 1.*

rational numbers that converges to the given number y; then define

$$b^y = \lim_{n \to \infty} b^{p/q}$$

which defines, that is, gives a meaning to, the expression b^y for all y, and therefore to the logarithm function as well. Indeed, given x, $\log_b x$ is *that* number y such that $b^y = x$.

Many important properties are easy consequences of the definition of the logarithm and the equation (7.4). First of all, if $x = y = 1$, then by (7.4)

$$\log_b 1 \cdot 1 = \log_b 1 + \log_b 1$$

so $\log_b 1 = 0$; the logarithm of 1 in any base is 0. Also, if s is any number, then

$$\log_b (x^s) = s \log_b x$$

because $b^y = x$ implies that

$$x^s = (b^y)^s = b^{ys}$$

whence $ys = \log_b x^s$; but, from the definition, $y = \log_b x$. This provides a quick method for calculating the value of any power or root of x.

A graph of the exponential curve $x = b^y$, wherein b is a fixed number greater than 1, is shown in Figure 7.2 with some sample points picked out. If the axes are rotated about the dotted line, which is the same as flipping the plane over along the dotted line, the graph of the function $y = \log_b x$ results (see Figure 7.3).

For historical reasons $b = 10$ has been used most often for computational pur-

FIGURE 7.2

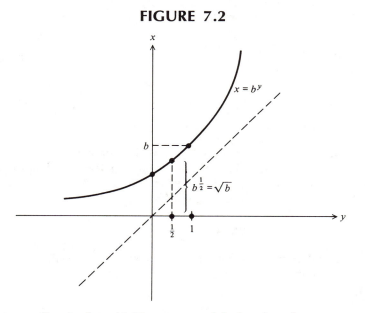

Graph of $x = b^y$. Note permuted designation of axes.

FIGURE 7.3

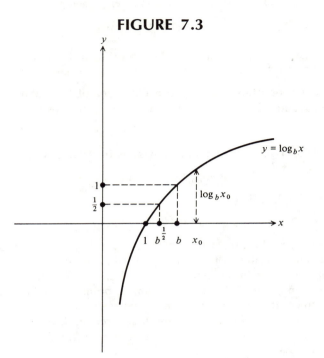

Graph of $y = \log_b x$ is the mirror image relative to the
dotted line of $x = b^y$ in Figure 7.2.

poses. From the point of view of calculus (see Chapter 9) and general theoretical
considerations the number[3]

$$e = 1 + \frac{1}{1} + \frac{1}{1 \cdot 2} + \frac{1}{1 \cdot 2 \cdot 3} + \frac{1}{1 \cdot 2 \cdot 3 \cdot 4} + \cdots$$

$$\approx 2.7182818$$

first used by Leonhard Euler (1707–1783), is the *natural base* for logarithms. It
turns out that e is an irrational number (indeed, it is transcendental) that plays a
role in mathematics of comparable importance to π.

In fact, the power e^y can be expressed by

$$e^y = 1 + \frac{y}{1} + \frac{y^2}{1 \cdot 2} + \frac{y^3}{1 \cdot 2 \cdot 3} + \cdots$$

so this "infinite polynomial" defines a number for each value of y. Define the natural logarithm of x by

(7.6) $\log_e x = y$ if and only if $e^y = x$

---

[3] *Using the calculus, it can be shown that another expression for* e *is* $\lim\limits_{N \to \infty} (1 + 1/N)^N$.

Suppose that $e^v = u$ for two other numbers u and v. Then the definition in (7.6) means that

(7.7) $$v = \log_e u$$

Now e is just a number, although it has a special "name" so the usual law of exponents holds for it:

(7.8) $$e^y \cdot e^v = e^{y+v}$$

Substitute the definition of $\log_e x$ and $\log_e u$ from (7.6) and (7.7) into (7.8) to find

(7.9) $$e^{\log_e x} \cdot e^{\log_e u} = e^{\log_e x + \log_e u}$$

Since $x = e^y = e^{\log_e x}$ and $u = e^v = e^{\log_e u}$, the left side of (7.9) is $x \cdot u$; but *by definition*

$$x \cdot u = e^{\log_e(x \cdot u)}$$

and from (7.9)

$$e^{\log_e(x \cdot u)} = e^{\log_e x + \log_e u}$$

If two powers of the same number are equal, then the exponents must also be equal. This means that

$$\log_e (x \cdot u) = \log_e x + \log_e u$$

which shows that the logarithm does indeed convert multiplication to addition.

§7.5. Napier proceeded in quite a different way. A logarithm is supposed to convert a product to a sum. A *geometric sequence*

$$a, \ ar, \ ar^2, \ ar^3, \ \ldots$$

is constructed of products. The initial term is a fixed number a, and each succeeding term is obtained from its predecessor by *multiplication* by a fixed quantity r. An *arithmetic sequence*

$$a_0, \ a_0 + s, \ a_0 + 2s, \ \ldots$$

involves sums. It has an initial term a_0 and each term is constructed from its predecessor by *addition* of a quantity s. Napier's definition of logarithms was as follows: to each term of the geometric sequence

(7.10) $$N, N\left(1 - \frac{1}{N}\right)^1, \ N\left(1 - \frac{1}{N}\right)^2, \ldots, N\left(1 - \frac{1}{N}\right)^N$$

he associated the corresponding term of the arithmetic sequence

(7.11) $$0, \ 1, \ 2, \ldots, N$$

and he called this correspondence the *logarithm* — in symbols

$$\text{Nap. log} \left[N\left(1 - \frac{1}{N}\right)^k \right] = k$$

Napier put it this way:

> *Logarithmi dici possunt numerorum proportionalium comites aequidifferentes.*
> (*Logarithms may be called equidifferent companions to proportional numbers.*)

Geometrically, Napier used the following description. Let P be a point on the line segment AB as in Figure 7.4. Suppose P is moving toward B at a rate proportional to its distance from B, while at the same time Q is moving at a constant speed along the segment CD; then Napier called the length $\overline{CQ}$ the logarithm of the length $\overline{AP}$. (This length was supposed to be the sine of an angle. Napier had the astronomical applications in mind.) In his words (from the English translation):

> *The logarithm of a given sine is that number which has increased arithmetically with the same velocity throughout as that with which radius began to decrease geometrically, and in the same time as radius has decreased to the given sine.*

FIGURE 7.4

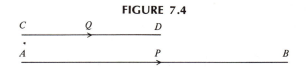

In the sequences (7.10) and (7.11) we see that the difference between successive terms is decreasing proportionally to the number of terms involved. This is the same idea illustrated in Figure 7.4.

Using these definitions, Napier was able to compute tables of his logarithms, for which he took N to be 10^7. Independently Jost Bürgi (1552–1632) developed a similar idea but published his results later. This particular definition was most suitable for computing $\log \sin \alpha$ for some angle α, and, as already mentioned, it is actually what Napier had in mind. His methods were refined by Henry Briggs (1561–1631), who published the first logarithms based on 10 which became the standard method for simplifying computation for some 300 years.

As is usual in the natural sciences, the innovative mathematician has usually assimilated many ideas from previous generations and remolded them into a new concept. To a certain extent this may also have been true in Napier's case, but logarithms did make a big splash when they were first presented. For instance, Briggs, visiting Napier in Scotland, is reputed to have said[4]:

> *My Lord, I have undertaken this long journey purposefully to see your person, and to know by what engine of wit or ingenuity you came first to think of this most excellent help in astronomy, viz., the logarithms.*

§7.6. How difficult is it to construct a useful table of logarithms, starting from scratch? Let us use our elementary definition of logarithms for base 10 given in Section 7.4 to approximate the logarithms of the numbers 1 through 10; that is, we shall find approximate values for $\log_{10} 1, \log_{10} 2, \ldots, \log_{10} 10$, which we denote

---

[4] F. Cajori, "*History of the exponential and logarithmic concepts,*" Amer. Math Monthly, **20**, (*1913*) p. 6.

by log 1, log 2, . . . , log 10. First, evidently $10^0 = 1$ and $10^1 = 10$, so immediately we have log $1 = 0$ and log $10 = 1$. Now let us find the others.

Consider log 2. The trick is to find a high power of 2 that is almost equal to a high power of 10 and then use the properties of logarithms. We see that

$$2^{10} = 1024 \simeq 1000 = 10^3$$

so we may assert that

$$\log 2^{10} \simeq \log 10^3$$

which implies

$$10 \log 2 \simeq 3 \log 10 = (3) \cdot (1) = 3$$

and yields

$$\log 2 \simeq \tfrac{3}{10} = 0.30$$

Now we can calculate log 5; for instance,

$$\log 5 = \log \left(\frac{10}{2}\right) = \log 10 - \log 2 \simeq 1 - 0.3 = 0.7$$

Let us try 3 and 7, the other primes[5] between 1 and 10. We have, for example,

$$3^7 = 2187 \simeq (2) \cdot (1093) \simeq 2 \times 10^3$$

Using this equation and our result for log 2, we obtain

$$\log 3 \simeq \tfrac{1}{7}(\log 2 + 3 \cdot \log 10)$$
$$\simeq \tfrac{1}{7} \cdot (0.3 + 3) \ = \frac{3.3}{7} = 0.47$$

For log 7 we use the simpler fact that

$$7^2 = 49 \simeq 50 = \frac{10^2}{2}$$

whence

$$\log 7 \simeq \tfrac{1}{2}(\log 10^2 - \log 2)$$
$$\simeq \tfrac{1}{2}(2 - 0.30) \simeq 0.85$$

We can also compute log 4, log 6, log 8, and log 9 by using $4 = 2^2$, $6 = 2 \cdot 3$, $8 = 2^3$, and $9 = 3^2$ and the values for log 2 and log 3. We thus obtain a primitive log table (see Table 7.1). The right-hand column gives the values of log x accurate to three places from a standard logarithm table; as we can see, there is a good deal of agreement. This is *not* the way logarithms are computed today, however. It is only an indication that in principle we can obtain some sort of approximation by elementary methods.

To compute logarithms in a systematic manner we once again use infinite series, unfortunately not available in modern form to Briggs and Napier, who had to use

[5] *An integer is* prime *if it is divisible (without a remainder) only by ±1 and ± itself.*

TABLE 7.1

x	$\log x$	$\log x$
1	0.00	0.000
2	0.30	0.301
3	0.47	0.477
4	0.60	0.602
5	0.70	0.699
6	0.77	0.778
7	0.85	0.845
8	0.90	0.903
9	0.94	0.954
10	1.00	1.000

laborious computations of roots to construct their tables. Today we know that[6]

$$\log_e (1 - x) = x + \frac{x^2}{2} + \frac{x^3}{3} + \cdots$$

as long as

$$-1 < x < +1$$

This will give the values of $\log_e x$ for $0 < x < 1$, which suffices, according to the rules for operating with logarithms, to determine the log of any number, in any base. For instance, there is the simple rule for changing base in a logarithm (which we shall not derive here) that relates base e and base 10 in the following way:

$$\log_{10} x = (\log_{10} e) \log_e x$$

Therefore

$$\log_{10} (1 - x) = (\log_{10} e) \left(x + \frac{x^2}{2} + \cdots \right)$$

and we can compute away. Also the example

$$\log_{10} (31.735) = \log_{10} (10^2 \times 0.31735) = 2 + \log_{10} (0.31735)$$

illustrates why we need only the logs of the numbers between 0 and 1 to find the log of any positive number.

Logarithms are, of course, still in use, and today they serve many purposes in addition to the simplification of multiplication and calculation of powers. These traditional applications are incorporated in the engineering slide rule[7] as well as in

---

[6] *This relation can be derived from Napier's definition of the logarithm in terms of rates of change of lengths of line segments by using* Taylor's theorem *as described in Chapter 10.*

[7] *The "C" and "D" scales of the typical slide rule are logarithmic, and addition of segments (parallel translation of the slide along the rule) corresponds to addition of the logarithms of the numbers etched on the segments, thus yielding the logarithm of the product (or quotient); the product (or quotient) can then be read off directly. The scales correspond to the tabulated values of the logarithm function and adjustment of the slide position to addition or subtraction.*

semilogarithmic and full logarithmic graph paper (used for Figures 0.1 and 7.1, respectively) but, with the advent of inexpensive electronic desk calculating machines, these uses have diminished in importance, whereas the use of the logarithm function as an intrinsic ingredient in the solution of problems continues to increase. Just as the vertical coordinate in the Mercator projection involves the logarithm function as the *result* of a practical calculation rather than as an intermediate simplifying technique, so too do these new applications treat the logarithm function on a par with the trigonometrical and other fundamental functions that occur constantly throughout all of mathematics.

EXERCISES

7.1. From the trigonometrical identity $\cos (x + y) = \cos x \cos y - \sin x \sin y$ and the facts that $\cos (-x) = \cos x$ and $\sin (-x) = -\sin x$ prove that

(*) $2 \cos x \cos y = \cos (x + y) + \cos (x - y)$

7.2. Use the identity labeled (*) in Problem 7.1 to calculate the following products (the "prosthaphaeretic rule") [*Hint for part* (*b*): recall that $\cos (\pi - x) = -\cos x$ if $0 \leqslant x \leqslant \pi/2$].
(a) 1576×998
(b) 242×358
(c) 242×0.358

7.3. Is the method of multiplying numbers in Problem 7.2 easier or more complicated than direct computation? Explain your answer. Give examples of situations in which it would be (a) useful; (b) not so useful (in fact a hindrance).

7.4. How many distinct operations are required to compute products by the prosthaphaeretic rule? How many are required by using logarithms?

7.5. Describe in general terms how a table of logarithms can be used to compute arbitrary (real) powers of a real number. In addition, give several nontrivial examples to illustrate your method.

7.6. Can the prosthaphaeretic rule be used to compute arbitrary powers of a given number? Explain.

7.7. (a) Compute the number e accurate to three decimal places by using the series

$$e = 2 + \frac{1}{2} + \frac{1}{2 \cdot 3} + \frac{1}{2 \cdot 3 \cdot 4} + \cdots$$

(that is, approximate by using enough terms of the infinite sum such that the first three decimal places do not change if further terms are added).
(b) Using this approximate value of e, find an approximate value for $\log_{10} e$ from the table of base 10 logarithms.

7.8. Use the series

$$\log_{10} (1 - x) = \left(x + \frac{x^2}{2} + \frac{x^3}{3} + \cdots \right) \log_{10} e$$

valid for $-1 < x < 1$ and the value of $\log_{10} e$ in Problem 7.7b to compute
(a) $\log_{10} 2$
(b) $\log_{10} 3$
(c) $\log_{10} 7$
and compare your values with those in Table 7.1 and in the logarithm table in the back of the book [*Hint:* Use $\log_{10} y = -\log_{10} (1/y)$].

7.9. By employing the "rough" technique in the text for approximating $\log_{10} 2$ without using series, can you improve on the value $\log_{10} 2 = 0.30$ given there?

7.10. Explain the advantages and disadvantages of the system of logarithms to
(a) base 2,   (b) base e,   (c) base 10.

7.11. Propose an explanation to account for the "excessive" time required by Zacharias Dase to mentally calculate the product of two 8-digit numbers, as shown in Figure 7.1.

7.12. On 1 July 1968 the estimated population of the metropolitan area of Houston, Texas, was 1,867,000 and was increasing at an annual rate of 3.4 percent. Eighteen months later the total population of the United States was estimated as 204,351,000 and was increasing at an annual rate of 1 percent. Use the values $\log_{10}(1.034) = 0.0145$, $\log_{10}(204,351,000) \simeq \log_{10}(204,000,000)$, $\log_{10}(1,867,000) \simeq \log_{10}(1,870,000)$ and Table 1.
(a) Assuming that the rates of increase will remain constant, in what year will the population of Houston equal the population of the United States?
(b) Does this mean that the rest of the United States will be empty?
(c) Does it mean that Houston will coincide with the United States?
(d) Is there any sensible conclusion that can be drawn from this calculation?

7.13. Can a table of values of the function $S(x) = x^2/4$ be used to convert division to addition and/or subtraction? Give a reason for your answer.

7.14. Find a formula for computing x^4 from a table of the function $S(x)$.

7.15. The purpose of this problem is to compare your ability to calculate with that of the prodigy Zacharias Dase and also with the theoretically optimum ability. In the text we showed that the time in seconds required to compute the product of two n-digit numbers should vary approximately according to the equation

$$T(n) = c \cdot n^2 \qquad c = p + s$$

where p stands for the time in seconds required to multiply two 1-digit numbers and s, the time in seconds required to add two 1-digit numbers. When $T(n)$ versus n^2 is graphed on full logarithmic graph paper, such as that used for Figure 7.1, the result should be a straight line of slope equal to 2. Recall that for Zacharias Dase the required time is proportional to $n^{2.72}$; that is, there is a constant c_D such that

$$T(n) = c_D n^{2.72}$$

and from the data given we can estimate $c_D = 0.102. \ldots$ Here is a short list of products whose factors have been chosen from a table of random numbers. Have someone read the numbered pairs to you and record the time in seconds that you required to calculate the product. Calculate the average time in seconds for products of 1 digit $\times$ 1 digit, 1 digit $\times$ 2 digits, 2 digits $\times$ 2 digits etc., as far as you can go.

Random Products

	Random Pairs	Time to Calculate (in seconds)	Average Time for Set
1 digit × 1 digit $n = 1$	7 × 5 2 × 6 4 × 5 9 × 5 3 × 7		
1 digit × 2 digits	9 × 63 7 × 57 3 × 74 3 × 19 6 × 53 3 × 37 6 × 13 2 × 32 8 × 56 6 × 68		
2 digits × 2 digits	66 × 15 88 × 38 61 × 12 22 × 25 71 × 46 98 × 18 65 × 68 98 × 11 31 × 74 17 × 88		
2 digits × 3 digits	31 × 617 39 × 941 72 × 979 44 × 164 91 × 497 11 × 936 44 × 864 86 × 978 34 × 538 52 × 147		
3 digits × 3 digits	172 × 133 385 × 145 144 × 569 629 × 935 128 × 231 594 × 845 352 × 825 976 × 892 934 × 167 747 × 864		

(a) Plot the required time in seconds for calculating the product of a k digit $\times$ m digit number on the graph paper provided at the point $n = \sqrt{km}$: *i.e.*, at $n = 1$ for 1 digit $\times$ 1 digit; at $n = 1.41 \ldots = \sqrt{1 \times 2}$ for 1 digit $\times$ 2 digits, etc.

(b) If the points on the graph lie nearly along a straight line, estimate the slope of that line, that is, the exponent E in the equation $T(n) = c_{you}n^E$. Compare your exponent with Dase's and with the theoretical minimum $E = 2$.

(c) Estimate the value of $(p + s) = c_{you}$ for you. Compare this with the estimate $c_D = 0.102$ for Dase.

(d) Propose an explanation for the differences in time required theoretically by Dase and by you.

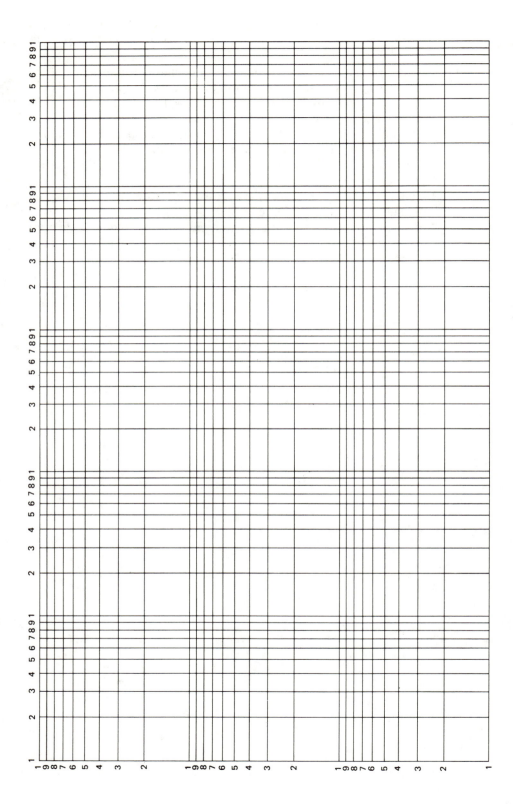

CHAPTER 8

The Algebrization of Geometry

As the classes of curves and other objects considered in geometry grew more extensive, it became necessary to find a method for converting geometrical problems into an algebraic (and later, an analytic) form. The advantage of "algebrizing" geometrical problems is that the solutions of algebraic problems usually follow an algorithm when they are possible at all, whereas the solutions of geometrical problems by geometrical methods usually require imaginative acts unique to each problem. In this sense the urge to algebrize is one of economy of effort. The present chapter discusses this problem and presents the solution developed by Descartes; it leads naturally to the theme of Part III, which is the application of analytical techniques, such as the calculus, to the solution of still more complicated classes of geometrical problems and ultimately to the description of the cosmos.

In addition to a theoretical and historical discussion we have included a "sampler" of famous and useful curves, the earliest dating from c. −200 and the most recent a product of the nineteenth century, to illustrate how the geometrical character of a curve becomes more and more complicated as the means for defining it become increasingly subtle, passing from synthetic geometrical definitions through definition by algebraic and transcendental equations to definitions that demand the power and formalism of the calculus.

§8.1. The idea of putting a coordinate grid on a geometric surface for purposes of measurement and location has a long history. Perhaps the most famous example from ancient times is Ptolemy's map of the world (see Figure 6.14) in which he represented points on the globe by pairs of angles that measure the number of degrees north and east of a fixed reference point. This system of latitude and longitude coordinates played a crucial role in the theory of navigation discussed in Chapter 5.

The chapter title "algebrization of geometry" specifically refers to the invention of *analytic geometry* by René Descartes (1596–1650) and Pierre Fermat (1601–1665). These mathematicians combined the notation and problem-solving ability of the algebraist, which originated with the Babylonians, with the geometry of the plane and space developed by the Greeks. The systematic transition from one to another is achieved by means of a *system of coordinates,* a combination that has enriched both algebra and geometry and indeed all of mathematics.

There were three major ingredients in the invention of analytic geometry:

1. The development of algebraic notation and algebraic problem-solving ability.
2. The classical synthetic geometry of the Greeks (as formulated, for instance, by Euclid).
3. The concept of a coordinate system.

§8.2. To begin the discussion of the "algebrization of geometry" it is appropriate to consider some aspects of the development of algebra as a branch of mathematics. There was progress along two fronts concerned respectively with problem-solving ability and algebraic notation. The ability to solve problems involved procedures such as factoring, reduction of complicated equations to a simpler form, methods of substitution of new variables, the use of square roots and exponents, etc., all of which are familiar to the average high school student today.

Algebraic notation passed through three stages of expression: *rhetoric* (grammatical sentences), *syncoptic* (shorthand abbreviations and words), and *symbolistic* (abstract symbols). The Babylonians wrote out their problems and solutions fully in words (*cp.* Chapter 3), and with few exceptions this was the accepted procedure in western societies until the time of Regiomontanus (1436–1476); for example, Leonardo di Pisa (Fibonacci, 1180–1250) describes the multiplication of two linear expressions, $(10 - x) \cdot (24 + x)$, as follows:

> *Ex 10 in 24 veniunt denarii 240; et ex 10 in re addita veniunt decem res additae; et ex 24 in re diminuta veniunt 24 res diminutae; a quibus si auferantur 10 res additae, remanebunt 14 res diminutae; et ex re addita in rem diminutam provenit census diminutus et sic habentur pro dicta multiplicatione denarii 240, censo diminutis et rebus 14.*

> Translation[1]: *From 10 in 24 come 240 known units; from 10 in an additive* x *come 10 additive* x*; and from 24 in a subtractive* x *come 24 subtractive* x*. If one subtracts the 10 additive* x *from this, then there remain 14 subtractive* x*. And from the additive* x *in the subtractive* x *results a subtractive* x²*. Therefore for the desired multiplication problem we have* 240 − x² − 14x.

Note that the + and − signs, parentheses, and the abstract symbol x for the unknown had not yet been invented. How difficult it must have been to try to express in a similar way a formula for the multiplication

$$(ax + b)(cx + d)(ex + f)$$

which is a triviality for any contemporary high school student. Less *thought* is necessary when modern formulas are used and therefore we can accomplish more; operations are automatic, once learned and understood.

The rhetorical mode of expressing algebraic relationships was transmitted to western Europe by the Arabs. The first mathematicians in Italy to take up Arabic algebra in the thirteenth century faithfully followed this tradition. The transition to the syncoptic period was encouraged by the desire to eliminate some of the excessive writing necessary for the rhetorical expression of mathematical relationships. The basic syncoptic form was still the grammatical sentence, but various shorthand notations represented often repeated words or concepts. The earliest known example of syncoptic notation is probably Diophantus' (of Alexandria, *fl. circa* +250) use of the particular Greek symbol sequence ϛ for the unknown quantity (from the final *s*-sound in the Greek word ἀριθμός – arithmetic). For example, with μ^δ denoting a constant quantity,

$$\bar{\iota}\,\varsigma = 10x \qquad \bar{\iota\alpha}\,\varsigma = 11x \qquad \bar{\lambda}\mu^\delta = 30$$

(*cp.* the discussion of the Greek number system in Chapter 1). Using these abbreviations, Diophantus wrote the "syncoptic sentence"[2]

---

[1] *Taken from Tropfke* [69] *p. 121, Vol. 3, whose German has been translated into English.*

[2] *Taken from Tropfke* [69] *and* [51], *as are the other examples in this section. The superscripts play a role similar to the "th" in "8th" (English) or "ieme" in "2$^{i\grave{e}me}$" (French) in designating ordinal numbers. We have reversed the order of the factors in the translation to conform to contemporary usage.*

$$\varsigma\varsigma^{\text{οι}} \ \overset{\text{''}}{\alpha}\rho\alpha \ \bar{\iota} \ \mu^{\text{ο}} \ \lambda \ \overset{\text{'}}{\iota}\sigma\text{οι} \ \epsilon\overset{\text{'}}{\iota}\sigma\overset{\text{`}}{\iota}\nu \ \varsigma\varsigma^{\text{οῖς}} \ \overline{\iota\alpha} \ \mu\text{ο}\nu\overset{\text{'}}{\alpha}\sigma\iota \ \overline{\iota\epsilon}$$
$$10x + 30 = 11x + 15$$

which represented a shorthand version of the sentence in mind:

Therefore 10 numbers (and) 30 units are equal to 11 numbers (and) 15 units.

Diophantus had shorthand notation for x^2 ($\delta^{\bar{v}}$, from the Greek word for square), x^3 ($\chi^{\bar{v}}$, from the Greek word for dice), etc.

Syncoptic notational systems became common in early Renaissance Europe. Regiomontanus wrote algebraic expressions in the following manner (modern translations are on the right-hand side):

$$\frac{2\alpha \ et \ 100 \ \tilde{m} \ 20\text{e}}{10\text{e} \ \tilde{m} \ 1\alpha} - 25 \qquad \frac{2x^3 + 100 - 20x}{10x - x^3} = 25$$

and

$$5 \ \tilde{m} \ \text{Radice de } 21\tfrac{8}{27}, \text{ ecce valor rei} \qquad 5 - \sqrt{21\tfrac{8}{27}} = x$$

Note the symbols for the different powers of the unknown x and also the mixture of symbols and words. A further example, from Michael Stifel (1486–1567), is

$$\frac{9\,\text{з} + 8\text{ı}}{6\alpha} \ \text{per} \ \frac{3\,\text{e}}{2} \ \text{facit} \ \frac{27\,\text{ı}\,s + 24\alpha}{12\,\alpha}$$

that is,

$$\frac{9x^4 + 8x^2}{6x^3} \cdot \frac{3x}{2} = \frac{27x^5 + 24x^3}{12x^3}$$

Note that x^4 is denoted by repeating the symbol for x^2 but that x^5 ($\text{ı}\,s$) requires a new pair of symbols altogether. Thus the relations among the various symbols representing different powers of the variable x had to be kept in mind. This is similar to the difference between the Egyptian (or Roman) and the Babylonian and modern (positional) number systems: the number of distinct symbols increased in the Egyptian system as the magnitude of the represented quantity increased, and the relations among symbols had to be kept in mind when computations were performed.

Such arbitrary systems of notation for the powers of a quantity x obscured their natural interrelation; there was no sensible way to express the important "law of exponents" $x^m \cdot x^n = x^{m+n}$ in such notations, and consequently the concept and properties of logarithms, which, after all, are nothing more than exponents relative to a fixed "base" quantity, seemed miraculous and subtle to early seventeenth-century mathematicians.

A further development can be seen in the notations of Thomas Harriot (1560–1621),

$$aaa - 3.bba ===== + 2.ccc$$
$$(x^3 - 3b^2x = 2c^3)$$

and William Oughtred (1574–1660),

$$Aqq + 4AcE + 6AqEq + 4AEc + Eqq$$
$$(a^4 + 4a^3b + 6a^2b^2 + 4ab^3 + b^4)$$

In this example q stands for "quadratic" or "squared," and c for "cubed"; the capital letters represent variable numbers, but there is still no exponential notation. Note also that this expression is a *formula*, free of specific values for A, B, . . . , which obviously represents the expansion of $(a + b)^4$ and is a special case of the Binomial Theorem now taught in high school; the concept of a *literal coefficient* was first introduced by the French mathematician Viète (1540–1603). In René Descartes' *La Géométrie* we find an expression that looks like modern notation:

$$x^3 - \sqrt{3}\, xx + \frac{26}{27}\, x - \frac{8}{27\sqrt{3}} \infty\ 0$$

(where ∞ means $=$). Earlier he had written

$$6\mathfrak{C} + 11_{\mathfrak{z}} + 6\varkappa + 1 \text{ ductum per } \sqrt[3]{\tfrac{3}{4}}$$
$$[\,(6x^3 + 11x^2 + 6x + 1) \cdot \sqrt[3]{\tfrac{3}{4}}\,]$$

whence it appears that the transition to modern symbolic notation was essentially completed in his time. Indeed, modern exponential notation was invented by Descartes and first appeared in *La Géométrie* in 1637.

Stifel's use of repetition to obtain the symbol denoting x^4 from that denoting x^2 contains the seed of the concept of exponent notation invented by Descartes almost 100 years later. The reader may wish to compare the coexistence of superfluous pictographic hieroglyphs and an essentially self-sufficient system of syllabic hieroglyphs in the Egyptian writing scheme with the awkward Renaissance syncoptic notations for powers of an unknown quantity which were often supplemented by instances of the future and incomparably more powerful exponential notation.

Algebraic notation has been standardized since the time of Leibniz (1646–1716). With a common and widely understood notation the ability to communicate mathematical information is greatly increased just as it is for speakers of the same language when compared with communication through an interpreter or with the necessity to learn a new language in order to establish communication. It is perhaps no accident that the development of a standard algebraic notation immediately led to the useful application of algebra to problems of geometry.

§8.3. In the elementary study of geometry the main questions of interest concern the properties of geometric objects in the usual Euclidean plane and in the three-dimensional space of perception. The objects usually considered first include points, lines, planes, curves such as circles, ellipses, parabolas, and hyperbolas, and surfaces such as spheres and cones.

All of these can be defined in natural and simple terms as the locus of the set of points having certain properties which depend only on the notions of *angle* and *distance*. For instance, a circle in the plane is the locus of points at a fixed distance from a given point, whereas a cone in space is the locus of points lying on straight

FIGURE 8.1

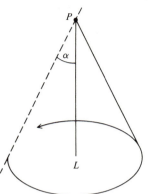

Generation of a cone by a line rotating about a fixed line L at a fixed angle α.

lines, all of which pass through a fixed point P and make a fixed angle α with a given line L through P (see Figure 8.1).

Geometric objects that can be defined by using properties related to angles and distances were essentially the only ones studied by the Greeks, for this technique of defining geometric objects was the only one known to them. It is unfortunately quite limited, for many curves and surfaces, some of which have important practical applications as well as interesting mathematical properties, cannot be defined by such essentially elementary geometrical means.

An important consequence of the algebrization of geometry is that the inventory of curves and surfaces that can be explicitly defined is vastly enriched and the variety of potential mathematical models available for the description of physical reality is greatly extended and their subtlety and refinement marvelously enhanced.

It will be worthwhile to recall the definition of a Cartesian coordinate system on the plane. Let L_1 and L_2 denote two perpendicular lines in the plane E, as illustrated in Figure 8.2. Suppose that a fixed notion of distance is given. At any point P in the plane E the perpendicular distances from P to L_1 and to L_2 are well defined; denote them by $\overline{PL_1}$ and $\overline{PL_2}$. Thus an ordered pair of signed numbers $(\pm\overline{PL_2}, \pm\overline{PL_1})$ can be associated with point P, where the sign of each number is determined by the quadrant in which P lies, as shown in Figure 8.2. Some sample points are plotted in Figure 8.3. Distances measured in the horizontal direction are conventionally denoted by x, those in the vertical direction by y, following a long-standing tradition. Thus to any point P in the plane there corresponds, according to this rule, a unique ordered pair of signed numbers. Conversely, given an ordered pair of signed numbers, there corresponds, relative to this particular coordinate system in the plane, a unique point P; for example, the pair (2, 1) is an ordered pair of numbers. Both numbers are positive, and in order to find the corresponding point in E, we move, starting at the point $0 = (0, 0)$, two distance units to the right (+ direction) and then one distance unit up (+ direction). If, instead, we

FIGURE 8.2

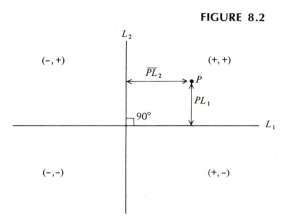

Cartesian coordinate system.

considered $(2, -1)$, the corresponding point would lie two units to the right $(+)$ and one down $(-)$, as in Figure 8.3.

Newton appears to have been the first to make use of negative numbers for the description of points in a coordinate system; before his time artificial ways of designating "direction" had been used, some of which persist to this day. The measurement of longitude "east" or "west" of the Greenwich meridian is one example—"+" or "−" would be more convenient for mathematical purposes—and the affixes A.D. and B.C. in the designation of calendrical time is another traditional example to which we have not adhered in this book.

FIGURE 8.3

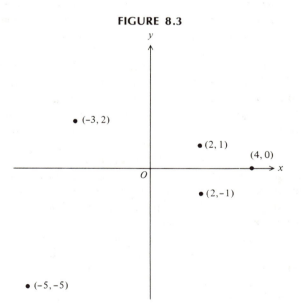

This correspondence between points of the plane and ordered pairs of numbers is called a *Cartesian coordinate system* for the plane. The point with coordinates (0, 0), called the *origin* of the coordinate system, is usually denoted by O and the ordered pair of numbers (x,y) associated with the geometric point P are called the *coordinates* of P. We have already seen that the notion of a coordinate system for points in a plane was known to Ptolemy; therefore it cannot be this correspondence between number pairs and points that alone constitutes a great achievement, certainly not of the seventeenth century, so many hundreds of years after Ptolemy. The significant seventeenth-century contribution to the algebrization of geometry was made by Descartes and Fermat who observed that a coordinate system for points in a plane provides a mechanism for introducing the calculational methods of algebra and arithmetic and thereby a method for describing and defining geometrical objects in an algebraic way.

Curiously enough, the Greeks could, in principle at least, have welded together Euclidean geometry and Babylonian algebra to anticipate the algebrization of geometry by more than 1000 years, but, in one of the classic instances in which the study of philosophy has had a practical, though unfortunately a negative, effect they rejected an identification of geometrical points in a plane with ordered number pairs. They argued thus: "numbers" are, in the first place, integers, used for counting, but by means of elementary arithmetic operations—addition and subtraction, multiplication and division—it is possible to construct the *rational numbers,* that is, fractions that are the ratios of integers. The collection of rational numbers suffices for all ordinary and practical purposes of calculation. Because they could not construct a coherent and complete theory of numbers of a more complicated nature, the Greeks in general refused to admit that any but the rationals were "real" numbers; there was just no philosophical justification for the introduction of other numbers and no practical need. This attitude led to a peculiar distinction between points in the plane and number pairs, for if an identification of points and number pairs were admitted it would automatically lead to the introduction of nonrational numbers as well. For instance, suppose that an isosceles right triangle ABC is drawn in a plane, as indicated in Figure 8.4. There should be no difficulty in using ordinary compass and straight edge methods of elementary geometry to construct such a triangle. Suppose that its sides AC and BC have length 1 in some convenient unit of measurement. The theorem of Pythagoras shows that the length of the hypotenuse, if this number exists, is $\sqrt{2}$. It is per-

FIGURE 8.4

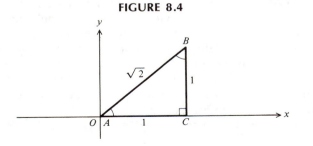

fectly clear from the construction of the triangle that the hypotenuse exists in a geometrical sense. Does $\sqrt{2}$ exist in the Greek sense? That is, is $\sqrt{2}$ a rational number? As in §1.5, we show that $\sqrt{2}$ is not rational.

Suppose that it is. Let its expression be $\sqrt{2} = m/n$, where m and n are positive integers without any common integral factors greater than 1. Then, squaring, we see that $2 = m^2/n^2$, so $2n^2 = m^2$. This means that m^2 is even. However, m itself must be even or odd. If even, then its square is also even, but if m is odd, say $m = 2k + 1$, then $m^2 = 4k^2 + 4k + 1$, so m^2 is also odd. We saw above that m^2 is even in our case; therefore m is also even. Say $m = 2p$. Then $2n^2 = m^2 = (2p)^2 = 4p^2$ and $n^2 = 2p^2$, which means that n^2 is even. Just as above, this implies that n is even. Both m and n are even, which contradicts the assumption that they have no common divisors greater than 1. This contradiction means that the assumption that $\sqrt{2}$ can be represented as a fraction is false: $\sqrt{2}$ is *not rational*.

From the Greek viewpoint the result asserts that $\sqrt{2}$ *is not a number,* but the **line segment, which is the hypotenuse of the isosceles right triangle, certainly** exists, whence the correspondence between numbers — that is, between *rational* (Greek) numbers — and points on a line is not exact. There are more points on a line than rational numbers. Since a Cartesian coordinate system is built from the correspondence of points on two axes with ordered pairs of numbers, it necessarily follows that there are more points in the plane than there are rational number pairs. This observation destroyed any possibility that the Greeks would invent analytic geometry. It set the studies of arithmetic and geometry apart from each other in a way that would probably have been irreconcilable and eternal were it not that the memory of man is short. The careful and subtle Greek philosophical investigations of the nature of number and the contradictions inherent in attempting to make planar points and (rational) number pairs correspond were forgotten, and the more pragmatic and simplistic Babylonian view of approximating all arithmetic quantities by rational numbers led ultimately to the notion that nonrational numbers could be defined by the collection of rational approximations to them. Said in other, and in some ways simpler, terms, nonrational numbers were considered by seventeenth-century mathematicians as the collection of decimal expressions that approximate them. This notion was formalized in the late nineteenth century and our current concept of real numbers was finally provided with an impeccable philosophical foundation. At this point the reader may profitably reinspect §§1.4–1.6.

Returning to the correspondence between number pairs and points of the plane, if by "number" we agree to mean infinite decimal expressions, then there is indeed a correspondence between number pairs and points such that every point is represented by a pair of numbers. The next step is to introduce the calculational methods of algebra and arithmetic into the study of geometry by using this correspondence. As has already been remarked, it is *this step* that constitutes the essential and significant achievement of the algebrization of geometry.

In order to illustrate how algebra can be introduced to study geometrical questions, let us consider some simple examples.

Associate with a general point P in the plane its coordinates, $P = (x, y)$. How can a circle $C(r)$ centered at the origin and of radius r be expressed in terms of the coordinate system? By definition, the circle $C(r)$ is the set of points whose distance from the origin is precisely r. So the problem is equivalent to that of determining which points $P = (x, y)$ are at a distance r from $O = (0, 0)$. To solve it we must know how to express the distance from (x, y) to $(0, 0)$. Consider the right triangle in Figure 8.5 whose sides are of length x and y and whose hypotenuse is of length PO. From the theorem of Pythagoras it is clear that

$$\overline{PO} = \sqrt{x^2 + y^2}$$

and that the set of points (x, y) satisfying $\overline{PO} = r$ is the same as the set of points whose coordinates (x, y) satisfy

$$\sqrt{x^2 + y^2} = r$$

or, what amounts to the same thing,

$$(8.1) \qquad\qquad x^2 + y^2 = r^2$$

This equation (8.1) is an *algebraic relation* that connects the two variable numbers x and y. Each pair of numbers (x, y) satisfying (8.1) corresponds to a point on the circle $C(r)$ and, conversely, each point on the circle has coordinates that must satisfy the equation. We say that (8.1) is the "equation of the circle" or the "equation defining the circle." Thus, given a circle, an equation in two unknowns x and y which represents this geometric object can be associated to it by means of a coordinate system. Other plane curves can be represented similarly by an algebraic equation in two variables.

FIGURE 8.5

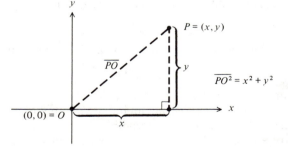

Pythagoras' theorem.

Conversely, any algebraic equation in two variables corresponds to some geometric object. A simple example is given by the following equation

$$(8.2) \qquad\qquad 3y - x = 0$$

Consider the set of all points in the plane which have coordinates (x, y) that satisfy (8.2). Again, by constructing a diagram (Figure 8.6) and plotting points that are solutions to (8.2), it appears that the points corresponding to numbers that satisfy the equation all lie on a straight line; for example, points $(3,1)$, $(6,2)$, $(9,3)$ satisfy (8.2) and all lie on the same straight line L, determined by any two of them. Using

FIGURE 8.6

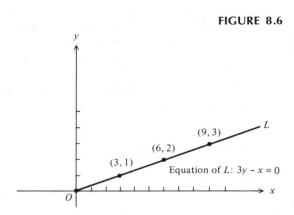

Equation of L: $3y - x = 0$

Linear equations correspond to straight lines.

the plane geometry of Euclid (proportionality of sides of right triangles), it is easy to prove that any solution (x, y) of (8.2) must lie on L. The conclusion we can draw here is that (8.2) determines, via its family of solutions, a straight line L in the plane with the coordinate system laid out on it. We say that (8.2) is the "equation of the straight line L." As in the case of the circle, had we started with L, (8.2) could have been derived. So the process is reversible. Certain geometric objects correspond to algebraic equations, and all algebraic equations correspond to geometric objects.

A geometric object may correspond to many different algebraic equations. This contrasts with the one-to-one correspondence between points of the plane and ordered pairs of numbers. Consider, for example, point O with coordinates $(0, 0)$. This point certainly corresponds to the algebraic equation

$$x^2 + y^2 = 0$$

because the equation has only $x = 0$ and $y = 0$ as solutions; but the equation

$$x^2 + 2y^2 = 0$$

also has only the pair $(x, y) = (0, 0)$ as solutions, and so it describes the same geometrical point object. It is easy to construct an infinite number of distinct algebraic equations that describe the same geometric point.

This multiplicity of algebraic descriptions of one geometric object has nothing to do with the simplicity of the particular object chosen for the example. There are infinitely many distinct equations that correspond to other more complex geometrical objects; for instance, the circle $C(r)$ discussed above can be described by the solutions of

$$(x^2 + y^2)^n = r^{2n}$$

where n denotes any positive integer. The reader will notice, however, that all these equations are related to one another in a simple way. This is not an accident, but reflects certain relatively deep properties of polynomials which we do not require for our further work. It suffices to remark that although the relation between geometric objects and algebraic equations is not one-to-one it nevertheless is susceptible to study, and a complete overview of the relations between the various equations that represent one object can be had.

§8.4. All the classical curves studied by the Greeks can be represented as solutions of (not necessarily algebraic) equations. The first fundamental observation made by both Fermat and Descartes was that straight lines in the plane correspond precisely to equations in two unknowns (x and y) which are linear, that is, equations in which x and y occur to the first power. The most general equation of this type is

$$ax + by + c = 0$$

where a, b, and c are fixed numbers. Note that the modern designation *linear* arises from the word *line*, which suggests the influence of the correspondence on mathematical nomenclature. A later development in the algebrization of geometry showed that all the classical conic sections studied by the Greeks — the ellipse, hyperbola, and the parabola (also including circles and straight lines as degenerate cases) — are precisely the set of curves in the plane corresponding to equations of the *second degree*,

(8.3) $$ax^2 + bxy + cy^2 + dx + ey + f = 0$$

where once again $a,b,c, \ldots$ are constant numbers. For example, if $a = c = 1$, $f = -r^2$, and the other constants are zero, then (8.3) reduces to (8.1), the circle of radius r centered at the origin.

The fundamental principle of analytic geometry is this correspondence between curves in the plane and equations in two unknowns. It was expressed by Fermat in the following manner

> *Whenever in a final equation two unknown quantities are found, we have a locus, the extremity of one of these describing a line, straight or curved.*

What is the significance of this correspondence? If the Greeks had their geometric way of defining and representing curves and there is also an algebraic way of describing the same curves, what is the advantage of the latter? One advantage is that for certain problems in geometry equations corresponding to the geometric problem can be found whose solution is equivalent to the solution of the geometric problem. For the solution process we can appeal to the many rules of algebra that have been developed over the centuries. Many of these rules are straightforward and such that once a problem is stated in the form of equations it is easy and nearly automatic to find its solution by algebraic manipulations of the quantities involved. The important point is that once the equations are known we can, for the time being, forget about their geometric derivation and concentrate on the purely algebraic problem at hand.

A more significant consequence of the correspondence between geometric objects and algebraic equations comes from application of the correspondence in the reverse direction: by using equations to *define* curves in the plane and in higher dimensional spaces the variety and quantity of describable curves and other geometric objects is vastly increased. An example will help to demonstrate this fact. The Greeks defined a parabola as the intersection of a plane with a right circular cone as shown in Figure 8.7. By using suitable Cartesian coordinates on the plane E it can be shown that the equation corresponding to the parabola is $y = x^2$, as shown in Figure 8.8. Now it is natural to ask what the curves corresponding to $y = x^3$, $y = x^4$, . . . , and so on look like. By plotting the solutions of these equations we obtain their geometric picture as illustrated in Figure 8.9.

There is no evident way of defining these curves by Greek methods. Moreover, the algebraic definition makes it clear that they are related and suggests that they may have certain properties in common, which is true and not difficult to show by using algebra and the "analytical" techniques developed in Chapter 9. Geometrical definitions expressed in terms of angle, distance, and intersections of surfaces would not exhibit the interrelationships of these curves and certainly would not have the simplicity of the algebraic definition

$$y = x^n$$

which, as n varies through the positive integers, efficiently includes in one formula the definitions of all these curves.

It may be possible, by means of complicated and ingenious arguments, to find a geometrical definition of the curve corresponding to $y = x^n$, where n is a positive integer. It may even be possible that such definitions could be extended to those cases in which n is assumed to be a not necessarily integral *rational* number, and that, for instance, the curves corresponding to the equations $y = x^{1/20}$ and $y = x^{3/4}$ would be definable by Greek-like geometrical methods. It would not be possible, however, to provide intuitively meaningful geometrical definitions for the curves defined by $y = x^n$, where x is *not* a rational number; there is no mechanism, for instance, for describing the curve corresponding to $y = x^{\sqrt{2}}$ in synthetic geometrical terms.

FIGURE 8.7

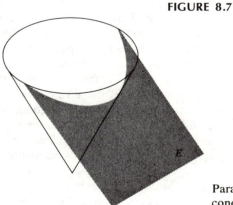

Parabola as the intersection of a cone with a plane.

FIGURE 8.8

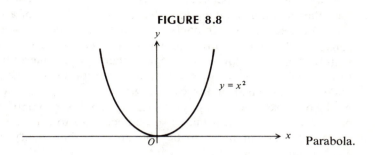

Parabola.

You may wonder of what interest such a curve may be. One of the central problems of theoretical physics is to find mathematical "models" that describe aspects of the real (physical) world. These mathematical descriptions are essentially sets of equations relating variables that stand for directly measurable physical quantities. Now it may happen that the equations that accurately describe some aspect of the physical world will turn out to be simple polynomial equations with integer exponents for all the variables; (8.3), the equation that defines the general conic section is of this kind. If this is the case, then it *may* be possible to provide a completely "Greek" geometrical definition of the equations involved — that is, to provide a completely "Greek," or synthetic *geometrical model* of the physical aspect of the world under consideration. What this means is, simply stated, that the Greeks could have discovered the model and have thereby attained a scientific un-

FIGURE 8.9

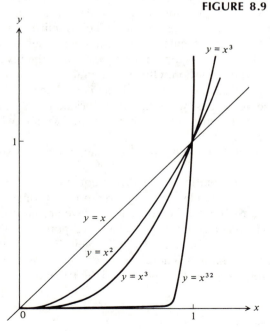

Graphs of power functions for $0 \leqslant x \leqslant 1$.

derstanding of the process as well as the ability to make accurate predictions. On the other hand, if the equations describing reality are of a more complicated nature, then geometrical models are out of the question—at least in the Greek sense—and scientific progress will be related to the state of algebraic and ultimately "analytic" knowledge. Since there is an inherently geometrical aspect to physical problems such as the determination of the motion of bodies in space under the influence of forces like gravity, the construction of a mathematical model of these aspects of physical reality requires geometry to enter in some way; but, as we have seen, it may also be necessary to have the apparatus of algebra and even more in order to be able to represent reality adequately. Thus the algebrization of geometry was intimately related to the progress of theoretical physics. We return to the problem of constructing models of physical reality in Chapter 12 in which still more powerful analytical tools are made available to aid in the description and comprehension of the structure of the universe.

§8.5.   The Greeks systematically studied the *conic sections,* that is, curves corresponding to equations of second degree of the form (8.3), from the viewpoint of synthetic geometry. They also invented various special curves for which synthetic geometric definitions are possible, but they were not able to extend their systematic studies of classes of related curves beyond the conic sections. Even *cubic curves,* corresponding to general equations of degree 3 in the variables x and y had to wait for algebrized geometry and Newton in the seventeenth century for comprehensive and illuminating study.

The remainder of this chapter presents a "sampler" of curves known to the Greeks, more complex curves whose definition and study require the mechanisms of algebrized geometry, and the still more complicated, interesting, and useful curves, whose study transcends the power of algebra, and which require as well the subtle and refined analytical tools of the calculus for their very definition.

Certain problems exercised Greek geometers for centuries and led to the invention of many special curves to aid in their solution. One was the famous "duplication of the cube": suppose a cube of side s is given; its volume is s^3. The problem is to find the side t of a cube whose volume is twice that of the original cube, that is, $2s^3$. This amounts to solving the equation $t^3 = 2s^3$, whence $t = 2^{1/3}s$; thus what is desired is a *geometrical construction* of $2^{1/3}$.

A second problem is the "trisection of a given angle," whose meaning is self-evident.

Neither problem can be solved by the solution of quadratic equations and both must therefore necessarily lead (by the principle of the algebrization of geometry!) to curves other than the conic sections. In order to construct $2^{1/3}$ we are naturally led to the cubic equation $y^3 - 2x = 0$ and consideration of the corresponding cubic curve. Trisection of the angle θ amounts to finding $\theta/3$ if θ is given. From Figure 8.10 θ is known if and only if $x = \cos\theta$ is known. Therefore trisection of θ is equivalent to the determination of $\cos(\theta/3)$ from $\cos\theta$. Now turn to the trigonometrical identity

$$\cos\theta = 4\cos^3\frac{\theta}{3} - 3\cos\frac{\theta}{3}$$

FIGURE 8.10

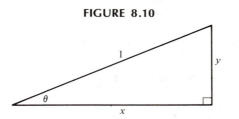

(which can easily be derived by repeated application of the formulas for $\cos(a+b)$ and $\sin(a+b)$) and put $x = \cos\theta/3$, $y = \cos\theta$. Then this identity is just the cubic equation

$$y = 4x^3 - 3x$$

and the value of $x = \cos\theta/3$ is determined by the value of $y = \cos\theta$. Thus it turns out that both famous problems amount to the study of special cubic equations and therefore of special cubic curves.

For instance, the *cissoid of Diocles* (Diocles, between -250 and -100), shown in Figure 8.11, corresponds to the cubic equation

(8.4) $$x^2 = \frac{(2r-y)^3}{y}$$

where $r > 0$ is a fixed number. Let C be determined as the intersection of $A'B$ with the cissoid, and drop the perpendicular CD to the y-axis. Let E denote the intersection of CD with the circle. Then it can be shown that $\overline{ED} = 2^{1/3}\overline{A'D}$, which solves the problem of duplication of the cube.

FIGURE 8.11

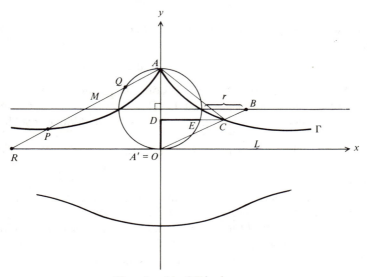

The cissoid of Diocles.

Indeed C is the intersection of the cissoid Γ and the line $A'B$ whose equation is

$$y = \tfrac{1}{2}x$$

Substitution of this value of y in (8.4), some further algebra, and application of Pythagoras' theorem yield the desired result.

This cissoid can also be described in a synthetic way which made it attractive to the Greek viewpoint.

In fact, let the circle of radius r be tangent to the line L at A' in Figure 8.11, and let A be the other endpoint of the diameter through A'. Draw any line M from A to line L and suppose that M meets the circle at Q and line L at R. Determine a point P on line M such that $\overline{AP} = \overline{RQ}$, then P lies on the branch of the cissoid of Diocles that is above the x-axis.

The *conchoid of Nicomedes* (Nicomedes, between -250 and -150), shown in Figure 8.12 with corresponding equation

(8.5) $$(x - a)^2(x^2 + y^2) - bx^2 = 0$$

of degree 4 can be used to solve both classical problems; it also has a synthetic geometric description.

FIGURE 8.12

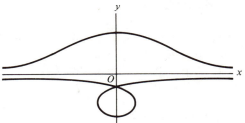

The conchoid of Nicomedes.

The variations in appearance of curves of degree 4 is remarkable; consider, for instance,

(8.6) $$a^2y^2 + b^2x^2 - x^2y^2 = 0 \qquad a,b \text{ constants}$$

whose graph is shown in Figure 8.13; it consists of four separate symmetrical curves as well as the isolated point (0,0). This curve is the special case of the family of Lamé curves (Gabriel Lamé, 1795–1870) defined by the equation

(8.7) $$\left(\frac{x}{a}\right)^m + \left(\frac{y}{b}\right)^m = 1 \qquad a,b \text{ constant}, m \text{ rational}$$

corresponding to $m = -2$.

If $m = 2$, the Lamé curve is an ellipse; more generally, if m is a positive even integer, the curve is closed. Figure 8.14 shows the case $m = 4$.

If m is a rational number of the form p/q and p is even, the curve is closed; if q is odd, it is star shaped as in Figure 8.15 for which $m = \tfrac{2}{5}$. Figures 8.16–8.18 show some other interesting forms corresponding to $m = \tfrac{3}{2}$, 5, and $-\tfrac{3}{2}$.

FIGURE 8.13

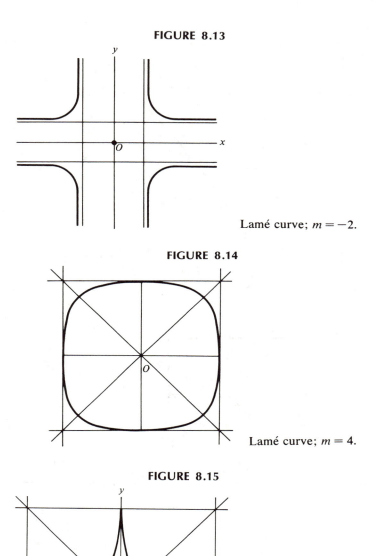

Lamé curve; $m = -2$.

FIGURE 8.14

Lamé curve; $m = 4$.

FIGURE 8.15

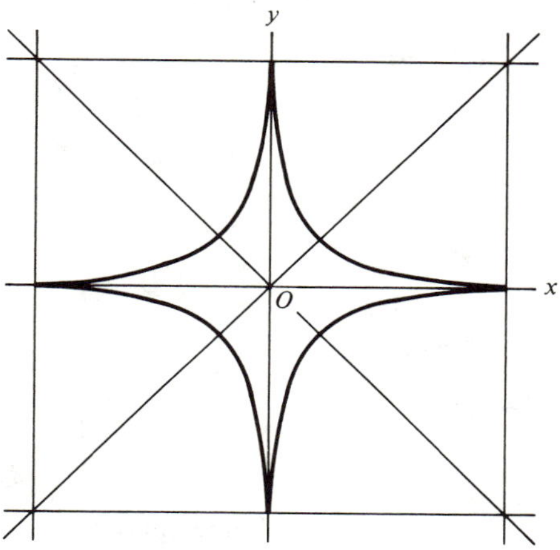

Lamé curve; $m = \frac{2}{5}$.

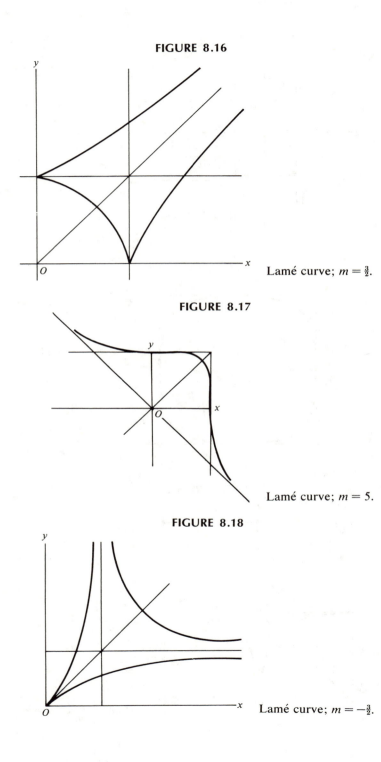

FIGURE 8.16

Lamé curve; $m = \frac{3}{2}$.

FIGURE 8.17

Lamé curve; $m = 5$.

FIGURE 8.18

Lamé curve; $m = -\frac{3}{2}$.

FIGURE 8.19

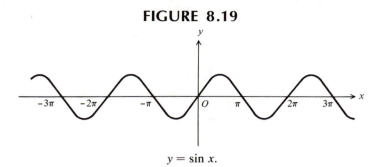

$y = \sin x.$

By combining algebraic equations with more complicated functions we can describe new geometrical objects. For instance, the trigonometrical functions naturally lead to periodic curves such as that corresponding to

(8.8) $y = \sin x$

in Figure 8.19 and the unbounded curve consisting of infinitely many disjoint pieces described by

(8.9) $y = \tan x$

in Figure 8.20.

FIGURE 8.20

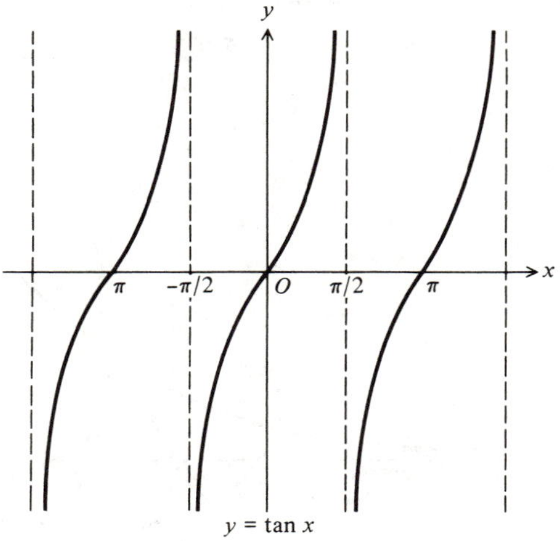

$y = \tan x$ $y = \tan x.$

The exponential function

(8.10) $y = a^x$ $a > 0$ constant

corresponds to the curve shown in Figure 8.21 and is there compared with various powers of x to show its rapid growth, for the case $a = e = 2.718. \ . \ . \ .$

FIGURE 8.21

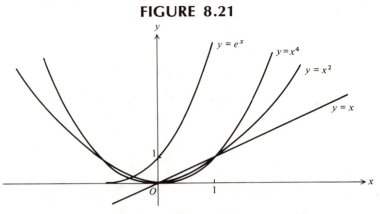

$y = e^x$ and $y = x^n$ for various n.

Combination of these functions in various ways leads for instance to

$$y = a^x \sin x$$

which is shown in Figure 8.22.
The curve corresponding to

(8.11)
$$y = \frac{a}{2} \left(e^{x/a} + e^{-x/a} \right) \qquad a > 0$$

FIGURE 8.22

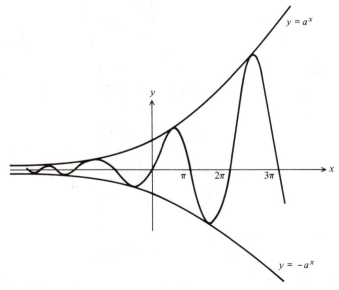

$y = a^x \sin x$.

called the *catenary*, is particularly interesting. It is the curve assumed by a heavy cable suspended under the action of gravity and consequently is well known to bridge engineers (*cp.* §9.11). The catenary is similar in appearance to a parabola, so much so that Galileo thought that the latter was the curve of a suspended cable; the true shape, described by (8.11), was discovered in 1690–1691 by Huygens, Leibniz, and Johann Bernoulli in their (separate) response to a challenge by James Bernoulli.

The catenary is related to the *tractrix,* a curve corresponding to the difficult equation[3]

$$(8.12) \qquad y \exp\left(\frac{x - \sqrt{a^2 - y^2}}{a}\right) = a - \sqrt{a^2 - y^2}$$

shown in Figure 8.23.

FIGURE 8.23

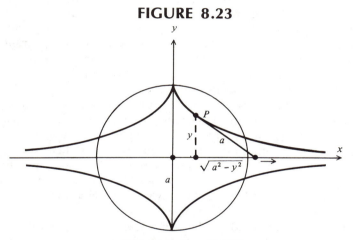

A Tractrix [see (8.12)].

The tractrix is the path followed by a child's pull toy if the child walks along a line and the initial location of the toy is at a distance from the line equal to the length of the pull string (Figure 8.24). If the tractrix is rotated about the *x*-axis, a surface is formed which is the ideal form (because it minimizes wear) for a bearing that supports a revolving shaft which exerts substantial loads on the bearing in the direction of the *x*-axis. This situation occurs, for instance, in oil-drilling rigs (in which, of course, the *x*-axis is taken to point toward the earth's center).

The (ordinary) *cycloid,* one arch of which is shown in Figure 8.25, corresponds to the equation

$$(8.13) \qquad \cos\left(\frac{x + \sqrt{2ay - y^2}}{a}\right) = 1 - \frac{y}{a} \qquad a > 0 \text{ constant}$$

---

[3] *It has become traditional to write exp z in place of e^z to aid the compositor when z is a complicated expression.*

FIGURE 8.24

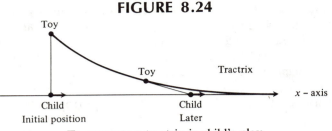

To generate a tractrix is child's play.

but it has a much simpler synthetic definition: let *P* be a fixed point on a circle of radius *a* and let the circle roll along a line without slipping. The curve traced out by *P* is the cycloid. Although the cycloid and a family of curves related to it were known in antiquity — *epicycloids,* formed by rolling a circle along another circle, were fundamental constituents in the Ptolemaic model of the apparent motion of the sun about the earth — the deeper and more interesting properties of cycloids had to wait for the invention of calculus for their study.

Turn one arch of the cycloid upside down to form a bowl. In this form the cycloid is the *brachystochrone,* that is, the curve along which a smooth particle will most quickly move under the influence of a (constant) gravitational field from one point to another, as James Bernoulli discovered. It is also the *tautochrone,* which is that curve such that the time of descent of a particle sliding smoothly along it under gravity's influence to its lowest point is independent of how high on the curve the particle begins its descent; finally, the cycloid is the *isochrone:* if a pendulum is constrained to swing between two cycloidal jaws as shown in Figure 5.9, then the period of the pendulum, that is, the duration of a complete swing, does not depend on the amplitude of the swing. Consequently the effect of friction, which diminishes the amplitude of the swing, will not affect the timekeeping properties of the pendulum. Huygens discovered both the tautochrone and isochrone properties of the cycloid and, as described in Chapter 5, unsuccessfully applied the latter to the problem of constructing a seaworthy navigational clock.

FIGURE 8.25

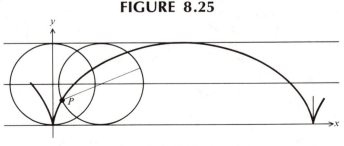

A cycloid.

A curve that has assumed ever increasing practical importance in the twentieth century is the *normal probability curve,*

(8.14)
$$y = \frac{1}{\sqrt{2\pi}} \, e^{-x^2/2}$$

shown in Figure 8.26, which describes the statistical distribution of quantities subject to additive independent variations, such as college grades.

FIGURE 8.26

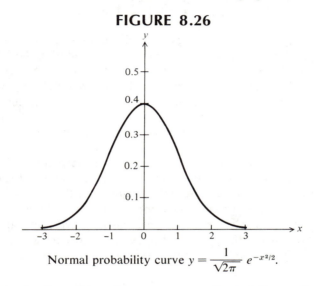

Normal probability curve $y = \dfrac{1}{\sqrt{2\pi}} \, e^{-x^2/2}$.

With the introduction of functions more complicated than polynomials, trigonometrical, and exponential functions, a larger inventory of curves becomes available. Here are several examples for which the definition itself requires a knowledge of the calculus, and no synthetic geometrical description can be hoped for. Figure 8.27 shows the curve corresponding to a certain *elliptic function,* which

FIGURE 8.27

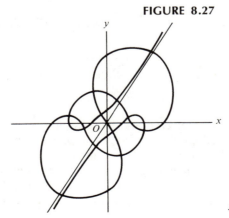

$x + iy = \text{sn}(2 + 3i)\,u.$

generalizes the ordinary trigonometrical functions; its complexity forbids its description here.

The *spiral of Cornu* occurs naturally in the theory of diffraction of light. This attractive curve is determined by the pair of equations

(8.15*a*)
$$x = \int_0^t \cos\left(\frac{\pi t^2}{2}\right) dt$$

(8.15*b*)
$$y = \int_0^t \sin\left(\frac{\pi t^2}{2}\right) dt$$

and is shown in Figure 8.28. The long Latin "S" denotes the *integral* (see §9.5); *t* is a parameter, and this pair of equations must be understood in the following sense: for each value assigned to *t* the right sides of (8.15*a*) and (8.15*b*) are numbers, respectively equal to *x* and *y*, and thereby defining a point $P = (x, y)$ in the plane. For example, if $t = 0$, then (as we shall learn in Chapter 9) necessarily

$$\int_0^0 \cos\ (\pi t^2/2)\ dt = \int_0^0 \sin\ (\pi t^2/2)\ dt = 0,$$

whence $x = y = 0$, and the corresponding point of the plane is the origin of the coordinate system. As *t* runs through the positive real numbers, the corresponding point with coordinates (x, y) runs from the origin along the part of the spiral above the *x*-axis, winds about $(\frac{1}{2}, \frac{1}{2})$ an infinite number of times, and approaches that point arbitrarily closely; similarly, as *t* runs through negative real values, the corresponding point traces out the portion of the spiral below the *x*-axis, winds an infinite number of times about the point $(-\frac{1}{2}, -\frac{1}{2})$, and approaches it arbitrarily closely.

FIGURE 8.28

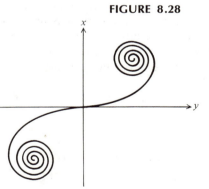

Cornu's spiral.

Once imagination is unfettered and increasingly refined ways of defining geometrical curves are adopted, the range of phenomena they exhibit is correspondingly increased. Unusual counterintuitive properties make their appearance, initially spreading confusion and concern about the logical and philosophical foundations of mathematics. Later, when they and their implications are understood and integrated into the continually growing mathematical structure, it becomes possible to recognize that a new level of sophistication and subtlety has been achieved, and from this higher and more rarified vantage point imagination once again begins its heady ascent, admitting neither restraint nor guide along its path but for the sure and steady staff of logic.

In the twentieth century a new plateau of sophistication was reached exemplified by the study of curves such as the one whose construction is described in Figure 4.28; it is contained entirely within a square, never crosses itself, fills the square but for a set of points whose "area" is 0, has infinite length, and does not have a tangent line at any of its points.

It is fitting to close this chapter with a description of the *logarithmic spiral* discovered by Descartes in 1638. This elegant curve, illustrated in Figure 8.29, corresponds to the equation

(8.16) $$y = x \tan (\log (x^2 + y^2))$$

If we write $r^2 = x^2 + y^2$, $\tan \theta = y/x$, it becomes

(8.17) $$r = e^{\theta/2}$$

This spiral has properties that have fascinated many mathematicians. If P is any point on the spiral other than the origin, then the part of the spiral from the origin to P is an *unending* curve of *finite* length. Any line through the origin meets the curve at points whose distance from the origin forms a geometric progression. If

FIGURE 8.29

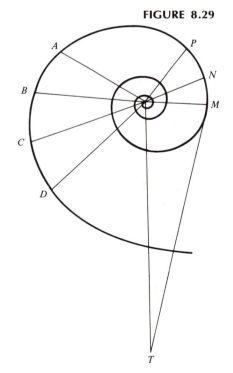

Logarithmic spiral.

any part of this curve is enlarged or reduced by any factor whatsoever, the result is congruent to another part of the same spiral.

Every line through the origin meets the spiral at the same angle. From this fact and the property (proved in §6.2; *cp*. Figure 6.9) that stereographic projection of a sphere on a plane preserves angles it follows that the logarithmic spiral is exactly the stereographic projection of Mercator's loxodromes on the earth's surface (*cp*. Figure 6.12 and §6.3), as Johann Bernoulli discovered. He was so entranced by this remarkable curve that he ordered it carved on his tomb with the inscription

Eadem mutata resurgo,

"Though changed I rise unchanged," a fitting epigraph for mathematics itself.

EXERCISES

8.1. Translate into modern notation:
(a) 1℞m̃ 1 per 1℞et 1 = 1ʒ m̃ 1
(b) 1℞m̃ 1 per 1ʒ et 1℞et 1 = ?
Note that the right side of (b) requires a new symbol.

8.2. Using Stifel's notation, translate the following:
(a) 1℞m̃ 1 per 1ᴄ + 1ʒ + 1℞ + 1
      faut 1ʒʒ m̃ 1
(b) 1℞m̃ 1 per 1ʒʒ + 1ᴄ + 1ʒ + 1℞ + 1
      faut 1ſsm̃ 1
(c) 1℞m̃ 1 per 1ſs + 1ʒʒ + 1ᴄ + 1ʒ
       + 1℞ + 1 = ?
Note that the right side of (c) requires a new symbol.

8.3. The equations in the last two problems are special cases of a simple algebraic identity. Write the identity in modern notation and *prove* that it is true.

8.4. Will your proof work so readily when expressed in the older notations? Why not?

8.5. Remember the summation notation $\sum\limits_{k=1}^{n} X_k$ which stands for $X_1 + X_2 + X_3 + \cdots + X_n$. Write the identity you found in Problem 8.3, using the summation notation.

8.6. Problems 8.1*a,b* have geometrical interpretations. What are they? (*cf.* Problems in Chapter 3.) What can you say about geometrical interpretations for Problem 8.3?

8.7. Is it necessary that the coordinate axes be perpendicular in order to have an effective coordinate system for the plane? Explain. What is the main advantage of the perpendicular coordinate system?

8.8. Suppose $y = x^2$ is the equation for a parabola. Write another equation to represent the same curve in the plane [that is, the same set of points (x, y) satisfies the equation].

8.9. Plot the points corresponding to the equation $x^2 - y^2 = 1$ on graph paper. This type of curve is called a *hyperbola*. Show in a diagram how such a curve occurs as the intersection of a cone and a plane.

8.10. What does Fermat mean by "line, straight or curved" in the quotation on p. 213? How would we express this today?

8.11. Does the equation $x^2 + y^2 = -1$ correspond to any curve in the plane? Explain.

8.12. Consider the set of points S in the plane that lies on either of the two perpendicular straight lines in Figure E8.12. Find an appropriate coordinate system for the plane with variable coordinates x and y and find a simple polynomial $P(x, y)$ with the property that $P(x, y) = 0$ corresponds to the point set S.

FIGURE E8.12

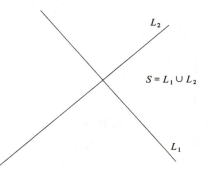

8.13. Assume that the equation $ax + by + c = 0$ is the most general equation of a straight line in the plane. Find a specific equation corresponding to the straight line that passes through points $(1, 1)$ and $(2, 1)$.

8.14. Consider the curve shown in Figure E8.15. Can a curve such as C have an equation of the form $y = ax^2 + bx + c$, where a, b, and c are constants? Why? (*Hint*: How many solutions are there to a quadratic equation?)

FIGURE E8.15

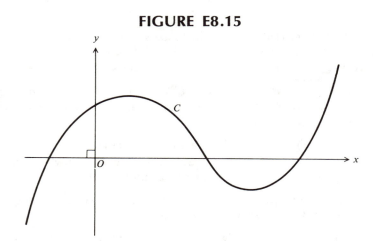

8.15.   Find an example of an equation whose associated curve has the form of the curve C in Problem 8.14.

8.16.   Under what conditions on a and b does the equation $ax^2 + by^2 = 1$ represent a circle?

8.17.   Give at least three reasons why the algebrization of geometry was a significant mathematical invention.

8.18.   If all the "information" about curves in the plane is contained in the equations that describe them, what is the utility of considering them as curves and not simply as equations, forgetting about the geometry?

*8.19.   Figure 8.9 appears to show that for $0 < x < 1$ the curve $y = x^n$ approaches the two straight-line segments consisting of the x-axis between $x = 0$ and $x = 1$ and the vertical line segment connecting $(1, 0)$ to $(1, 1)$. Give a precise statement that will describe this phenomenon.

*8.20. $x^2 + y^2 = 1$ describes a circle of radius 1 centered at the origin. This circle is tangent to the square of side 2 which is centered at the origin and has sides parallel to the coordinate axes (see Figure E8.21). Describe the curves that correspond to the equations $x^n + y^n = 1$, where n is an integer greater than 1. What happens if n is allowed to become large?

FIGURE E8.21

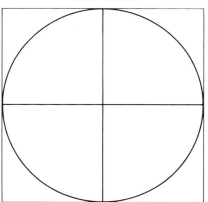

PART THREE

The Rise of Geometrical Analysis

CHAPTER 9

Infinitesimal Calculus

Here the fundamental ideas underlying the mathematical theory known as the infinitesimal calculus *are introduced. This powerful tool, invented by Newton and Leibniz at the end of the seventeenth century, has had an impact on science and technology greater than any other single advance. One hundred years before the invention of calculus the description of the motion of an object freely falling under the influence of gravity or of a ball rolling down an inclined plane presented a problem that taxed and indeed defeated the greatest scientists of the time. One hundred years after its creation the calculus had enabled Laplace, the French mathematician and physicist, to account for all but the most insignificant aspects of the heavenly motion of the moon and planets and had provided him with the mathematical foundations for the first scientific speculations into the origin of the solar system.*

This chapter begins by introducing the notion of limit, *which is basic to all that follows and indeed is the foundation of the definition of the real numbers given in Chapter 1. This concept appeared in embryonic form in many earlier periods of mathematics. It underlies positional notation itself, is inherent in the Babylonian technique for approximation of square roots, appears in geometrical form in Archimedes' estimate of* π *obtained by comparing the circumference of a circle with the perimeters of inscribed and circumscribed regular polygons, and again in his procedure for finding the area bounded by a parabola and a chord. The hypothesis that the radius of the earth is so small compared with the radius of the celestial sphere that the horizon and the "artificial horizon" need not be distinguished for the purposes of calculating the altitude of a star is another instance of an appeal to properties of limits. The extension of the concept of logarithm for rational powers of the base to a logarithm for any* arbitrary *power of the base, glossed over in Chapter 7, demands, for its correct and comprehensible statement, knowledge of the properties of limits. This recurrent theme, unobtrusively ever present throughout the preceding work, from this point on assumes a central and visible role in the advancement of mathematics and its application to the understanding and mastery of the physical world.*

Following the introduction of arithmetical *and* geometrical *definitions of limit, a special limit, the* derivative, *is introduced, methods for calculating its value for simple functions are developed, and its geometrical significance as the slope of a tangent line to a curve at a given point is explored. The importance of an algebrized geometry becomes evident from this example. Next another, and apparently independent, special limit, called the* integral, *is shown to have a geometrical interpretation as an* area *and indeed provides the first meaningful way to define the concept of area for geometrical objects whose boundaries do not consist of straight-line segments.*

A fundamental discovery by Newton and Leibniz was the duality of the notions of derivative and integral. They showed that the two processes are inverse to each other in a certain specific sense. The importance of this discovery is that it provides a method for calculating values of integrals (hence areas) by means of operating with derivatives; as the latter are usually much simpler, this technique is highly productive.

The chapter continues with some applications to the study of motion on a line and the calculation of the length of curves, the latter a problem of subtlety and complexity. It ends with a brief historical study of the application of calculus to the design of suspension bridges.

§**9.1.** In the preceding chapters we have encountered various methods for approximating arithmetic quantities or geometric objects, all of which are special cases of *limiting processes;* for instance, in Archimedes' estimate of π (§4.6) the length of a circle was approximated by the length of inscribed and circumscribed regular polygons which, as the number of their sides increased, appeared to approximate ever more closely the circle itself. It has also been mentioned that sometimes infinite series can be used to represent a quantity; for example,

$$\tfrac{\pi}{4} = 1 - \tfrac{1}{3} + \tfrac{1}{5} - \tfrac{1}{7} + \cdots$$

The three dots following the "plus" sign indicate that the series continues according to the obvious law of formation of terms — consecutive odd denominators and alternating signs — and that by taking enough terms it is possible to approximate $\pi/4$ as closely as desired; the reader may wish to look ahead to Figure 10.4 and the accompanying discussion to learn more about this example. The number of terms that may be required to attain an approximation of a prescribed degree of accuracy is another problem and often quite a difficult one.

The definition of real numbers given in §1.4 implicitly makes use of limiting processes, because the infinite decimal expansion $b_n b_{n-1} \cdots b_1 b_0 . b_{-1} b_{-2} b_{-3} \cdots b_{-k} \cdots$ of a real number stands for the "sum" of the particular infinite series

$$(9.1) \quad b_n 10^n + b_{n-1} 10^{n-1} + \cdots + b_1 10 + b_0$$

$$+ \frac{b_{-1}}{10} + \frac{b_{-2}}{10^2} + \frac{b_{-3}}{10^3} + \cdots + \frac{b_{-k}}{10^k} + \cdots$$

The fundamental notion that underlies both the geometrical and arithmetical approximations given in (9.1) is that of a *limit*. In order to avoid too many technicalities, we treat limits in an intuitive way but everything said can be justified. First we define an *arithmetical limit*. Suppose that

$$(9.2) \qquad a_1, a_2, \ldots, a_n, \ldots$$

is a sequence of numbers. An example is

(9.3) $\qquad\qquad 10^{-1}, 10^{-2}, 10^{-3}, \ldots, 10^{-n}, \ldots$

We denote the sequence (9.2) by $\{a_n\}$ and say that the real number a is the *limit* of this sequence if the decimal expansions of the numerical differences $(a - a_n)$ begin with more consecutive zeros as n gets larger. For instance, the sequence (9.3), which is more compactly denoted $\{10^{-n}\}$, has the limit 0 because the differences $(0 - 10^{-n}) = (-10^{-n})$ have the decimal expansions shown in Table 9.1.

TABLE 9.1

n	$a - a_n = 0 - 10^{-n}$
1	-0.1
2	-0.01
3	-0.001
4	-0.0001
.	.
.	.
.	.
n	$-0.\underbrace{000 \ldots 01}_{}$
	$(n-1)$ zeros
.	
.	

Therefore an increasing number of initial zeros appear in the decimal expansion of $(a - a_n)$ as n gets larger. If the numbers a_n are thought of as points on a line (this is an example of the algebrization of geometry!), the sequence of numbers a_n has the number a as a limit just in case the points corresponding to a_n approach the point corresponding to a in such a way that for large enough n the distance between a_n and a is smaller than any preassigned (positive) value. This is illustrated in Figure 9.1 for the sequence $\{10^{-n}\}$.

FIGURE 9.1

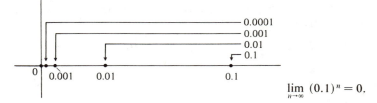

$$\lim_{n \to \infty} (0.1)^n = 0.$$

If a is the limit of the sequence a_n, it is customary to write

$$\lim_{n \to \infty} a_n = a$$

which is read "a_n approaches the limit a as n approaches infinity."

One consequence of this definition of limit is that the familiar nonterminating decimal expansions of fractions (rational numbers) can be thought of as sequences of numbers which have the corresponding fraction as limit. Consider $\frac{1}{3}$. It has the nonterminating expansion

$$\frac{1}{3} = 0.33333 \ldots$$

where just as before, the dots mean that the expansion continues on indefinitely according to the obvious rule. Now define a sequence associated with this decimal expansion; for instance, put

$$a_1 = 0.3$$
$$a_2 = 0.33$$
$$a_3 = 0.333$$
$$a_4 = 0.3333$$
.
.
.
$$a_n = 0.\underbrace{333 \ldots 3}_{n \text{ times}}$$
.
.
.

That is, the nth term of the sequence is the number obtained by placing n three's after the decimal point. It is easy to see that the sequence $\{a_n\}$ has a limit, and indeed

$$\lim_{n \to \infty} a_n = \frac{1}{3}$$

To show this consider the decimal expansions of the differences

$$(a - a_n) = (\tfrac{1}{3} - 0.\underbrace{333 \ldots 3}_{n \text{ times}})$$

listed in Table 9.2:

TABLE 9.2

n	$(a - a_n) = (\tfrac{1}{3} - 0.\underbrace{333 \ldots 3}_{n \text{ times}})$
1	0.0333 . . .
2	0.00333 . . .
3	0.000333 . . .
.	.
.	.
.	.
n	0.$\underbrace{000 \ldots 0}_{n \text{ zeros}}$333 . . .
.	
.	
.	

The expansions in this table are computed by expressing $\frac{1}{3}$ as 0.333 . . . so that

$$a - a_n = (0.333 \ . \ . \ .) - (0.\underbrace{333 \ . \ . \ . \ 3}_{n \ 3's})$$

$$= 0.\underbrace{000 \ . \ . \ . \ 0}_{n \ zeros}333 \ . \ . \ .$$

The number of initial zeros in the decimal expansion of $(a - a_n)$ clearly increases as n increases. In the same way you can show that $0.5 = 0.4999$ . . . , which really means

$$0.5 = \lim_{n \to \infty} a_n$$

with

$$a_n = 0.4\underbrace{9999 \ . \ . \ . \ 9}_{n - 1 \ times}$$

Another example is

$$1 + \tfrac{1}{2}, \ 1 - \tfrac{1}{3}, \ 1 + \tfrac{1}{4}, \ 1 - \tfrac{1}{5}, \ . \ . \ .$$

This sequence has the limit 1 because the differences of the terms of the sequence from 1 are

$$-\tfrac{1}{2}, \ +\tfrac{1}{3}, \ -\tfrac{1}{4}, \ . \ . \ .$$

and the decimal expansions of these numbers have increasingly many consecutive zeros after the decimal point.

Not all sequences of numbers have limits; for instance, the sequence

$$0, \ 1, \ 0, \ 1, \ 0, \ 1, \ . \ . \ .$$

alternates between 0 and 1. If it had a limit, say a, then the differences $(a - a_n)$ would be

$$a, \ a - 1, \ a, \ a - 1, \ a, \ a - 1, \ . \ . \ .$$

whose decimal expansions cannot have an increasing number of consecutive initial zeros as n increases. In this case we say that the limit of the sequence *does not exist* or that this sequence *has no limit*.

§9.2.　There is a corresponding notion of the *limit of a sequence of geometrical objects*. It is difficult to make this concept precise without using the relationship between ordered pairs of numbers and geometrical points in the plane (and the corresponding relationships for points on a line or in space), but, in its elementary form, it is so intuitively reasonable that it is worthwhile to present it, albeit in a philosophically unsatisfactory way. As an example we consider a tangent line to a given curve as the geometrical limit of secants. Let Γ be a curve, as illustrated in Figure 9.2, and consider a sequence of secants (a *secant* is a line segment joining two points on the curve) that tends closer and closer to a given point P on the

FIGURE 9.2

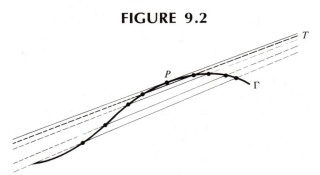

Tangent line as a limit of secant lines.

curve so that the distance of the endpoints of the secants from P has zero as a limit. For the curve in Figure 9.2 the *geometrical limit* of the lines determined by the sequence of secants is the line T tangent to the curve at P.

Of course, if the curve has a "corner," so that it is not "smooth" at the point P (see Figure 9.3), then the situation is much more complicated. There can be *two* positions for lines that are limits of sequences of secants, and there is *no* unique tangent line to the curve at P. In this case it is possible to find a sequence of secants that does not converge in the geometrical sense. One way is to choose a sequence of secant lines $S_1, S_2, S_3, \ldots$, that converges to the line T in Figure 9.3 and another secant sequence $S'_1, S'_2, S'_3, \ldots$, that converges to the line T'. Now construct a new sequence of secants by interlacing the terms of the preceding two sequences:

$$S_1, S'_1, S_2, S'_2, S_3, S'_3, \ldots$$

FIGURE 9.3

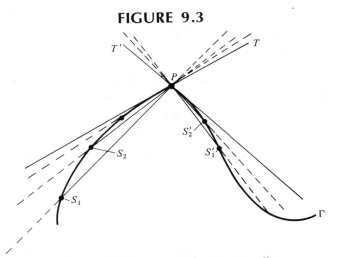

At a "corner" there is no unique tangent line.

The secant lines in this sequence alternately approach the limit lines T and T', and the sequence alternates but does not approach any fixed line as a limit. Compare this case with the alternating sequence 0, 1, 0, 1, 0, 1, . . . , examined in the preceding paragraph, which did not converge to a limit in the arithmetic sense.

§9.3. By using coordinates and equations to represent curves the two notions of limit can be combined. We apply this combination to the case of tangent lines. Suppose that a sequence of secant line segments S_n converges to the tangent line T to the curve Γ at point P as in Figure 9.4. This is equivalent to two conditions: first, the sequence of lines determined by the secants converges to a line through point P and, second, the *slopes* of the secant lines converge to the slope of the tangent line T.[1] Since a sequence of points P_1, P_2, P_3, . . . , has P as a limit just in case the sequence of distances between P and P_n has the limit zero, we have shown that the *geometrical* notion of a tangent line as the limit of a sequence of secant lines is equivalent to the *arithmetical* notion that the number sequence corresponding to the distance from P to the secants has zero as limit *and* the slopes of the secant lines have the slope of the tangent line as limit.

FIGURE 9.4

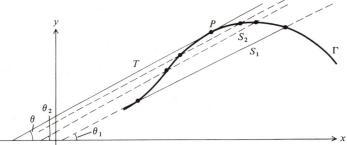

If a tangent line is the limit of a sequence of secants, its slope is the limit of the sequence of secant slopes.

Referring to Figure 9.4, the angle determined by the positive x-axis and the secant line S_n is θ_n. The slope of S_n is therefore $\tan \theta_n$. Let θ denote the corresponding angle made by the tangent line T. Then the sequence of angles

$$\theta_1, \theta_2, \ . \ . \ . \ ,\theta_n, \ . \ . \ .$$

must converge to the angle θ; but this is the same as saying that the sequence of numbers

$$\tan \theta_1, \tan \theta_2, \ . \ . \ . \ ,\tan \theta_n, \ . \ . \ .$$

converges to $\tan \theta$.

---

[1] *If a line corresponds to the Cartesian coordinate equation $y = mx + b$, then the* slope *of the line is the number m. Recall that $m = \tan \theta$ where θ is the angle made by the positive x-axis and that part of the line which lies above the x-axis.*

We shall try to compute the slope of the tangent to a curve at a particular point by using this idea. Consider, for example, the parabola Γ defined by $y = x^2$, whose graph is drawn in Figure 9.5. Let us compute the slopes of a sequence of secants and see if we can find their limit. We can try, for instance, the sequence of secants defined by the sequence of pairs of points on Γ which is determined by the sequence of pairs of x-coordinates $\{(x_n', x_n'')\}$ given by

$$(1 - \tfrac{1}{2}, 1 + \tfrac{1}{2}), \ (1 - \tfrac{1}{3}, 1 + \tfrac{1}{3}), \ \ldots \ , (1 - \frac{1}{n}, 1 + \frac{1}{n}), \ \ldots$$

FIGURE 9.5

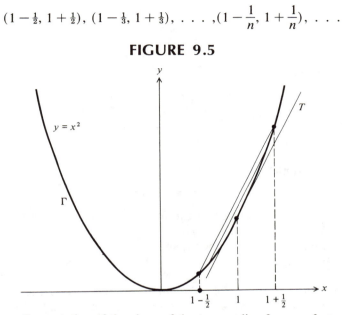

Computation of the slope of the tangent line for $y = x^2$ at $x = 1$.

The slope of the first secant is (*cp.* Figure 9.6)

$$m_1 = \frac{(\tfrac{3}{2})^2 - (\tfrac{1}{2})^2}{1} = \tfrac{8}{4} = 2$$

of the second,

$$m_2 = \frac{(\tfrac{4}{3})^2 - (\tfrac{2}{3})^2}{(\tfrac{2}{3})} = \frac{2(\tfrac{2}{3})}{(\tfrac{2}{3})} = 2$$

In general, the slope of the nth secant is

$$\begin{aligned} m_n &= \frac{(1 + 1/n)^2 - (1 - 1/n)^2}{(1 + 1/n) - (1 - 1/n)} \\ &= \frac{1 + 2/n + (1/n)^2 - 1 + 2/n - (1/n)^2}{2/n} \\ &= \frac{2(2/n)}{(2/n)} = 2 \end{aligned}$$

FIGURE 9.6

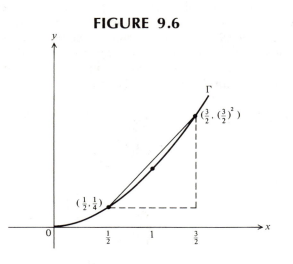

The sequence of slopes is therefore

$$m_1 = 2, \ m_2 = 2, \ m_3 = 2, \ \ldots$$

and it is clear that it converges to the limit 2, which must be the slope of the tangent T to the curve at point $(1, 1)$, as shown in Figure 9.7.

To examine this process in general, suppose a curve defined by an equation of the form $y = f(x)$ is given for some function f; for example, f might be $f(x) = \sin x$, or e^x, or $\log x$, or $x^3 + 2x + 1$, etc. Figure 9.8 exhibits part of the *graph* of the function $y = f(x)$, that is, of the set of points in the plane whose Cartesian coordinates (x, y) are related by $y = f(x)$. Let P be the particular point on the graph with coordinates $(x_0, f(x_0))$, and consider a fixed sequence of secants $S_1, S_2, \ldots$, which approaches the point P as shown. This sequence, as we have seen, may or may not have some line as a limit, but, for the time being, suppose it has. Then consider the sequence of corresponding slopes,

$$m_1, m_2, \ldots, m_n, \ldots$$

FIGURE 9.7

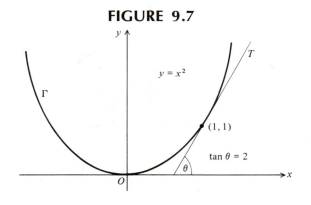

FIGURE 9.8

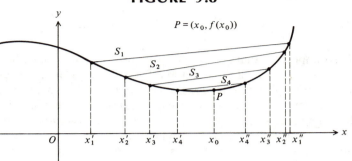

Computation of slopes in general: part 1.

Let us compute these slopes. To do so, names for the points on the x-axis determining the secants are needed. Let these pairs of points that determine S_1, S_2, ... be called (x'_1, x''_1), (x'_2, x''_2), ..., as shown in Figure 9.8. The slopes of the corresponding secants are (Figure 9.9)

$$m_1 = \frac{f(x''_1) - f(x'_1)}{x''_1 - x'_1}$$

$$m_2 = \frac{f(x''_2) - f(x'_2)}{x''_2 - x'_2}$$

.
.
.

$$m_n = \frac{f(x''_n) - f(x'_n)}{x''_n - x'_n}$$

etc.

FIGURE 9.9

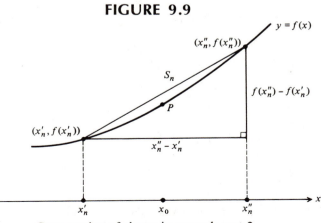

Computation of slopes in general: part 2.

The limiting slope is the slope m of the tangent line, if one exists. Its value is expressed in terms of the symbols we have introduced by

(9.4)
$$m = \lim_{n \to \infty} m_n = \lim_{n \to \infty} \frac{f(x_n'') - f(x_n')}{x_n'' - x_n'}$$

This limit is so important that it deserves a name. We have called it the slope of the tangent line to the curve at the point $(x_0, f(x_0))$, but it is more properly known as the *derivative of the function* $y = f(x)$ *at the point* x_0. More precisely, if the limit of the m_n's exists for every such sequence of secants and if this limit is always the same number, then *that number* is the *derivative of the function at that point*. As we have seen, the derivative of a function at a point does not always exist; for instance, a curve with a corner may have secant sequences that approach two distinct limiting lines.

§9.4.   Recall that a *function f* is a rule that assigns to each number x a number y, written $y = f(x)$. If the function f has a derivative at each point, which will be the case if the graph of the function has a tangent line at each of its points, that is, if it is a smooth curve with no corners, then a new function $\frac{df}{dx}(x)$ can be formed, defined as follows: $\frac{df}{dx}(x)$ is the value of the derivative of the function f at the point x.

For instance, if $f(x) = 3x$, then its graph is given by a straight line with slope equal to 3 (Figure 9.10). What is the graph of the derivative of $f(x) = 3x$? We see by inspection that for any two points (x_1, x_2) on the x-axis the secant joining the points on the graph (Figure 9.11) will have slope $= 3$, the same as the slope of the straight line itself. Therefore any sequence of secants tending to a specific point

FIGURE 9.10

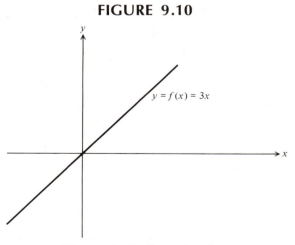

The graph of a linear function.

FIGURE 9.11

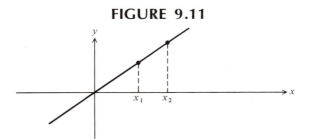

will also have slopes all equal to 3, and the limit of any such sequence of slopes *must* be the number 3 (Figure 9.12). Therefore the derivative of the function $f(x) = 3x$ has value 3 at each point x, and its graph (Figure 9.13) is a straight line parallel to the x-axis as shown in Figure 9.13. This procedure assigns to a given function f a new function called the *derivative* of f, customarily written $\frac{df}{dx}$. In our example

$$f(x) = 3x$$

and then

$$\frac{df}{dx}(x) = 3$$

Formulas for *differentiating* functions (that is, calculating $\frac{df}{dx}$) turn out to be of extreme importance in mathematics and the key to the usefulness of calculus. Given the unwieldy definition of a derivative, it would be impractical to check all possible sequences of secants to determine that a derivative of a function has a particular value at a specific point. Fortunately, it is possible to derive simple

FIGURE 9.12

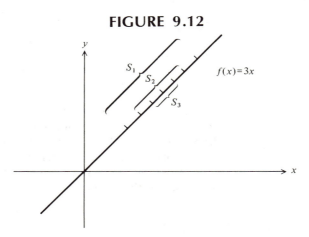

FIGURE 9.13

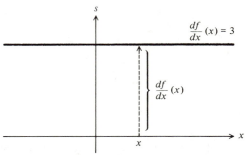

$\frac{df}{dx}(x) = 3$

$\frac{df}{dx}(x)$

The graph of the derivative of the linear function shown in Figure 9.10.

formulas for the calculation of the derivatives of commonly occurring functions and their combinations.

One such formula is

(9.5)
$$\frac{d(x^2)}{dx} = 2x$$

which means that the function $f(x) = x^2$ has as derivative the function $\frac{df}{dx}(x) = 2x$. In our original example the slope of the curve $y = x^2$ was computed at the point $P = (1, 1)$. By using formula (9.5) we immediately obtain

(9.6) $f(x) = x^2$ $\frac{df}{dx}(x) = 2x$ whence $\frac{df}{dx}(1) = 2 \cdot 1 = 2$

the derivative of f at the point $x = 1$, which (by *definition*) is the value of the slope of the tangent line at the point $(1, f(1))$. This agrees with our earlier calculation, but we can also find the slope at any point. For example, the slope of the tangent line to the curve $y = x^2$ at point $(0, 0)$ is

$$\frac{df}{dx}(0) = 2 \cdot 0 = 0$$

and at $(\frac{1}{2}, \frac{1}{4})$ it is

$$\frac{df}{dx}(\tfrac{1}{2}) = 2 \cdot \tfrac{1}{2} = 1$$

which agrees with intuition, since the slope of $y = x^2$ at $x = 0$ is clearly zero (the tangent line is horizontal; Figure 9.14) and the slope at $x = \frac{1}{2}$ should be somewhere between 0 and 2, since the slope appears to increase in passing from $x = 0$ to $x = 1$. If the derivative of $y = x^2$ is plotted as a function of x, a straight line is obtained; in fact, according to (9.5), it corresponds to the equation

$$\frac{df}{dx}(x) = 2x$$

FIGURE 9.14

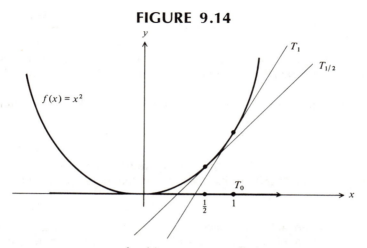

$y = x^2$, with some tangent lines.

As before, the curve representing the derivative is less complicated than the original. This is not always so, but it is true in general for polynomial functions, as the following important formula shows.

THEOREM: Let $a_0, a_1, \ldots, a_n$ be constant real numbers and let

$$f(x) = a_n x^n + a_{n-1} x^{n-1} + \cdots + a_0.$$

FIGURE 9.15

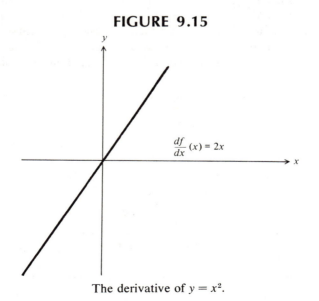

$$\frac{df}{dx}(x) = 2x$$

The derivative of $y = x^2$.

Then

$$\frac{df}{dx}(x) = na_nx^{n-1} + (n-1)a_{n-1}x^{n-2} + \cdots + a_1$$

This formula states how the derivative of a polynomial function can be calculated without the need for examining the properties of limits of secant lines. Its proof is not hard if one has some facility in handling limits.

As an example of how the theorem is used, let $f(x) = 4x^5 + 3x^2$; then

$$\frac{d(4x^5 + 3x^2)}{dx} = 20x^4 + 6x$$

The procedure specified by the theorem is obviously a great deal simpler to apply than the original limit definition; its drawback is that it applies only to *polynomial* functions.

§9.5. In §§4.8 and 4.10 the concept of area was examined and a method for finding the area of a plane figure as the limit of approximating areas was introduced and applied to a region bounded by a parabola and a straight line. This procedure, first used by Eudoxus and greatly amplified by Archimedes, is known as the *method of exhaustion*. It is the basic idea in *integration theory*, or *integral calculus*, to which we now turn our attention.

To simplify the discussion let us assume that we want to find the area under the graph of a function $y = f(x)$ between two vertical lines and above the x-axis, as in Figure 9.16. Recall that one way of applying the method of exhaustion consisted of approximating the desired area from within and from without by the sums of areas of rectangles. In the following diagram this process is illustrated by showing a set of circumscribed and inscribed rectangles. More precisely, subdivide the interval between $x = a$ and $x = b$ by points $x_0 < x_1 < x_2 \ldots < x_n$, where $x_0 = a$ and $x_n = b$. Then construct a rectangle above each of these smaller intervals formed by the subdivision (between x_i and x_{i+1} in general) whose height is the maximum (the R_i's) or minimum (the r_i's) of the curve in that interval. Thus we obtain rectangles

FIGURE 9.16

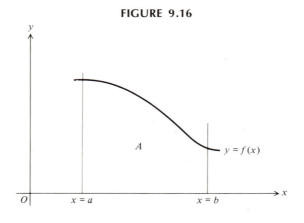

$R_1, R_2, \ldots, R_n$, and $r_1, r_2, \ldots r_n$, the *circumscribed* and *inscribed* rectangles, respectively, illustrated in Figure 9.17, where n has been chosen equal to 6. It is clear that the area A under the curve, understood in the sense of §4.10, should certainly be smaller than the sum of the areas of the rectangles $R_1, R_2, \ldots, R_n$ and certainly larger than the sum of the areas of the smaller rectangles $r_1, r_2, \ldots, r_n$;

FIGURE 9.17

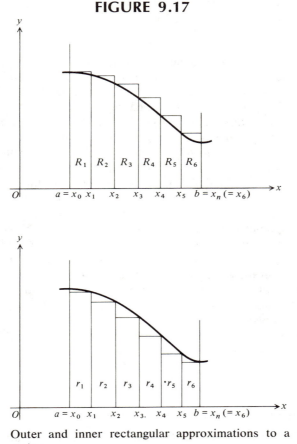

Outer and inner rectangular approximations to a region.

letting $A(R_1)$ denote the area of R_1, etc., we should have

$$A(r_1) + A(r_2) + \cdots + A(r_n) < A < A(R_1) + A(R_2) + \cdots + A(R_n)$$

Now suppose that the lengths of the intervals are made smaller and smaller so that the widths of the rectangles approach zero. It seems clear that the sum of the areas of the big rectangles will get smaller and the sum of the areas of the small rectangles will get larger.

For the sum of the circumscribed and inscribed rectangles to the particular parabola described in §4.8 we had (Figures 4.13, 4.14), for example,

$$\tfrac{4}{3} + \frac{1}{N} - \frac{1}{3N^2} > A > \tfrac{4}{3} - \frac{1}{N} - \frac{1}{3N^2} \cdot$$

where N denotes the number of intervals used. In order for the widths of the rectangles to tend to zero, the number of intervals must increase without bound; that is, N must become very large in this example. However, $1/N$ tends to zero as N becomes large, whence for a large number of intervals

$$\tfrac{4}{3} - \text{small} < A < \tfrac{4}{3} + \text{small}$$

where the terms denoted by "small" have the limit zero if the widths of the rectangles all tend to zero. Thus, both the "circumscribed sum" and "inscribed sum" tend to $\tfrac{4}{3}$ and the area under the parabola is squeezed in between. Therefore the area under the parabola must indeed be *equal* to $\tfrac{4}{3}$, as argued in §4.8.

Now use the idea in this example to *define* the area under a curve in general. Referring to Figure 9.17, *define* the area under the graph of the function $y = f(x)$ between the vertical lines $x = a$ and $x = b$ to be the *limit* of the sums of the areas of approximating circumscribed and inscribed rectangles as the widths of the rectangles tend to zero. For the definition to make sense this limit must exist and be the same for all different choices of sequences of inscribed and circumscribed rectangles (this is analogous to the definition of a derivative as the limit of slopes of an *arbitrary* sequence of secants tending to a given point). If we know, for some reason, that all such limits will be the same, then it suffices to find one convenient set of limiting rectangles for which the limit of the sums of their areas is computable. For instance, in the example of the parabola, either the circumscribed or the inscribed sequence of rectangle area sums gives the desired answer, $\tfrac{4}{3}$. This definition of area is the same as the one given in Chapter 4, but it is phrased in a more precise and formal way.

Let us agree to call this limit, if it exists, the *integral of the function $f(x)$ from the point a to the point b* and to denote it by

$$\int_a^b f(x)\ dx$$

This symbolic notation was invented by Leibniz about 1680. The symbol $\int$ is a Latin long S which here represents the *sum* of the areas of rectangles taken in the *limiting* sense already discussed. The symbol dx represents the "infinitesimal" width of one of the approximating rectangles in the limiting process, and $f(x)$ denotes the height of the infinitesimal rectangle whose base is at x.

How can the integral of a function f from a to b be computed? One way is to find a sequence of approximations by rectangles, the limit of whose summed area can be computed, as we did for a parabola in §4.8. It can be shown that the limits will exist and always be the same for any sequence of inscribed or circumscribed rectangles if the curve is *continuous*. This means that the curve can be drawn by some "continuous" motion of a point; for example, the point of a pencil moving in contact with a piece of paper. But for many frequently occurring and important functions there is a much simpler and better way, based on the justly named *Fundamental Theorem of Calculus,* independently discovered by Newton and Leibniz, which is presented in §9.7 after a brief historical digression.

§9.6. In 1647 Cavalieri found the following formula for the area A under the curve $y = x^n$, illustrated in Figure 9.18, where n is a nonnegative integer:

(9.7)
$$A = \int_0^b x^n \, dx = \frac{b^{n+1}}{n+1}$$

Cavalieri proved this formula for $n = 1, 2, \ldots , 9$, and then surmised that it was probably true in general, although he had no proof.

FIGURE 9.18

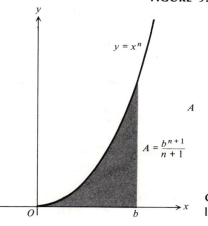

$$A = \frac{b^{n+1}}{n+1}$$

Geometric interpretation of Cavalieri's formula.

Cavalieri's method used the *theory of indivisibles,* which was in vogue at that time but has long since been replaced for the measurement of area by the *method of exhaustion.* Briefly, in the theory of indivisibles a plane area was considered as consisting of an infinite number of parallel line segments of finite length and of "indivisible" width. The area of such a region was considered to be the "sum" or "union" of these indivisible elements. To compute the area of a geometrical region R by this method the elements of R must be compared with the elements of a region whose area is known. For example, such reasoning can easily be used to find the area of a triangle, assuming that we know the area of a parallelogram: each of the indivisible line elements of the triangle appears twice in the corresponding parallelogram whence the area of the triangle must be half that of the parallelogram. A more complicated form of this kind of reasoning was employed by Cavalieri in his derivation of formula (9.7). See, for example, Boyer [8], p. 362, for a more complete discussion.

The significance of Cavalieri's work is not in its method but rather that it was the first published *formula* which showed how to integrate a class of functions. As we have already remarked, it is such general formulae that give calculus a great deal of its power and practical importance.

Fermat (1601–1665) made numerous contributions to the development of differential and integral calculus as well as to the foundations of analytic geometry, as mentioned in Chap. 8. He introduced the notion of the "difference" quotient [which is used in (9.4) to define the derivative] and used it in the study of problems of maxima and minima. Fermat intuitively understood the concept of derivative of a function at a point but did not have control of the limit concept, which was introduced much later. He followed the method in §9.3 to derive the formula for the derivative of a polynomial function.

The achievements of Cavalieri in integral calculus and Fermat in differential calculus were coupled by Newton and Leibniz in their formulation of the Fundamental Theorem of Calculus.

§9.7. In the preceding sections independent developments of differential and integral calculus were presented. About 1680, and independently of one another, Newton and Leibniz discovered the relation between these two apparently distinct concepts, which can be thought of geometrically as finding tangents and measuring areas, respectively. This relationship is now known as the *Fundamental Theorem of Calculus*. It can be formulated as follows:

Suppose $y = f(x)$ is a function having a derivative at each x such that $a \leqslant x \leqslant b$. Then

$$(9.8) \qquad \int_a^b \frac{df}{dx}(x)\ dx = f(b) - f(a)$$

Note that the derivative and integral concepts both appear in the left-hand side of the equation. Recall that each represents limiting processes, applied to the function concerned. What the relation states is that these two limiting processes are *inverses* of each other, somewhat similar to the way extraction of a square root is inverse to squaring a number.

Consider a particular example. Suppose that we want to prove Cavalieri's formula (9.7), assuming that we know the simpler formula for calculating the derivative of a polynomial. We then choose

$$f(x) = \frac{x^{n+1}}{n+1}$$

From the differentiation formula in §9.4

$$\frac{df}{dx}(x) = x^n$$

Then, using the Fundamental Theorem (9.8), with $a = 0$, find

$$\int_0^b \frac{df}{dx}(x)\ dx = \int_0^b x^n\ dx = f(b) - f(0)$$

$$= \frac{b^{n+1}}{n+1} - \frac{0^{n+1}}{n+1}$$

$$= \frac{b^{n+1}}{n+1}$$

which is precisely Cavalieri's formula (9.7).

Of course this applies immediately to the problem of the quadrature of the parabola (as does Cavalieri's formula). If we want to find the area under the curve $y = 1 - x^2$, between $x = -1$ and $x = 1$, as in Figure 4.12, we find a function $f(x)$ such that

(9.9) $$\frac{df}{dx}(x) = 1 - x^2$$

For instance, we take

$$f(x) = x - \frac{x^3}{3}$$

and check that (9.9) is indeed satisfied. Notice that, by symmetry, the area from $x = -1$ to $x = 0$ is the same as that between $x = 0$ and $x = 1$. Now use the Fundamental Theorem of Calculus to find

$$\text{area} = 2\int_0^{+1}(1 - x^2)\ dx = 2\int_0^{+1}\frac{df}{dx}(x)\ dx$$
$$= 2\,(f(1) - f(0))$$

where $f(x) = x - x^3/3$. Therefore

$$\text{area} = 2\left(+1 - \frac{(+1)^3}{3}\right) - 2\left(0 - \frac{0}{3}\right)$$
$$= 2(1 - \tfrac{1}{3})$$
$$= \tfrac{4}{3}$$

which is the result previously, and much more laboriously, obtained. It also provides a verification of the Fundamental Theorem for this special case since the previous derivation of the result $\tfrac{4}{3}$ is independent of this computation.

Any formula known for integration becomes, via the Fundamental Theorem, a formula for differentiation and vice versa. In practice it is easier to derive formulas for differentiating functions, so the Fundamental Theorem provides numerous formulae for integrals of specific functions. Portions of two pages from an early compilation of integrals are shown in Figure 9.19; note that this compilation ran at least 567 pages.

§9.8. The simplest really useful application of the differential calculus is to the solution of *extremal problems,* which call for either the maximization or minimization of some function of one or more variables. They occur throughout all aspects of civilized life. Some extremal problems require the use of integral as well as differential calculus, but the simpler types can be dealt with by the latter alone.

Here are some typical extremal problems:

1. A manufacturer of cans wishes to produce a cylindrical can with a pre-assigned volume but using a *minimal* amount of material.

2. The Council of Economic Advisers wishes to *maximize* employment while *minimizing* inflation. This problem may not, from the mathematical viewpoint, have a solution.

FIGURE 9.19

F. Alg. rat. ent.   TABLE 1.   Lim. 0 et 1.

1) $\int (1-x^2)^a dx = \frac{(2^{a/2})^2}{1^2 a+1/1}$ (VIII, 239).   2) $\int (1-x)^{p-1} x\, dx = \frac{1}{p(p+1)}$ (VIII, 319).

3) $\int (1-x)^p x^{1-p} dx = \frac{1}{2} p\pi (1-p)\, Cosec\, p\pi =$ 4) $\int (1-x)^{1-p} x^p\, dx\, [p^2 < 1]$ (IV, 27).

5) $\int (1-x)^{p-1} x^{q-1} dx = \frac{\Gamma(p)\,\Gamma(q)}{\Gamma(p+q)} = \frac{1^{p-1/1}}{q^{p/1}} = \begin{bmatrix} p \\ q \end{bmatrix} = B(p,q),$ l' intégrale Eulérienne de pre- mière espèce (VIII, 262).

6) $\int (1-x)^{q+b-1} x^{p+a-1} dx = \frac{p^{a/1} q^{b/1}}{(p+q)^{a+b/1}} \cdot \frac{\Gamma(p)\,\Gamma(q)}{\Gamma(p+q)}$ (VIII, 262).

7) $\int (1-x)^{b-p} x^{p+c} dx = \frac{(1+p)^{c/1}(1-p)^{b/1}}{1^{b+c+1/1}} \cdot \frac{p\pi}{Sin\, p\pi} =$ 8) $\int (1-x)^{p+c} x^{b-p} dx$ (IV, 28).

9) $\int (1-x)^{b-p} x^{p-c} dx = \frac{(1-p)^{b/1}}{p^{c/1-1} 1^{b-c+1/1}} \cdot \frac{p\pi}{Sin\, p\pi} =$ 10) $\int (1-x)^{p-c} x^{b-p} dx$ (IV, 28).

11) $\int (1-x^2)^q x^{2a-1} dx = \frac{1^{a-1/1}}{2 \cdot (q+1)^{a/1}}$ (VIII, 238).

12) $\int (1-x^2)^q x^{2a} dx = \frac{2^{q/2}}{(2a+1)^{q+1/2}}$ (VIII, 238).

13) $\int (1-x^r)^{p-1} x^{q-1} dx = r^{p-1} \frac{1^{p-1/1}}{q^{p/r}} = \frac{1}{pr} \cdot \frac{pr+q}{(p+1)q} \cdot \frac{2(pr+q+r)}{(p+2)(q+r)} \cdot \frac{3(pr+q+2r)}{(p+3)(q+2r)} \cdots$
                                                                (VIII, 233, 234).

14) $\int (1-x)^{a-1} (1+q x^b)^c x^{p-1} dx = 1^{a-1/1} \sum_0^\infty \binom{c}{n} \frac{q^n}{(p+nb)^{a/1}} \quad [q^2 < 1]$ (VIII, 475).

15) $\int [(1+x)^{p-1}(1-x)^{q-1} + (1+x)^{q-1}(1-x)^{p-1}] dx = 2^{p+q-1} \frac{\Gamma(q)\,\Gamma(q)}{\Gamma(p+q)}$ (VIII, 631).

16) $\int [p^r x^{r-1}(1-px)^{q-1} + (1-p)^q x^{q-1}\{1-(1-p)x\}^{r-1}] dx = \frac{\Gamma(q)\,\Gamma(r)}{\Gamma(q+r)}$ (VIII, 631).

Page. 27.

F. Algébrique;
Exponentielle;   TABLE 396.   Lim. 0 et $\frac{\pi}{2}$.
Circulaire Directe.

1) $\int e^{-px} Sin x . x\, dx = \frac{1}{(1+p^2)^2} \left[\left\{ 1-p^2-\frac{1}{2}p\pi(1+p^2) \right\} e^{-\frac{1}{2}p\pi} + 2p \right]$ (VIII, 566).

2) $\int e^{-px} Cos x . x\, dx = \frac{1}{(1+p^2)^2} \left[p^2-1+\left\{ \frac{\pi}{2}(1+p^2)+2p \right\} e^{-\frac{1}{2}p\pi} \right]$ (VIII, 566).

3) $\int e^{-q\, Tg\, x} \frac{x\, dx}{Cos^2 x} = \frac{1}{q} \left[Ci(q).Sin q + Cos q.\left\{ \frac{\pi}{2} - Si(q) \right\} \right]$ V. T. 271, N. 2.

4) $\int e^{-q\, Tg\, x} \frac{Sin x + Cos x}{Cos^2 x} x\, dx = Sin q.\left\{ \frac{\pi}{2} - Si(q) \right\} - Ci(q).Cos q$ V. T. 271, N. 3.

5) $\int e^{-Tg^2 x} Sin 4x \frac{x\, dx}{Cos^6 x} = -\frac{3}{2} \sqrt{\pi}$ V. T. 272, N. 9.

6) $\int e^{-Tg^2 x} Sin^3 2x \frac{x\, dx}{Cos^3 x} = 2\sqrt{\pi}$ V. T. 272, N. 9.

7) $\int e^{-q\, Tg^2 x} \frac{q - Cos^2 x}{Cos^4 x . Cot x} x\, dx = \frac{1}{4} \sqrt{\frac{\pi}{q}}$ V. T. 272, N. 9.

8) $\int e^{-q\, Tg^2 x} \frac{q - 2 Cos^2 x}{Cos^6 x . Cot x} x\, dx = \frac{1+2q}{8} \sqrt{\frac{\pi}{q}}$ V. T. 272, N. 11.

Page 567.

Early integral table by Bierens de Haan, 1867.

3. An engineer wishes to determine the curve of a suspension bridge cable that will carry the *maximal* load for a given amount of structural material used; see §9.11.

4. A corporation executive wishes to *maximize* profits.

5. NASA wishes to determine space flight paths that *minimize* fuel consumption while permitting completion of the flight within a predetermined time interval (*cp.* Chapter 12 and especially §12.8).

6. Describe a curve of *minimal* length that joins two prescribed points on a given surface. Note that if the surface is a plane the solution is a straight line; if it is a sphere, then the curve is a great circle. See Chapter 11 for an extensive discussion of this problem and Chapter 12 for its application to the creation of models of the universe.

7. Extremal problems occur in less abstract and less pleasant contexts. For instance, determine the angle of a cannon such that a ballistic shell will have *maximal* range. If air resistance is neglected, this is a simple question. Otherwise it is not. Napoleon's interest in mathematics and mathematicians may have been due in part to his training as a commissioned artillery officer.

You can easily devise a large number of problems related to your own field of interest which involve — perhaps in some not yet mathematized way — the notion of maximization or minimization of some variable quantity.

To demonstrate how the differential calculus can be used to solve extremal problems consider the following one, which is simple but characteristic.

A farmer has a given amount of fence to enclose a rectangular area bounded on one side by a river. What should the dimensions of the rectangle be to maximize the enclosed area? Figure 9.20 displays the various elements of the problem. As-

FIGURE 9.20

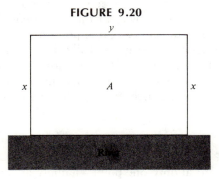

sume that the length of fence available is 200 yards. Then

(9.10) $\qquad 2x + y = 200 \qquad$ (= length of fence in yards)

(9.11) $\qquad xy = A \qquad$ (= area of region in square yards)

For a given value of x, y is determined by the first equation; hence the area A is determined by the second equation. Then A can be considered as a function of the single variable x. In fact, solving (9.10) for y and substituting the result in (9.11) produces

$$A(x) = x \cdot y = x(200 - 2x)$$

Figure 9.21 exhibits the curve corresponding to the function $A(x)$. Notice that it appears to have a *maximum* at the point above $x = 50$ and conclude therefore that the area of the field must be a maximum if $x = 50$, which solves the problem. How can the precise point on the curve corresponding to the maximum be determined?

FIGURE 9.21

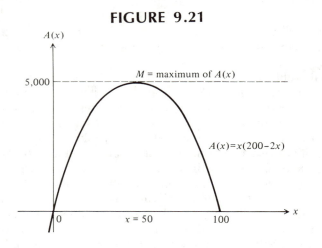

If we were to plot many points, the location of the maximum could be fixed to any desired degree of precision, but this procedure is tedious and in principle inexact. From a theoretical point of view it is clear that the maximum point has the property that the tangent line to the curve at that point must be *parallel to the x-axis,* which means that the slope of this tangent line is equal to 0. According to the theory of derivatives, the derivative of the function $A(x)$ which defines the curve must be 0 at the value of x above which the maximum lies. What is the derivative of the function $A(x)$? Using the theorem for differentiating a polynomial in §9.4, we easily find

$$\frac{dA}{dx}(x) = 200 - 4x$$

This function of x is zero at only one point — namely at $x = 50$. We conclude that $x = 50$ is precisely where the function $A(x)$ takes on its maximum value.

This simple problem gives some indication of the power of the differential calculus for treating such matters.

§9.9. Newton thought of the derivative as the rate of change of the position of a particle moving in space. This enabled him to introduce the essentially geometrical outlook of the calculus into physics and, as a consequence, to formulate general "laws of nature" in mathematical terms.

Indeed, if $x(t)$ represents the position of a particle moving along the x-axis as a function of time t (a rather simplified view of the three-dimensional world but one not devoid of interesting applications) and if a certain "force" F is exerted on the particle, such as a constant force like that of gravity near the earth's surface (in which case we think of the x-axis as perpendicular to the earth's surface), then Newton postulated that the *second derivative* of $x(t)$ [the derivative of the derivative of $x(t)$] is proportional to the force F and the constant of proportionality is the *mass* m of the particle. In symbols,

$$(9.12) \qquad F = m \, \frac{d^2x}{dt^2}$$

where $\dfrac{d^2x}{dt^2}$ is the traditional notation for the second derivative of $x(t)$.

A typical problem of that branch of physics concerned with the motion of particles is to determine where the particle will be at some future time t if it is known to have been at point $x(t_0)$ at time t_0 and was then traveling with velocity $v(t_0)$. It is assumed, of course, that the nature of the force F is known, and in this event the future motion of the particle will be found by solving the *differential equation* (9.12) for the unknown function $x(t)$. For instance, if F is a constant force, then the second derivative of $x(t)$ will equal the constant ratio F/m. According to the theorem on the differentiation of a polynomial in §9.4, the derivative of a polynomial of degree 2 is a polynomial of degree 1, and the derivative of a polynomial of degree 1 is constant; combining these statements and the definition of *second derivative*, we find that the second derivative of a polynomial of degree 2 is a constant. This means that there is some polynomial

$$(9.13) \qquad x(t) = at^2 + bt + c$$

that satisfies Newton's equation (9.12). By differentiating (9.13) twice we find

$$(9.14) \qquad \frac{dx}{dt} = 2at + b$$

and

$$\frac{d^2x}{dt^2} = 2a$$

Comparison of the last equation with (9.12) shows that the constant a must have the value

$$a = \frac{F}{2m}$$

Furthermore, the constant c is the position of the particle at $t = 0$, as is shown by substituting $t = 0$ in (9.13). It is not hard to convince ourselves that dx/dt represents[2] the velocity of the particle at time t, so we have shown that for a constant force Newton's differential equation is satisfied by a parabola (9.13) whose coefficients are completely determined by the initial position and velocity of the particle.

If the position of the particle at $t = 0$ is the origin, so that $x(0) = 0$, and if its velocity then is v, the solution to Newton's equation simplifies to

$$(9.15) \qquad x(t) = \frac{F}{2m} \, t^2 + vt$$

---

[2] *We sometimes write dx/dt in place of $\dfrac{dx}{dt}$, d^2x/dt^2 in place of $\dfrac{d^2x}{dt^2}$, etc., for the compositor's convenience.*

This precise equation gives the position of the particle at any time t. It represents, for instance, the position of an object falling under the influence of the *constant force* of the earth's gravitation (near the surface of the earth and neglecting wind resistance). The position of the object does not vary at a constant rate but at an ever-increasing rate, quadratic in time t. This agrees with experience which shows that any falling object picks up speed as time progresses.

In general, calculus is used to express the fundamental equations of the paths and motion of "particles" such as spaceships, planets, molecules in chemical reactions and nuclear explosions, fluids in pipes, air flowing past aircraft wings, and electrons in electrical wiring systems — almost any type of dynamical system. Observe that the laws of motion are embodied in *differential equations,* that is, in equations involving unknown functions and their derivatives, of which (9.12) is a simple example. Equations of this kind are generalizations of algebraic equations; their study follows the study of elementary calculus and is mandatory for engineers, physicists, economists, chemists, demographers, psychologists, and others attempting to describe the various aspects of orderly change in the world.

§9.10. The preceding discussion of differential and integral calculus was motivated by the geometric problems of finding the tangent lines to a curve and measuring the area of a region. There is a third geometric problem whose importance will become clearer when we consider differential geometry in Chapter 11 but which is introduced here since it, too, is "solvable" by means of the calculus. This is the problem of measuring the *length* of a portion of a curve ("arc length"). For instance, the measurement of the length of a circle of diameter equal to 1 is really the same thing as the measurement of π, which, as shown by Archimedes' computations, is not a simple matter. Archimedes' method provides a clue to a general principle; the definition of arc length given below is similar to that used by him for the circle and will also have points of contact with the definition of derivative and integral in that it will be the limit of a sequence of approximations.

Let Γ be a curve in the plane joining points A and B as in Figure 9.22. Select a sequence of $(n + 1)$ consecutive points on Γ, say $A = P_0, P_1, \ldots , P_n = B$, and join them by line segments as in Figure 9.23, where n is equal to 4. It seems clear that the length of Γ, which we

FIGURE 9.22

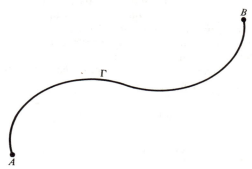

FIGURE 9.23

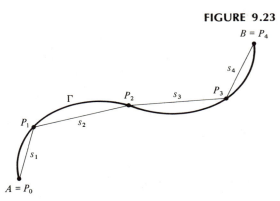

Polygonal line approximating a curve.

denote by $l(\Gamma)$, is greater than or equal to the sum of the lengths of the straight-line segments $s_1, \ldots, s_n$; that is,

$$l(\Gamma) \geq l(s_1) + \cdots + l(s_n)$$

We could try to proceed as in the measurement of area to determine whether there is any quantity similar to the right side of this inequality that naturally bounds the length $l(\Gamma)$ from above, *i.e.*, is $\geq l(\Gamma)$. For the circle Archimedes used the perimeter of a circumscribed polygon, but without further assumptions it is not clear what figure to choose; therefore we ignore an upper bound and proceed anyway. Without being completely rigorous, define $l(\Gamma)$ to be the *limit* of the sum of the lengths of the segments joining sequences of points, as the number of points becomes increasingly large (more precisely: tends to infinity) and the distance between successive points on the curve is made very small. This corresponds to taking finer and finer subdivisions of the curve and taking the limit of the lengths of the corresponding "polygonal approximations" to the curve Γ. This limit will not always exist, but it will for most smooth curves. It turns out to be quite a difficult problem to compute the *value* of this limit explicitly and exactly.

Suppose that the curve is the graph of a function $f(x)$ that has a derivative; for instance, $f(x)$ might be a polynomial. A *formula* for the length of the curve can be derived. Consider the situation in Figure 9.24. For a given approximation Pythagoras' theorem shows that

$$(9.16) \quad l(s_1) + \cdots + l(s_n) = \sqrt{[f(x_1) - f(x_0)]^2 + (x_1 - x_0)^2} + \cdots +$$
$$\sqrt{[f(x_n) - f(x_{n-1})]^2 + (x_n - x_{n-1})^2}$$

since the distance between two points with coordinates (a, b) and (c, d) is (according to Pythagoras)

$$\sqrt{(d - b)^2 + (c - a)^2}$$

Rewriting (9.16), we obtain

$$(9.17) \quad l(s_1) + \cdots + l(s_n) = \left\{ 1 + \left[\frac{f(x_1) - f(x_0)}{x_1 - x_0} \right]^2 \right\}^{1/2} (x_1 - x_0) + \cdots$$

where the ellipsis represents similar terms in $x_2 - x_1$, $x_3 - x_2$, etc. Now consider the function

$$g(x) = \left\{ 1 + \left[\frac{df}{dx}(x) \right]^2 \right\}^{1/2}$$

FIGURE 9.24

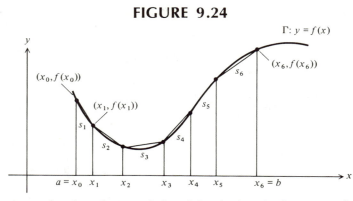

Approximation of a curve's length by the length of an approximating polygonal line.

and assume that $df/dx\,(x)$ exists everywhere; that is, the curve in Figure 9.24 has a tangent at each point. Then for very small $(x_1 - x_0)$ we have

$$\frac{df}{dx}(x_1) \sim \frac{f(x_1) - f(x_0)}{x_1 - x_0}$$

(where $\sim$ means "approximately equal to") because the difference quotient, which is the approximating secant slope, is close to the tangent slope, with similar results for the other x_j's. It is intuitively reasonable to expect that

$$(9.18)\quad l(s_1) + \cdots + l(x_n) \sim g(x_1)(x_1 - x_0) + \cdots + g(x_n)(x_n - x_{n-1})$$

where the expression on the right is a sum of the areas of rectangles, as shown in Figure 9.25. Suppose the subdivision of the curve is made very fine; that is, the largest difference $(x_k - x_{k-1})$ is made very small. Then the sum on the right in (9.18) must tend to

$$\int_a^b \left\{1 + \left[\frac{df}{dx}(x)\right]^2\right\}^{1/2} dx$$

the area under the curve in Figure 9.25. On the other hand, the expression on the left must tend to the arc length of the curve in Figure 9.24.

The length of the curve whose equation is $y = f(x)$ from $x = a$ to $x = b$ is therefore given by the integral

$$(9.19)\qquad l(\Gamma) = \int_a^b \left\{1 + \left[\frac{df}{dx}(x)\right]^2\right\}^{1/2} dx$$

This is a *formula* for the length of a portion of a curve represented in terms of the function that, according to the principle of algebrization of geometry, corresponds to it. Our intuitive proof of its validity admittedly contains several "gaps" but it can be made rigorous by a careful appeal to the formal theory of limits and some auxiliary results concerning the differentiation of functions.

Notice how complicated formula (9.19) is. There are three limiting processes: the derivative of the function f, the square root function, and the integral, a third

FIGURE 9.25

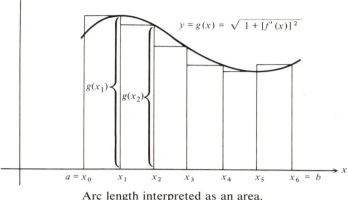

Arc length interpreted as an area.

limiting process superimposed on the other two. Considerable information is required to measure the arc length of a curve.

Formulae such as (9.19) were obtained by Van Heuraet (1633–1660) and Neil (1637–1670), who were contemporaries of Fermat. At that time the problem of determining arc lengths was at the frontier of mathematical research.

The availability of the arc length formula does not ensure that exact numerical answers will be easy to find unless the integrals are computable (but see Chapter 10 for approximations). The square root in (9.19) makes evaluation of the integral formidable and presented mathematicians with a host of research problems for two centuries as they tried to evaluate this integral for common and important functions $f(x)$. Even if $f(x)$ is a polynomial, the integral[3] will *not* be, which is a principal part of the difficulty.

In Fermat's only paper published during his lifetime (1660) there is a solution to the arc length problem for the *semicubical parabola*

$$y = x^{3/2} \qquad 0 \leqslant x \leqslant a$$

shown in Figure 9.26. The part below the x-axis corresponds to the choice of the negative square root of x^3.

The arc length integral for the semicubical parabola can be evaluated in terms of elementary functions, which came as a great surprise in the seventeenth century because Descartes believed it was impossible to measure the length of a curve defined by polynomial equations. As he put it in *La Géométrie:*

> *Geometry should not include lines (or curves) that are like strings, in that they are sometimes straight and sometimes curved, since the ratios between straight and curved lines are not known, and I believe cannot be discovered by human minds, and therefore no conclusion based upon such ratios can be accepted as rigorous and exact.*

[3] *Considered as a function of its upper limit b.*

FIGURE 9.26

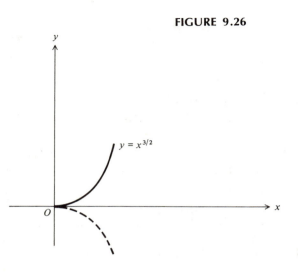

$y = x^{3/2}$

Semicubical parabola.

History did not bear out Descartes' dogmatic and groundless assertion, nor fortunately did his great prestige suppress the research of younger and more fertile minds, for in fact it is easy to show that

$$\frac{d}{dx}\left[\tfrac{8}{27}(1 + \tfrac{9}{4}x)^{3/2}\right] = (1 + \tfrac{9}{4}x)^{1/2}$$

which implies that the length $l(\Gamma)$ of one branch of the semicubical parabola from $x = a$ to $x = b$ is explicitly given by the straightforward formula

$$l(\Gamma) = \tfrac{8}{27}\left[(1 + \tfrac{9}{4}b)^{3/2} - (1 + \tfrac{9}{4}a)^{3/2}\right]$$

For instance, its length from $x = 0$ to $x = 1$ is simply $(13^{3/2} - 8)/27 \cong 1.4397$. . . .

§9.11.   Amongst the common engineering applications of the calculus few are more romantic and impressive than the design and construction of great suspension bridges — thin ribbons of steel strung from lofty towers in arcs at once beautiful and purposeful. From the 70-foot span over the River Tees constructed in 1741 to the exquisite 4200-foot span of the Golden Gate completed in 1937 (Figure 9.27) there has been a general progression of ever longer and increasingly slender suspension bridges. Second only to this impressive sequence of triumphs of the engineers' art is the unremitting succession of bridge failures including the 260-foot span of the River Tweed footbridge at Drybourgh Abbey in Berwick County, Scotland, in a gale six months after its completion in 1817; the 449-foot Union Bridge over the Tweed near Berwick in 1820, also six months after completion, in a violent wind; the Brighton Chain Pier in Sussex County, England, which consisted of 255-foot spans of a 12⅔-foot roadway, in a storm; and the Wheeling, West Virginia, 1010-foot span over the Ohio River in 1854, six years after completion, in a hurricane. A newspaper reporter, witness to this

FIGURE 9.27

The Golden Gate Bridge: an exercise in engineering mathematics.

event, wrote:

> *With feelings of unutterable sorrow, we announce that the noble and world-renowned structure, the Wheeling Suspension Bridge, has been swept from its strongholds by a terrific storm, and now lies a mass of ruins. Yesterday morning thousands beheld this stupendous structure, a mighty pathway spanning the beautiful waters of the Ohio, and looked upon it as one of the proudest monuments of the enterprise of our citizens. Now, nothing remains of it but the dismantled towers looming above the sorrowful wreck that lies beneath them.*
>
> *About 3 o'clock yesterday we walked toward the Suspension Bridge and went upon it, as we have frequently done, enjoying the cool breeze and the undulating motion of the bridge. . . . We had been off the flooring only two minutes, and were on Main street when we saw persons running toward the river bank; we followed just in time to see the whole structure heaving and dashing with tremendous force.*
>
> *For a few moments we watched it with breathless anxiety, lunging like a ship in a storm; at one time it rose to nearly the height of the tower, then fell, and twisted and writhed, and was dashed almost bottom upward. At last there seemed to be a determined twist along the entire span, about one half of the flooring being nearly reversed, and down went the immense structure from its dizzy height to the stream below, with an appalling crash and roar.*
>
> *For a mechanical solution of the unexpected fall of this stupendous structure, we must await further developments. We witnessed the terrific scene. The great body of the flooring and the suspenders, forming something like a basket swung between the towers, was swayed to and fro like the motion of a pendulum. Each vibration giving it increased momentum, the cables, which sustained the whole structure, were unable to resist a force operating on them in so many different directions, and were literally twisted and wrenched from their fastenings. . . .*
>
> *We believe the enterprise and public spirit of our citizens will repair the loss as speedily as any community could possibly do. It is a source of gratulation that no lives were lost by the disaster.*

In 1864, 14 years after its construction, the 1043-foot span over the Niagara River connecting Lewiston, New York, and Queenston, Canada, failed in a heavy wind after its stabilizing guy cables had been temporarily removed to permit the passage of an ice jam during the spring thaw; in 1886, a year after an inspection, a 216-foot span over the Ostrawitza River at Ostraw, Czechoslovakia, collapsed from the vibrations created when a small troop of Uhlan cavalry consisting of 26 soldiers, 16 horses, and 2 carriages, a very light load, charged over it.

By far the most spectacular and carefully documented suspension-bridge failure was that of the 2800-foot span Tacoma Narrows Bridge, which connected the Olympic Peninsula with the rest of the State of Washington, a mere four months after its completion in 1940, in a 40-mile-an-hour wind.

> *Dynamic oscillations of suspension bridges reached their climax in the disaster at Tacoma Narrows. . . . From the beginning this structure showed signs of instability. During erection moderate vertical oscillations were observed from time to time. . . . Sometimes these movements took place in very light winds (3 or 4 miles per hour), and at other times much higher winds failed to produce motion.*
>
> *During the early morning of November 7, 1940, there was a fairly strong wind which caused some movement of the bridge. . . . The main span was vibrating in vertical motion with moderate amplitude at a frequency of about 36 cycles per*

*minute. This motion continued regular until about 10:00 a.m., the wind increasing in
the meantime to a velocity of 42 miles per hour by the anemometer.*

*At this time the entire motion of the structure changed violently. The cause of this
change is not certain, but there is some evidence to support the belief that one of the
bands holding the center cable ties slipped. At any rate, the previous vertical motion
immediately changed from 36 cycles per minute to 12. The resulting motion was tor-
sional and was very violent, the angular distortions reaching an amplitude of approximately 45 degrees each way from the static position. . . .*

*The bridge tenaciously withstood this violent rolling and twisting motion for about
an hour with relatively minor damage, but finally it yielded and a section of slab and
laterals near the center of the main span dropped [into the water 208 feet below].*

*Nature had destroyed this great structure by subjecting it to forces which it was
not designed to withstand. It was the most notable and most spectacular failure in
suspension bridge history.*

This quotation is from *The Mathematical Theory of Vibration in Suspension
Bridges* [5], which proceeds to show that the failure of the Tacoma Narrows
Bridge, as well as the others mentioned above, was due to the reinforcing and
automatically synchronizing effect of wind-induced vibrations of the structure
which resulted in progressive amplification of these motions until the bridge was
destroyed.

Suspension-bridge design has always been a highly mathematical subject. Consider the suspending cable C in Figure 9.28. Its shape is determined by giving y as
a function of x, say $y = y(x)$. If $w(x)$ is the weight per horizontal unit length at

FIGURE 9.28

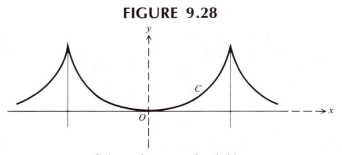

Schematic suspension bridge.

point x, then the shape of the curve C is determined by the differential equation

$$(9.20) \qquad \frac{d^2y}{dx^2} = -\frac{w(x)}{H}$$

where H is a constant (equal to the horizontal tension in the cable). If the cable
has a fixed weight w per foot of length, which, of course, it will if it is made from a
uniform material of constant diameter, then

$$(9.21) \qquad w(x) = w\left[1 + \left(\frac{dy}{dx}\right)^2\right]^{1/2}$$

(*cp.* Figure 9.29, which shows how the length of the cable from P to Q is related to the coordinates x and y by Pythagoras' Theorem, and recognize the expression for arc length in (9.21)), and it can be shown that the solution of (9.20) is

$$y = \frac{H}{2w}\left(e^{(w/H)x} + e^{-(w/H)x} - 2\right)$$

where e is the base of the natural logarithms introduced in Chapter 7. This curve, which is naturally assumed by a freely suspended uniform cable under the influence of gravity, is called a *catenary* (*cp.* §8.5).

FIGURE 9.29

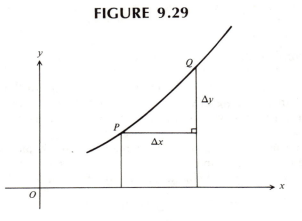

In fact, a suspension bridge consists not only of the suspending cables but also of the suspended roadway; the vehicles that move across it must also be taken into account. The weight per unit length of the roadway alone is essentially constant, say $= a$; were it be suspended by weightless cables, (9.20) would simplify to

$$\frac{d^2y}{dx^2} = -\frac{a}{H}$$

The right-hand side is constant, that is, a polynomial of degree 0; therefore, by the theorem on differentiation in §9.4

$$y = -\frac{a}{2H}x^2 + bx + c$$

for some constants b and c; the cable curve is a *parabola*.

The actual state of affairs lies somewhere between the weightless roadway (for which the cable curve is a catenary) and the weightless cable (for which it is a parabola) (*cp.* Figure 9.27). The differential equation must be solved by using the actual weights per unit length contemplated by the designer, taking into account the anticipated vehicle load, the strength of the materials to be used for construction, changes in size due to temperature variation, and the weight exerted on the structure by winds of various speeds.

None of these loads and forces was responsible for the failures described above.

The effect of synchronized vibration cannot be determined without studying the variation with time of the deflection of points on the bridge under the influence of changing wind forces. Unfortunately, these studies were not undertaken until prompted by the motion pictures that showed the tortured twisting movements of the Tacoma Narrows Bridge: the possibility of such motion had remained unsuspected by most bridge designers until then, but 10 years later an essentially complete mathematical solution to the complicated problem of suspension bridge vibration was presented in [5]. It is unlikely that there will ever again be a bridge failure due to this cause.

EXERCISES

9.1. Decide whether the following sequences of real numbers have some number as a limit. Where a limit does exist, write down the limiting value.

(a) $1, \frac{1}{2}, \frac{1}{4}, \frac{1}{8}, \cdots, \frac{1}{2n}, \cdots$

(b) $1, 2, 3, 4, \ldots$

(c) $1, \frac{1}{2}, 0, \frac{1}{3}, 0, \frac{1}{4}, 0, \frac{1}{5}, \cdots$

(d) $1, \frac{1}{2}, 1, \frac{1}{3}, 1, \frac{1}{4}, 1, \frac{1}{5}, \cdots$

(e) $2, \frac{3}{2}, \frac{4}{3}, \frac{5}{4}, \cdots, \frac{n+1}{n}, \cdots$

9.2. Give an intuitive definition of a tangent line to a smooth curve. Can you give a precise definition without using the concept of limit?

9.3. Using a Cartesian coordinate system, write the equations of a sequence of lines in the plane (a) which *do not* converge to some limiting line; (b) which *do* converge to some limiting line. Can you find a curve so that your sequence of lines in part (a) will be a sequence of secants to the curve?

9.4. In the graph of the parabola $y = x^2$ in Figure 9.5, show that the sequence of secants determined by the points on the parabola whose x-coordinates are the pairs of numbers

(a) $\left\{1 - \frac{1}{2}, 1 + \frac{1}{2}\right\}, \left\{1 - \frac{1}{4}, 1 + \frac{1}{3}\right\},$

$$\left\{1 - \frac{1}{8}, 1 + \frac{1}{5}\right\}, \ldots, \left\{1 - \frac{1}{2n}, 1 + \frac{1}{n}\right\}, \ldots$$

(b) $\left\{\frac{1}{2}, 1\right\}, \left\{\frac{3}{4}, 1\right\}, \left\{\frac{7}{8}, 1\right\}, \ldots, \left\{1 - \frac{1}{2n}, 1\right\}, \ldots$

both have the line passing through $(1, 1)$ with slope equal to 2 as a limiting line.

9.5. Consider the curve defined by $y = x^3$ and determine the limit of the secants to the curve determined by the points whose x-coordinates are the pairs of numbers

$$\left\{1 - \frac{1}{2}, 1 + \frac{1}{2}\right\}, \left\{1 - \frac{1}{3}, 1 + \frac{1}{3}\right\}, \ldots, \left\{1 - \frac{1}{n}, 1 + \frac{1}{n}\right\}, \ldots$$

(*cp.* Problem 9.4). Do you think there will be only one limiting line for all such sequences? Why?

9.6. Replace the number 1 in Problem 9.5 with an arbitrary x-coordinate x and derive a formula for the slope of the tangent line at any such point (x, x^3)

on the curve by (a) using a particular sequence to compute the limit, as in Problems 9.4 and 9.5, and assuming that all such limits are the same or (b) using any limiting sequence of secants for a point (x, x^3) on the curve.

9.7. Draw the graphs of the functions

$$y = f(x) \qquad y = \frac{df}{dx}(x) \qquad y = \frac{d^2f}{dx^2}(x) \qquad \text{(2nd derivative)}$$

for

(a) $f(x) = x$ (c) $f(x) = x^2$
(b) $f(x) = x + 2$ (d) $f(x) = x^3$

9.8. Use the basic theorem concerning the derivative of a polynomial to compute the following derivatives:

(a) $\dfrac{d}{dx}(1 + 2x + 3x^2)$

(b) $\dfrac{d}{dx}(\sqrt{2}x^4 + \left(\dfrac{\pi}{e}\right)x^2 + \pi^2)$

(c) $\dfrac{d}{dx}[(ax + b)(cx + d)]$, a, b, c, d, constants

(d) $\dfrac{d}{dx}(x - 1)^3$

9.9. Consider the function $f(x)$ defined by

$$f(x) = \begin{cases} x & \text{if } x \geq 0 \\ -x & \text{if } x < 0 \end{cases}$$

(a) Draw the graph of $f(x)$.
(b) Determine all the points at which $f(x)$ has a derivative.
(c) Compute the derivative of $f(x)$ at all points at which it exists and draw a graph of the derivative function.

9.10. Prove that the function $f(x)$ in Problem 9.9 has no derivative at $x = 0$. (*Hint:* See Figure 9.3.)

9.11. Consider the four curves in Figure E9.11. Find the area of the shaded region for each of these curves by using as many of the following three methods as are applicable. If one of the techniques is not (readily) applicable, explain why and what kind of information would be necessary to make it work.
Method A: Elementary plane geometry.
Method B: Approximation from the inside by inscribed rectangles (as in Figure 9.17).
Method C: By the Fundamental Theorem of Calculus and the basic theorem for calculating the derivatives of a polynomial.

9.12. Which method of measuring area in Problem 9.11 (A, B, or C) seems most powerful? Which is weakest? Explain.

FIGURE E9.11

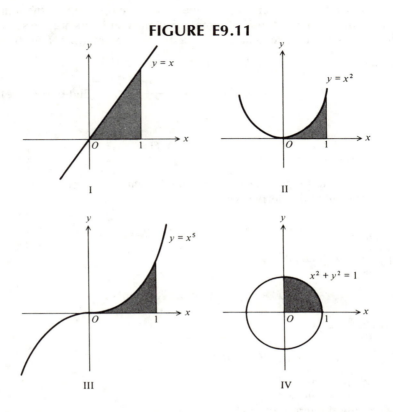

I

II

III

IV

9.13. Suppose that a sheet-metal worker has a square sheet of metal whose sides are 2 feet long and that he wants to remove squares from the corners so that the remainder can be made into a container by folding the edges up along the dotted lines as shown in Figure E9.13 (the squares to be removed are shaded). What is the length x of the side ab of the square that should be cut out in order that the container will have maximum volume? (*Hint:* Compare the example in the text and Figure 9.18 and use the fact

FIGURE E9.13

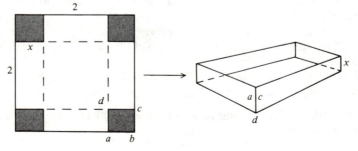

that the derivative of a function will vanish at a maximum or minimum value. The function to be *maximized* is the volume.)

9.14. Compute the following integrals by applying the Fundamental Theorem of Calculus:

(a) $\int_0^1 (2x - 3x^2) \, dx$

(c) $\int_0^2 \sqrt{2} \, x^2 \, dx$

(b) $\int_2^3 (1 - x^2) \, dx$

(d) $\int_0^1 dx$

9.15. Give an intuitive reason why the length of the curve Γ in Figure 9.23 should be $\geq l(S_1) + \cdots + l(S_n)$, as indicated there.

9.16. The arc length of a segment of a circle in Figure E9.16 is defined by

FIGURE E9.16

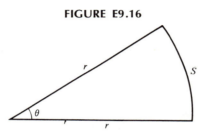

$S = r \cdot \theta$, where r is the radius and θ is the subtended angle (measured in radians). Does this definition agree with the more general one given in §9.10? Give reasons (but not a detailed proof) for your answer. [*Hint:* Consider Archimedes' work on the measurement of π (Chapter 4) and what *radian measure* of angles means.]

CHAPTER 10

The Calculus and Calculation

Chapter 10 is devoted to the study of Taylor's theorem. *This beautiful and remarkably useful result exhibits the value of a function at a point in terms of its value and the values of all its derivatives at some nearby point in the form of an infinite series. By means of this relation, values of functions can be calculated by man or machine with as much accuracy as desired; indeed, almost all modern tables of function values — logarithms, trigonometrical functions, and others that are much more complicated, are calculated from series expressions derived from Taylor's theorem or from a variant form of it. Taylor's Theorem also has an important geometrical interpretation. It provides the solution to the problem of constructing the "best" approximation by polynomial curves to the graph of a function near any of its points, which generalizes the evident geometrical fact that the tangent line at a point is the "best" approximating first degree polynomial curve.*

§**10.1.** From one standpoint the progress of mathematics appears as a repetitive process of oscillation from periods of computational ascendency to periods of formal and theoretical advances. For instance, the Babylonian introduction of positional notation led to great achievements in astronomical calculating capabilities, which in their turn led to advances in the conception of the geometrical nature of the universe as well as to progress in the study of trigonometrical functions. The later development of exponent notation and algorithms for algebra led to the invention of logarithms and the creation of algebrized geometry. These in their turn paved the way for the invention of calculus.

In this section we present an amalgamation of the techniques of algebrized geometry and calculus which yields remarkably powerful tools for the calculation of function values and the deeper understanding of the nature of functions. The earliest and still most potent representative of this class of results is known as *Taylor's theorem,* after the English mathematician Brook Taylor (1685–1731); it is still very "modern" and in various generalized guises is used again and again in current mathematical research as well as in applications of mathematics to the sciences and engineering.

The problem that Taylor's theorem solves is most intuitively posed in geometrical terms. Suppose that a portion of a curve, say Γ, is given, as in Figure 10.1. As a matter of convenience, suppose also that the curve passes through the origin O of a Cartesian coordinate system. If the curve is smooth enough, it will have a well-defined tangent line at O which meets the curve at O and lies close to it for points close to O. Expressed in other terms, the tangent line to Γ at O *approximates* Γ near the point of tangency. One consequence of this intuitively obvious approximation of curve by line is that the ordinate y of a point (x, y) lying on the curve can be approximated by the corresponding ordinate y_L of the point (x, y_L) lying on the tangent line L; for example, if the slope[1] of the tangent line is m, then the line corresponds to the equation

$$y_L = mx$$

---

[1] *Recall that the* slope *of a line is the tangent of the angle between the positive x-axis and the part of the line lying above the x-axis.*

FIGURE 10.1

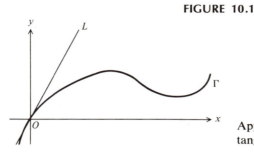

Approximation of a curve by its tangent line at one point.

and for x near 0 the value y for the point (x, y) lying on Γ must be approximately

$$y \sim mx$$

also.[2]

So far we have used the most elementary principles of algebrized geometry. Now we introduce the concept of derivative from calculus. Assume that the curve Γ corresponds to some equation, say $y = f(x)$. Then we know (from §9.3) that the slope m of the tangent line to the curve at the origin is given by

$$m = \frac{df}{dx}(0) = \frac{df}{dx} \quad \text{evaluated at } x = 0$$

This means that values of $y = f(x)$ for x near zero are approximately given by

$$f(x) \sim \left(\frac{df}{dx}(0) \right) x$$

In other words, the *straight line* through the origin which "best" approximates the curve corresponding to $y = f(x)$ is the line whose slope is the same as the slope of the curve at O.

The algebrization of geometry introduced a myriad of new curves into geometrical considerations. In particular, those curves that correspond to equations of the form $y = P_n(x)$, where P_n stands for a polynomial of degree n, are now on a common footing with simple straight lines (which correspond to P_1, first-degree polynomials) and parabolas (corresponding to P_2, second-degree polynomials); therefore it is entirely natural to ask for that curve of degree n that most closely approximates $y = f(x)$ at the origin. For instance, what is the equation of a parabola that "best" fits $y = f(x)$ at O? To answer this question in an impeccable way, "best" must be precisely defined; intuitively, it ought to mean that the approximating parabola has as much contact with the curve Γ as it possibly can, just as the tangent line has more contact with Γ than any other line meeting the curve at O. This condition, when expressed in mathematical terms, leads to the conclusion that the parabola must correspond to the equation

$$y = ax + bx^2$$

[2] "$\sim$" *means "approximately equal to."*

where

$$a = \frac{df}{dx}(0)$$

and

$$b = \frac{1}{2}\frac{d^2f}{dx^2}(0)$$

This means that the parabola and the curve Γ have the same tangent line at O; moreover, the curves corresponding to their derivatives also have the same tangent line at O.

Carrying this process further, we can attempt to approximate $y = f(x)$ by a polynomial of degree n as closely as possible at the origin; if this is done, the coefficients of the polynomial are related to the function f as shown in (10.1), in which we use the convenient shorthand notation $f^{(k)}(0)$ for the kth derivative of $f(x)$ evaluated at $x = 0$; the polynomial T_n is called the *Taylor polynomial* of f:

$$(10.1) \qquad T_n = f^{(1)}(0)x + \frac{f^{(2)}(0)}{2!}x^2 + \cdots + \frac{f^{(n)}(0)}{n!}x^n$$

The numbers[3]

$$f^{(1)}(0), \ \ldots, \ \frac{f^{(k)}(0)}{k!}, \ \ldots, \ \frac{f^{(n)}(0)}{n!}$$

are constants that depend only on the function f.

The curve corresponding to this nth-degree polynomial equation approximates $y = f(x)$ near $x = 0$ with increasing fidelity as n is taken larger and larger. The content of *Taylor's theorem* is that this *approximation* becomes an *equality* as n grows without bound (*i.e.*, as n approaches infinity); that is, if $f(0) = 0$, then

$$(10.2) \qquad f(x) = f^{(1)}(0)x + \cdots + \frac{f^{(n)}(0)}{n!}x^n + \cdots$$

This theorem is *not true* for all functions f. Certain conditions must be satisfied; some are obvious, others are more complicated to state and to check, but most of the functions that can be written down by ordinary means are perfectly acceptable candidates for the application of Taylor's theorem.

Figure 10.2 exhibits approximating *Taylor polynomials* of various degrees to the curve corresponding to $y = \sin x$; it is absolutely clear that the approximations of the sine curve by the polynomials become increasingly accurate as the degree of the polynomial is increased.

We have considered curves that pass through the origin for the sake of convenience. This means that the function f has the property that $f(0) = 0$. To remove this unnecessary and inconvenient restriction simply observe that if f is any function then $y = f(x) - f(0)$ corresponds to a curve passing through the origin; now the derivative of the constant $f(0)$ is zero, and the nth derivative of $f(x) - f(0)$ is

---

[3] $n! = 1 \cdot 2 \cdot 3 \cdots \cdots n$; $1! = 1$, $2! = 2$, $3! = 6$, $4! = 24$, $5! = 120$, *etc.*

FIGURE 10.2

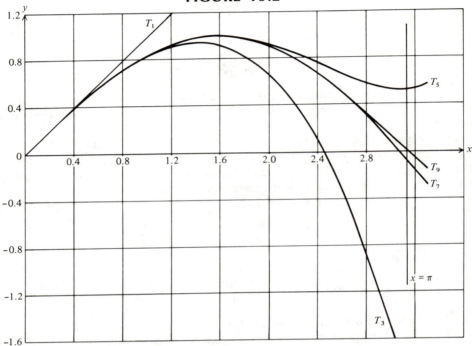

Taylor polynomial approximants: T_1, T_3, T_5, T_7, T_9 to $y = \sin x$.

the same as the nth derivative of $f(x)$ itself, which means that Taylor's theorem can be applied to $f(x) - f(0)$ for any reasonable function f to produce the following formula:

$$(10.3) \qquad f(x) - f(0) = f^{(1)}(0)x + \cdots + \frac{f^{(n)}(0)}{n!}x^n + \cdots$$

Finally, observe that we inquired about the nature of Γ near $x = 0$. We could have asked for its behavior near any point $x = a$; if we had done that and performed an analysis of the same kind, the result would have been the fully general *Taylor's theorem*:

$$(10.4) \quad f(x + a) = f(a) + f^{(1)}(a)x + \frac{f^{(2)}(a)}{2!} x^2 + \cdots + \frac{f^{(n)}(a)}{n!} x^n + \cdots$$

This version reduces to (10.3) if $a = 0$; it reduces to (10.2) if $a = 0$ and also $f(0) = 0$. The *infinite series* on the right side is called the *Taylor expansion of the function f about the point a.*

§10.2. Brook Taylor was not the first to discover the theorem named for him, but he was the first to publish the result, which attached his name to it forever.

The theorem was discovered and rediscovered in many slightly different forms near the turn of the seventeenth century by the Scotsmen James Gregory (1638–1675), dead at 36 years of age, and Colin Maclaurin (1698–1745), by Jean Bernoulli (1667–1748) of the celebrated Swiss family of mathematicians and physicists (see Figure 10.3), and by Taylor himself. The idea was important, had beautiful as well as immediately useful practical consequences, and was bound to occur to several researchers almost simultaneously. Its practical significance cannot be overestimated; even today it provides the most obvious and often the most successful method for approximating the solution of engineering equations which are too complicated to admit an exact analysis.

FIGURE 10.3

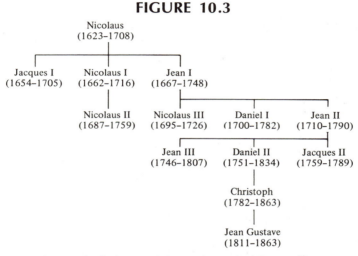

A genealogical tree of the mathematical Bernoullis.

§10.3. Some examples of the use of Taylor's theorem will help to make its effectiveness clear. Table 10.1 contains a list of formulas for derivatives that will be useful for these examples and also for the work in Chapter 11.

Consider, for example, the function

$$f(x) = \sqrt{1 + x} = (1 + x)^{1/2}$$

It is easy to calculate the derivatives of this function and, after seeing the first few, to understand their general law of formation. For the first, by formula 2 (or 3) in Table 10.1, with $s = \frac{1}{2}$,

$$\frac{df}{dx} = \frac{1}{2} (1 + x)^{-1/2}$$

this is the same kind of function; therefore its derivative, that is, the *second* derivative of f itself, can be calculated in a similar manner as (see Exercise 10.15)

$$\frac{d^2f}{dx^2} = \frac{1}{2} \left(-\frac{1}{2}\right) (1 + x)^{-3/2}$$

TABLE 10.1
Some Convenient Derivatives

1. $\dfrac{d}{dx}(a_0 + a_1 x + a_2 x^2 + \cdots + a_k x^k) = a_1 + 2a_2 x + 3a_3 x^2 + \cdots + ka_k x^{k-1}$

2. $\dfrac{d}{dx}(a + x)^s = s(a + x)^{s-1}$ for arbitrary s and a

3. $\dfrac{d^n}{dx^n}(a + x)^s = s(s - 1)(s - 2) \cdots (s - n + 1)(a + x)^{s-n}$ for arbitrary s and a

4. $\dfrac{d}{dx}\left(\dfrac{1}{1 - x}\right) = \dfrac{1}{(1 - x)^2}$

5. $\dfrac{d^n}{dx^n}\left(\dfrac{1}{1 - x}\right) = \dfrac{n!}{(1 - x)^{n+1}}$ where $n! = 1 \cdot 2 \cdot 3 \cdots n$

6. $\dfrac{d}{dx}\log_e(1 + x) = \dfrac{1}{1 + x}$

7. $\dfrac{d^n}{dx^n}\log_e(1 + x) = \dfrac{(-1)^{n-1}(n - 1)!}{(1 + x)^n}$

8. $\dfrac{d}{dx}\sin x = \cos x$

9. $\dfrac{d^n}{dx^n}\sin x = \begin{cases}(-1)^k \cos x & \text{if } n = 2k + 1 \\ (-1)^k \sin x & \text{if } n = 2k\end{cases}$

10. $\dfrac{d}{dx}\cos x = -\sin x$

11. $\dfrac{d^n}{dx^n}\cos x = \begin{cases}-(-1)^k \sin x & \text{if } n = 2k + 1 \\ (-1)^k \cos x & \text{if } n = 2k\end{cases}$

12. $\dfrac{de^x}{dx} = e^x$ where e is the base of the system of natural logarithms

13. $\dfrac{d^n e^x}{dx^n} = e^x$

Again the result has the form to which the differentiation theorem applies, and

$$\frac{d^3 f}{dx^3} = \frac{1}{2}\left(-\frac{1}{2}\right)\left(-\frac{3}{2}\right)(1 + x)^{-5/2}$$

It should now be evident how the process will continue and that the nth derivative will have the expression

$$(10.5) \qquad \frac{d^n f}{dx^n} = \frac{1}{2}\left(-\frac{1}{2}\right)\left(-\frac{3}{2}\right)\left(-\frac{5}{2}\right) \cdots \left(-\frac{(2n - 3)}{2}\right)(1 + x)^{-(2n-1)/2}$$

Now let us choose a convenient value for the parameter a in Taylor's theorem. Since the values of the various derivatives of f must be computed for the value a chosen, we should select a to make these calculations as simple as possible. Examination of (10.5) shows that it will be easy to evaluate the derivatives at $a = 0$, and we make this choice. Therefore from (10.5) with $x = 0$

(10.6) $$\frac{d^n f}{dx^n}(0) = (-1)^{n+1} \frac{3 \cdot 5 \cdot 7 \cdot \cdots \cdot (2n - 3)}{2n} \text{ for } n > 1$$

Substitution of this result in the expression for $f(x)$, given by Taylor's theorem, leads to

(10.7) $$\sqrt{1 + x} = 1 + \tfrac{1}{2}x - \tfrac{1}{8}x^2 + \tfrac{1}{16}x^3 + \cdots$$

For instance, if $x = 0.03$, then $\sqrt{1.03}$ is given by

$$\sqrt{1.03} = 1 + \frac{0.03}{2} - \frac{0.0009}{8} + \frac{0.000027}{16} + \cdots$$
$$= 1 + 0.015 - 0.0001125 + \cdots$$
$$= 1.01489$$

correct to four decimal places. This is obviously a convenient and powerful tool for determining square roots to as many places of accuracy as desired. It makes use of no more than the calculational methods of elementary algebra.

In addition to providing a convenient and accurate way of computing square roots, Taylor's theorem also sheds some light on the Babylonian method of approximating square roots examined in §3.3. Recall that when searching for $\sqrt{x}$ the Babylonians would start with some approximation x_1 and then generate a sequence of increasingly better approximations by computing

$$x_2 = \frac{x_1 + (x/x_1)}{2}$$

$$x_3 = \frac{x_2 + (x/x_2)}{2}$$

$$\cdot$$
$$\cdot$$
$$\cdot$$

$$x_{n+1} = \frac{x_n + (x/x_n)}{2}$$

Let us apply their technique to estimate $\sqrt{1 + x}$, starting with the "approximate" value 1. Then the second, better, approximation is $\tfrac{1}{2}[1 + (1 + x)/1] = \tfrac{1}{2}(2 + x) = 1 + x/2$; but this is just the Taylor polynomial T_2 — the first two terms of the Taylor series shown in (10.7). The next Babylonian approximation is already more difficult to compute:

(10.8) $$\frac{1}{2}\left[\left(1 + \frac{x}{2}\right) + \frac{1 + x}{1 + x/2}\right]$$

If x is smaller than 1, then, as you recall from your work in high school and can verify by long division,

$$\frac{1}{1 + x/2} = 1 - \frac{x}{2} + \left(\frac{x}{2}\right)^2 - \left(\frac{x}{2}\right)^3 + \cdots$$

Substitute this expression[4] into (10.8) to find the expression for the third Babylonian approximation *carried only through terms that are quadratic in x*

$$\frac{1}{2}\left[1 + \frac{x}{2} + (1 + x)\left(1 - \frac{x}{2} + \frac{x^2}{4} - \cdots\right)\right]$$

$$= \frac{1}{2}\left[1 + \frac{x}{2} + \left(1 + x - \frac{x}{2} - \frac{x^2}{2} + \frac{x^2}{4} + \cdots\right)\right]$$

$$= 1 + \frac{x}{2} - \frac{x^2}{8} + \cdots$$

which is the Taylor polynomial of degree 3. Continuation of this process will indeed lead to the Taylor polynomials of higher and higher degrees. For those readers who are concerned about the delicate questions of mathematics note that we introduced a *second* kind of approximation into our calculations by replacing the quotient $(1 + x/2)^{-1}$ with a nonterminating polynomial and then chopping off most of that polynomial, retaining only the first three terms. This procedure will work only if x is between -1 and $+1$, but the Babylonian approximations work for all x and are even better than Taylor's theorem for the special case of calculating square roots.

For any particular function there may well be a special algorithm that is more effective for its computation than the Taylor series; the power and significance of Taylor's theorem is that it provides a *uniform method* for finding approximations for a marvelously broad and important class of functions; it permits the user to dispense with creative thought, which is, as we have insisted all along, one of the primary utilitarian features of mathematics. To illustrate how thought can indeed be dispensed with by utilizing Taylor's theorem consider the problem of calculating powers — not necessarily integral powers — of numbers close to 1, that is, of calculating the expression

$$(1 + x)^s$$

for small x and arbitrary s. The square root previously considered is the special case $s = \frac{1}{2}$. This more general problem is important to banks and insurance companies, amongst others, for it is the formula for computing compound interest: if the interest rate is x percent per unit of time, then one dollar invested for s time units will grow to $(1 + x)^s$ dollars. It would be nice to know how to compute this number for various choices of s and x; x, unfortunately, always too small and s too large (from the viewpoint of bank depositors, at least).

Using formula (3) in Table 10.1 and Taylor's theorem, we *automatically and thoughtlessly* (but carefully) find

$$(10.9) \quad (1 + x)^s = 1 + sx + \frac{s(s-1)}{1 \cdot 2}x^2 + \frac{s(s-1)(s-2)}{1 \cdot 2 \cdot 3}x^3$$
$$+ \cdots + \frac{s(s-1)\ \cdots\ (s-n+1)}{1 \cdot 2 \cdot 3 \cdots \cdot n}x^n + \cdots$$

---

[4] *The expression on the right is just a geometric series.*

For instance,

$$(1.01)^{1/10} = 1 + (0.1)(0.01) + \frac{(0.1)(0.1-1)(0.01)^2}{2} + \cdots$$

$$= 1 + 0.001 - 0.0000045 + \cdots$$

$$= 1.0009955 \ldots$$

nearly.

§**10.4.** In Chapter 7 the elementary properties and applications of logarithms were discussed and the problem of calculating tables of logarithms touched on. It should be clear that the crude technique used there to illustrate how we might conceivably construct a table of logarithms is indeed much too crude to permit a truly useful table to be produced within a lifetime. Furthermore, that method made it difficult to assess the amount of error present in any of the calculated values. Taylor's theorem can easily be applied to $f(x) = \log_e (1 + x)$ to yield a simple Taylor series that reduces the preparation of a table of logarithms to a rote procedure best performed by a machine, which is indeed the way such tables are now prepared.

From formulae 6 and 7 in Table 10.1 and taking $a = 0$ in Taylor's theorem, find the following series (which, after changing x to $-x$, is the same as that on p. 194):

$$(10.10) \qquad \log_e (1 + x) = x - \frac{x^2}{2} + \frac{x^3}{3} - \frac{x^4}{4} + \cdots$$

This series works as an approximation to the logarithm only if $-1 < x < 1$, but, according to the laws of logarithms, this suffices to find the logarithm of *any* number; for instance, to find $\log_e 2$, we can proceed in this way: since $2 = 1 + 1$, the series cannot be used, but we know that $\log (2) = -\log (\frac{1}{2})$. Now $\frac{1}{2} = 1 - \frac{1}{2}$, which is a usable form. We have found

$$\log_e 2 = -\log_e (\tfrac{1}{2}) = -\log_e (1 - \tfrac{1}{2})$$

$$= -\left[\left(-\frac{1}{2}\right) - \frac{(-1/2)^2}{2} + \frac{(-1/2)^3}{3} - \cdots\right]$$

$$= \tfrac{1}{2} + \tfrac{1}{8} + \tfrac{1}{24} + \cdots$$

$$= 0.5000 \ldots + 0.12500 \ldots + 0.0416666 \ldots$$

$$+ 0.0156250 \ldots + 0.0062500 \ldots$$

$$+ 0.0026041 \ldots + \cdots$$

$$= 0.6911 \ldots$$

by summing the first six terms. The correct value of $\log_e 2$ (obtained by summing sufficiently many terms of the series) is 0.6913. . . .

The base e of the natural logarithm system is easier to evaluate. The series for the function $f(x) = e^x$ is, according to formulae 12 and 13 in Table 10.1 and Taylor's theorem,

$$(10.11) \qquad e^x = 1 + x + \frac{x^2}{2!} + \frac{x^3}{3!} + \cdots + \frac{x^n}{n!} + \cdots$$

Setting $x = 1$ yields

(10.12)
$$e = 1 + 1 + \tfrac{1}{2} + \tfrac{1}{6} + \tfrac{1}{24} + \tfrac{1}{120} + \tfrac{1}{720} + \cdots$$
$$= 1 + 1 + 0.5 + 0.166666 \ldots + 0.0416666 \ldots$$
$$+ 0.0083333 \ldots + 0.0013888 \ldots + \cdots$$
$$= 2.718 \ldots$$

The value of e correct to 12 decimals is actually

$$e = 2.718281828459 \ldots$$

As two final examples the Taylor series for the sine and cosine functions are certainly worthy of attention. By the completely standard process that we have gone through several times already, this time using formulae 8 to 11 in Table 10.1, find (*cp.* Figure 10.2):

(10.13)
$$\sin x = x - \frac{x^3}{3!} + \frac{x^5}{5!} - \frac{x^7}{7!} + \cdots$$

and

(10.14)
$$\cos x = 1 - \frac{x^2}{2!} + \frac{x^4}{4!} - \frac{x^6}{6!} + \cdots$$

From this point of view Pythagoras' theorem about the sides of a right triangle, which, of course, is the same thing as the basic equation

(10.15)
$$(\sin x)^2 + (\cos x)^2 = 1$$

becomes an *equation in infinite series*. Squaring the series for $\sin x$ and $\cos x$ leads to series that begin as follows:

$$(\sin x)^2 = x^2 - \frac{2x^4}{3!} + \cdots$$

and

$$(\cos x)^2 = 1 - \frac{2x^2}{2!} + \frac{x^4}{(2!)^2} + \frac{2x^4}{4!} + \cdots$$

Add these equations to find

$$(\sin x)^2 + (\cos x)^2 = 1$$

correct at least through the fourth power in x. With care and conviction it can be shown that all the powers of x cancel on the right side; consequently the Taylor series provides a nontrigonometrical method of proving the basic identity (10.15) of trigonometry.

At this point the reader should return to Aristarchus' ingenious device for bounding the value of sin 3°, discussed in §4.1, and try to obtain a better pair of inequalities from (10.13); remember that $3° = \pi/60 = 0.052 \ldots$ in radian measure, which *must* be used in applying (10.13). By using only the first *two* terms of the Taylor series and the above approximation for 3°, which amounts to knowing

π correct to only three figures, that is, 3.14, we do indeed improve on Aristarchus, but with rather more powerful tools.

§10.5. In the preceding subsections little has been said about how to determine the *accuracy* of an approximation made using Taylor's theorem. For instance, in the first example, calculation of $\sqrt{1.03}$ led to the value 1.0149 by taking the first three terms of the series. Looking at the rapid way in which the values of the successive series terms appear to decrease makes it seem plausible that the contributions from later terms in the series will not add enough to affect the fourth decimal place. This kind of "plausibility argument" can be misleading, and anyway it would inspire confidence to have a simple prescription available that would give an absolutely reliable measure of how great an error would be made by taking only the first one, or two, or three, or n terms of a given series. Part of Taylor's theorem, which we have suppressed because it is too technical to be worth including, provides such a prescription and makes the theorem a tool excellently adapted to practical application as well as to research.

There is one case for which it is easy to estimate the maximum error made by using only the first n terms of the Taylor series or indeed of *any* series. If the *sign* of successive terms of the series alternates, as, for instance, in

(10.16) $$1 - \tfrac{1}{3} + \tfrac{1}{5} - \tfrac{1}{7} + \cdots$$

(the denominators run through the odd integers), or in the series (10.7) for $\sqrt{1+x}$, then the error made by stopping after n terms of the series have been summed is always less (in absolute value) than the value of the next term (in absolute value)! For (10.16) this means that the error made by summing only the first three terms, $1 - \tfrac{1}{3} + \tfrac{1}{5} = 0.966666 \ldots$, must lie between $-\tfrac{1}{7}$ and $+\tfrac{1}{7}$, that is, between $-0.1428 \ldots$ and $+0.1428. \ldots$ This will mean more to you when you recall from §4.7 that the series (10.16) is *Gregory's series* and its sum is $\pi/4$. The error rule assures us that $\pi/4$ can be approximated as closely as desired by using Gregory's series and that to get an approximation accurate, say, to within one hundredth (0.01), it will do to sum terms until one that lies between -0.01 and $+0.01$ is reached. By examining the series it is plain that this will happen when the term $+1/101$ is reached. Therefore summation of the first 50 terms suffices to achieve to required accuracy. You now have a more sophisticated viewpoint from which to appreciate Archimedes' estimation of π, for it was not much more work for him to obtain a result of just this accuracy than it would be to sum the first 50 terms of the series. But the latter is a mechanical method and can be continued in an automatic way; if a result accurate to within one-ten-thousandth (0.0001) is desired (or required), simply sum the first 5000 terms of the series – or let a machine do it in less than a second.

Figure 10.4 shows how Gregory's series slowly approaches $\pi/4$ as its terms are summed. The sum of the first n terms is graphed (as the ordinate) against n and the resulting points have been connected by straight-line segments to make it easier to see what is happening. The resulting broken-line curve is alternately larger and then smaller than $\pi/4$ because of the alternation in sign of the terms in the series

FIGURE 10.4

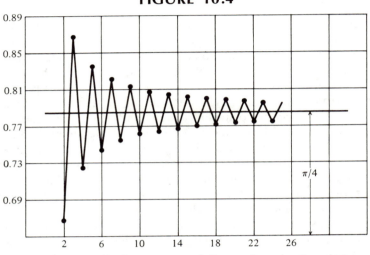

The error remaining after n terms of Gregory's series for $\pi/4$ have been summed.

(10.16). This is typical of alternating term series but not of the sums of other series.

For instance, the decimal expansion of a number represents a numerical series, as, say,

$$\frac{1}{3} = 0.33333 \ldots = \frac{3}{10} + \frac{3}{10^2} + \frac{3}{10^3} + \cdots$$

and, since all these terms are positive, each approximate sum is larger than all of those preceding that include fewer terms. Incidentally, this particular series is a special case of the geometric series, in which the first term is $\frac{3}{10}$ and the ratio, $\frac{1}{10}$. From the discussion in Chapter 1 you know how the approximate sums for the geometric series approach its value as increasing numbers of terms are summed; *cp.* pp. 33–35.

In the same way it can now be determined that the first three terms of the series (10.6) for $\sqrt{1.03}$ do indeed provide a result accurate to four decimal places because the fourth term of that *alternating series* is 0.000027/16 and the error made by summing only the first three terms must be less than that.

If a series is not alternating, that is, if its terms do not alternate in sign, it is harder to decide how much error is made by stopping after n terms have been summed. The usual prescriptions provided as part of Taylor's theorem are stated in terms of the size of the derivatives of the function expressed by the Taylor series or, in another useful form, by the size of an integral related to that function. It is usually possible to estimate these quantities without much trouble, and thereby to determine the error bounds for the Taylor series.

EXERCISES

10.1. Consider the quadratic function

$$f(x) = 5 + 3x^2$$

(a) Compute the *Taylor expansion* of the function $f(x)$ up to and including the second derivatives; that is, compute the Taylor polynomial T_2 of order 2 for this function f at the point $x = 0$.

(b) What is the difference between $f(x)$ and T_2; that is, what is the error in this approximation?

10.2. As in Problem 10.1, compute the Taylor polynomial of order 2 for the function

$$f(x) = 5 + 3x^2 + 2x^3$$

at $x = 0$. What is the error in the approximation of $f(x)$ by T_2?

10.3. Consider the function

$$f(x) = (1 + x)^5$$

(a) Write $f(x)$ in the form

$$f(x) = a_0 + a_1x + \cdots + a_5x^5$$

(giving the explicit values of the a_k) either by using the Binomial Theorem or by carrying out the indicated multiplication.

(b) Derive the same result as in (a) by using Taylor's theorem and formula (2) in Table 10.1.

10.4. Consider the function

$$f(x) = \sqrt{4 + x}$$

Find the Taylor expansion of $f(x)$ at $x = 0$ by finding, in particular, a formula for $(d^n f/dx^n)(0)$. [*cp.* (10.6)].

10.5. Using the fact (which we assume here) that the Taylor expansion for $f(x)$ in Problem 10.4 makes sense (converges) as long as $-4 < x < +4$, compute the value of $\sqrt{2}$ accurate to *four* decimal places.

10.6. By using the Taylor expansion of the function

$$f(x) = (1 + x)^s$$

compute the amount of interest earned in six months in a savings deposit of $100 with an annual rate of interest of 4 percent accurate to the nearest cent.

10.7. Let $i = \sqrt{-1}$, that is, $i^2 = -1$, be the imaginary unit. Using the series for sin x, cos x, and e^x show that

$$e^{ix} = \sin x + i \cos x$$

by writing out the series for e^{ix} and separating it into two series. (*Note:* One operates with imaginary quantities according to the rules $(ix)^2 = i^2x^2 = -x^2$, $(ix)^3 = i^3x^3 = i^2 \cdot ix^3 = -ix^3$, and so on, exactly as with real numbers).

10.8. On p. 286 the expression "squaring the series for sin x" was used. Define what this should mean and give several examples. (*cp.* the definition of multiplication of real numbers given in Chapter 1.)

10.9. Show that by using two terms for the Taylor expansion of sin x we can improve Aristarchus' estimate of sin 3° (recall 3° = $\pi/60$ and assume that π is approximated by 3.14).

10.10. (a) How many terms of Gregory's series must be used to obtain a value of π accurate to
  (i) one tenth,
  (ii) one hundredth,
  (iii) one millionth.
(b) Carry out this computation for part (i).
(c) How does this compare with the accuracy of Archimedes' computation of π?

10.11. Give a formula for the error in summing n terms of a geometric series and illustrate it with two examples. (*Hint:* See Chapter 1 for a discussion of geometric series.)

*10.12. Consider a series of terms

$$a_0 + a_1 + a_2 + \cdots + a_n + \cdots$$

in which a_n does not approach zero as n gets large. Do you think that such a series can converge to a number? Why? Give an example to illustrate your answer. (*Hint:* First look at examples of series that *do* converge and then construct examples of series that do not.)

10.13. Consider an alternating series

$$a_0 - a_1 + a_2 - a_4 + \cdots \pm a_n \cdots$$

in which the terms a_n go to zero as n gets large; for example,

$$1 - \tfrac{1}{2} + \tfrac{1}{3} - \tfrac{1}{4} + \cdots$$

Do you think that such a series will always converge? Why?

10.14. Give an example of an alternating series that does not converge to any real number.

10.15. Using the definition (9.4) of the derivative, show that

$$\frac{dcf}{dx} = c\,\frac{df}{dx}$$

if c is a constant and f is a function.

CHAPTER 11

Differential Geometry

Differential geometry *combines the concerns of geometry with the techniques of the calculus to provide a method for studying geometry on curved surfaces. This chapter introduces these methods by studying the geometry of the sphere, some of which was introduced in Chapters 5 and 6, and comparing it with Euclidean geometry. The intuitive concept of* curvature *is made precise, and the description of* geodesic curves *on the surface in terms of derivatives of certain functions is made plausible. This chapter is meant to provide the mechanism for studying relativity theory in Chapter 12 and the framework for comparing the physical theories of Newton and Einstein in an intuitive but still accurate way.*

§11.1.　Not long after the invention of the calculus by Newton and Leibniz that powerful tool was directed toward some of the oldest problems in mathematics, concerned with the geometry of curves, surfaces, and space in general. This application of the calculus to geometry became known as *differential geometry,* which, as an independent mathematical discipline, developed from infancy in 1800 to maturity at the beginning of the twentieth century. In 1916 it was used by Albert Einstein in his formulation of the General Theory of Relativity. The purpose of this chapter is to introduce the fundamental concepts of differential geometry with a discussion of their origin. They will be used in the description of General Relativity in Chapter 12.

Our consideration of differential geometry is consistent with one of the basic themes of this book: it is a major development of the mathematical ideas necessary to an understanding of the geometrical nature of space beyond man's immediate perception.

Many of the basic concepts of differential geometry were developed as a consequence of an attempt by Carl Friedrich Gauss (1777–1855) to describe the detailed geometric shape of the surface of the earth in a concrete manner. This problem requires mathematical tools that are a step beyond the *flat Euclidean geometry* of the ancients (corresponding to their "flat" world; *cp.* Figures 6.15 to 6.19) and the symmetrical *spherical geometry* of the followers of Ptolemy and the Renaissance navigators (*cp.* Chapters 5 and 6), for in order to make further progress it turns out to be necessary to introduce the limit notions underlying the calculus.

§11.2.　Differential geometry studies properties of curves and surfaces in space: more specifically properties of the *length* of curves and the *area* bounded by curves on surfaces as well as *curvature* and *distortions* under various types of constraint. Three fundamental mathematical concepts arise in this context and are basic to the study of differential geometry. Underlying all three, however, is the fundamental notion of the *space* in which the geometry of curves or surfaces is to be considered. This space might, for example, be the plane or the surface of a sphere or three-dimensional space or the surface of an ellipsoid or the surface of the earth flattened at the North and South poles. It is not easy to give a definition of the spaces that are admissible for the study of geometry. One very useful definition of geometrical space was contributed by Bernhard Riemann (1826–1866); it underlies most modern models of the universe and is touched on later. The earlier geometers, preceding 1850, had certain fixed ideas about the spaces that were admissible; the examples given above were included, but not many more. "Spaces" of more than three dimensions were, for instance, not comprehended within their framework. As shown later, however, this generalization is indispensable to relativity theory, in which the *geometrical space* underlying the theory is *four*-dimensional.

Suppose that we are agreed about what constitutes a "space." The three fundamental concepts of differential geometry relative to this space are a *metric, geodesic,* and *curvature.*

1. A *metric* is a rule, or formula, that determines (defines) the *distance* between two points in the space.

2. A *geodesic* is a curve joining two given points in the space which has a length equal to the distance between the points.[1]

3. *Curvature* is the amount of intrinsic bending or curvedness of the space at a given point in it.

To elaborate these informal definitions we illustrate them with some examples. Suppose that the space is the ordinary Euclidean plane E^2 or ordinary three-

FIGURE 11.1

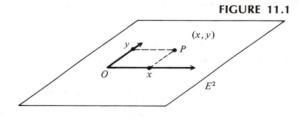

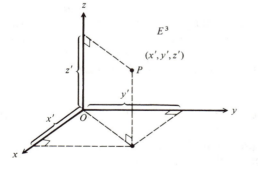

Cartesian coordinate systems in the plane and in three-dimensional space.

[1] *We use this definition of geodesic to simplify a difficult concept. It differs from the usual definition, but the two agree if the given points are sufficiently close, and in any event a geodesic in our sense is always a geodesic in the ordinary usage of differential geometry.*

dimensional Euclidean space E^3. By choosing a fixed point O and perpendicular coordinate axes, as in Figure 11.1, a point P can be represented by its coordinates (x, y) in E^2 and (x, y, z) in E^3. Recall that this is a description of the points of these geometric spaces by analytic geometry, as introduced in Chapter 8. The *distance* between two points P and P' is given by Pythagoras' theorem (Figure 11.2). Here it is intuitively clear that the shortest distance between two points is the length of the straight-line segment joining them. Thus the geodesics for E^2 and E^3 are segments of straight lines. At this stage the calculus is not needed and the concept of the *curvature* of E^2 and E^3 is not geometrically apparent, for these spaces turn out to be *flat;* their curvature is zero.

FIGURE 11.2

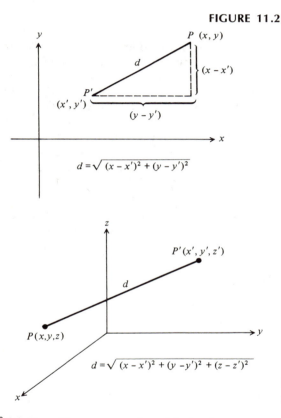

$$d = \sqrt{(x - x')^2 + (y - y')^2}$$

$$d = \sqrt{(x - x')^2 + (y - y')^2 + (z - z')^2}$$

Pythagoras' theorem in the plane and in three-dimensional space.

§11.3. There are various kinds of curvature, and the concept of the curvature of a given space is not a simple one to understand. First consider a simpler notion: the *curvature of a curve* in space and, as a particular case, of a curve in the plane. Consider a curve in E^2 which is *not* a straight line; then it is indeed *curved,* as is Γ in Figure 11.3.

The curvature of a curve is a measure of how "bent" or "curved" it is at each of its points; that is, it is a measure of the extent to which it deviates from being a straight line.

FIGURE 11.3

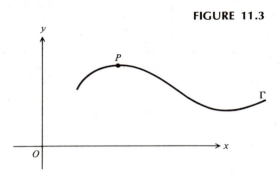

A curve which is curved at P.

Recall that one of the problems of the calculus was to find the *tangent line* to a given curve. An analogous problem is the following: find the radius of a circle tangent to Γ at P which approximates the curve Γ better at P than any other circle. For example, in Figure 11.4 there are three circles, S_0, S_1, S_2 (of radii R_0, R_1, and R_2, respectively), tangent to Γ at P. Clearly S_1 is a better approximation to the curve Γ at P than either of the circles S_0 and S_2. This is analogous to the problem of finding an approximation of a given function by a polynomial, as discussed in Chapter 10. It is, however, not obvious that there is a *best circle* but for most points of curves which occur naturally it can be proved that it does exist.

Consider a family of circles with a common tangent as in Figures 11.4 or 11.5. Clearly, the smaller the radius of one of the circles in Figure 11.5, the more bent or curved it is at the point P, in *relation* to the others in the family. This motivates us to define the *curvature* K_P of a circle as $K_P = 1/R$, where R is the radius of the circle under consideration. As R becomes large, K_P becomes small, and con-

FIGURE 11.4

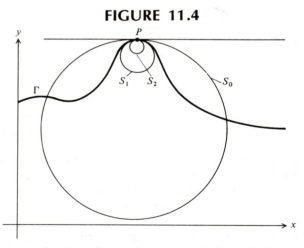

Approximation of curve by tangent circles at P. S_1 is a better approximant than either S_0 or S_2.

FIGURE 11.5

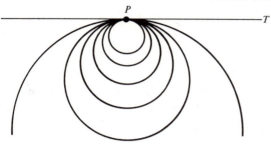

Curvature of circles of various radii.

versely, as R becomes small the curvature K_P becomes large. As the radius R increases without bound, the large circles approach the tangent line T, which will therefore have curvature $K_P = 0$ and is said to be *flat* at point P. More generally, the *curvature of Γ at P* in Figure 11.4 will be a number that tells how curved Γ is at that point. If R_0 and R_2 are the radii of S_0 and S_2, respectively (see Figure 11.4), then intuition suggests that

$$\frac{1}{R_0} < K_P < \frac{1}{R_2}$$

so we define the curvature of Γ at P to be $K_P = 1/R_1$ where S_1 is the best approximating circle at that point.

The problem of finding the curvature of Γ at P can be phrased as a minimization problem of sorts: find the circle S tangent to Γ at O for which the distance $d(P, P')$ between corresponding points on S and Γ (Figure 11.6) goes to zero *faster* (as P and P' both tend to O) than the corresponding distance for any other such circle. This sounds complicated and indeed it is. What is being minimized is a certain rate of change (of a distance as P and P' move toward O). In Chapter 9 it was stated that a rate of change of distance involves a *first derivative* (the *velocity*). A minimization problem also requires the computation of a derivative. Thus it seems that the curvature of a curve Γ at a point P is a number that could be computed by knowing the *first* and *second derivatives* (derivative of a derivative) of the functions defining the curve Γ. This is true, but the minimization problem is too complicated to be solved here although its result will be stated momentarily. A

FIGURE 11.6

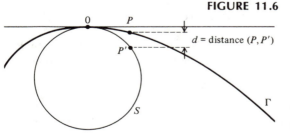

The best approximating circle tangent to a curve at 0.

consequence of the solution of this minimization problem is that the curvature defined heuristically by "best approximation" by circles makes sense.

The formula for the curvature of a curve defined as the graph of a function $y = f(x)$ in the (x, y) − plane is

$$K_P = \pm \frac{f''(x)}{[1 + [f'(x)]^2]^{3/2}}$$

(see, for example, Thomas [65], p. 406), in which the sign is chosen to make K_P a positive number, the reciprocal of the radius of the best approximating circle.

FIGURE 11.7

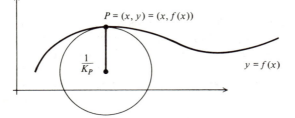

Curvature of curve at P.

For example, if we consider the parabola $y = x^2$, then

$$y = f(x) = x^2$$

and

$$f'(x) = 2x$$
$$f''(x) = 2$$

Hence according to the above formula

$$K_P = K_x = \frac{2}{[1 + (2x)^2]^{3/2}}$$

In particular $K_0 = 2$, whence the best approximating circle (called the *osculating circle*) at $x = 0$ has radius $\frac{1}{2}$. As x increases, the curvature at P_x decreases (Figure 11.8) which means that the radius of the approximating circle increases so the curve becomes flatter. This certainly agrees with our intuitive picture of a parabola.

The curvature of a curve in E_2 is *not yet* an example of the *curvature of the space* in which geometry is being considered, but it has bearing on this problem as we show later. In this example the curvature of the space E^2 is still *zero*.

§11.4. Consider a sphere S (that is, the surface of a ball) of radius R in E^3. In analogy with our discussion of circles in the plane it makes sense to say that one sphere S is *more curved* than a second sphere S if the former has a *smaller* radius and conversely. This definition immediately gives rise to examples of "spaces" − surfaces of spheres − with a nonzero "curvature" (Figure 11.9). A sphere S of radius R is an example of a two-dimensional space as are the surface

FIGURE 11.8

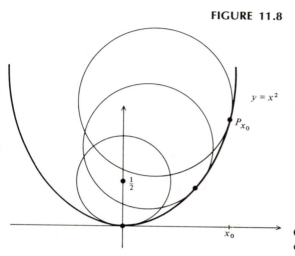

$y = x^2$

P_{x_0}

$\frac{1}{2}$

x_0

Osculating circles for selected points on $y = x^2$.

of the earth and the Euclidean plane. Concerning the sphere, we may ask what is the *metric*, what are *geodesics*, and what is the *curvature* of this space? In this section we discuss the first two questions. The more difficult problem of defining and computing the *curvature* of a general surface in E^3 is discussed in §11.5.

If P and P' are two points on a sphere, we can calculate the usual distance between them *on the sphere* (not in the three-dimensional space). For example, when we speak of the distance between New York and London, we mean distance on the surface of the earth and "as the crow flies," that is, the shortest possible distance, which is the same as the length of the shortest curve that *lies on the sphere* and joins P to P'. In §9.10 we saw how the length of a curve in the plane could be measured; formula (9.19) gives it as the value of a certain integral. Let us generalize this formula to curves in the three-dimensional space E^3. Suppose a curve in (x, y, z)-space is thought of as the trace of the motion of a point as a function of time t. Then the coordinates (x, y, z) of a point P on the curve become functions of time $(x(t), y(t), z(t))$, as illustrated in Figure 11.10. A formula for the length of such a path or curve can be derived as (9.19) was, and we find that the

FIGURE 11.9

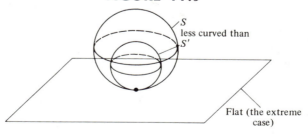

S
less curved than
S'

Flat (the extreme case)

Spheres of various radii.

FIGURE 11.10

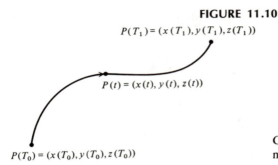

$P(T_1) = (x(T_1), y(T_1), z(T_1))$

$P(t) = (x(t), y(t), z(t))$

$P(T_0) = (x(T_0), y(T_0), z(T_0))$

Curve considered as the path of a moving point.

length of the curve measured from time $t = T_0$ to time $t = T_1$ is given by the formula

$$(11.1) \qquad l = \int_{T_0}^{T_1} \left[\left(\frac{dx}{dt} \right)^2 + \left(\frac{dy}{dt} \right)^2 + \left(\frac{dz}{dt} \right)^2 \right]^{1/2} dt$$

which is the limit of the lengths of a sequence of polygonal line segments which approximate the curve between $P(T_0)$ and $P(T_1)$; one such polygonal approx-imant is shown in Figure 11.11. This length depends on the usual metric of E^3 which finds its expression in the theorem of Pythagoras used to derive (9.19) and (11.1).

Suppose we consider the symbols dx, dy, dz, and dt as denoting formal "small numbers" and rewrite (11.1) in a suggestive (and purely formal) fashion by setting

$$ds = [(dx)^2 + (dy)^2 + (dz)^2]^{1/2}$$

$$= \left\{ \left[\left(\frac{dx}{dt} \right)^2 + \left(\frac{dy}{dt} \right)^2 + \left(\frac{dz}{dt} \right)^2 \right] \cdot (dt)^2 \right\}^{1/2}$$

$$= \left[\left(\frac{dx}{dt} \right)^2 + \left(\frac{dy}{dt} \right)^2 + \left(\frac{dz}{dt} \right)^2 \right]^{1/2} dt$$

FIGURE 11.11

$P(T_1)$

$P(t)$

$P(T_0)$

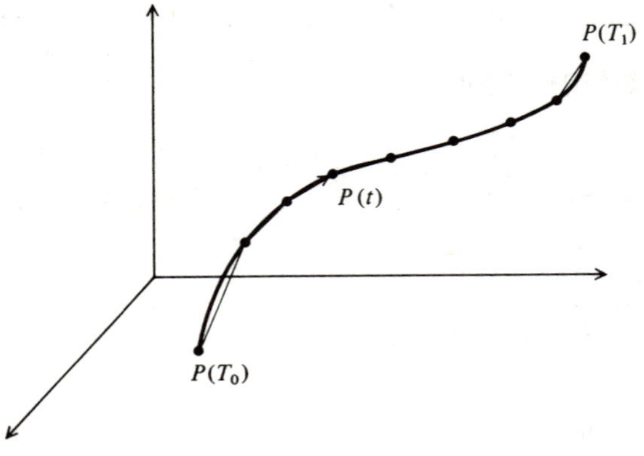

Space curve as the path of a moving point.

where the last formula is the square root of the sum of the squares of the derivatives of the functions $x(t)$, $y(t)$, $z(t)$ at time t and dt is a formal symbol representing an "infinitesimal time" in the integration process of formula (11.1). Comparison with (11.1) shows that

$$(11.2) \qquad l = \int_{T_0}^{T_1} ds = \int_{T_0}^{T_1} [(dx)^2 + (dy)^2 + (dz)^2]^{1/2}$$

The rigorous interpretation of formula (11.2) is given by (11.1). The symbol

$$(11.3) \qquad (ds)^2 = (dx)^2 + (dy)^2 + (dz)^2$$

or

$$ds = [(dx)^2 + (dy)^2 + (dz)^2]^{1/2}$$

is picturesquely called the *infinitesimal element of arc length* and intuitively corresponds to the length of a segment of the polygonal approximant in Figure 11.11. Integrating ds along a curve $(x(t), y(t), z(t))$ as in (11.2) yields the length of the curve. Therefore, given the curve as a function of time and the expression (11.3), the length can be obtained. Here, of course, x, y, and z are coordinates in E^3.

Now consider a curve on the sphere S expressed in the *coordinates of S* (as given, for example, by latitude and longitude for the two-dimensional sphere). How do we find a corresponding expression for the infinitesimal element of arc length? This is important for finding the lengths of curves on the earth, since these curves (for, example, the flight path of an airplane) are expressed as a function of time in the latitude-longitude coordinates (α, β) given in Figure 5.2 (see Figure 11.12). By expressing α and β as functions of x, y, and z we obtain the following formula for the length of a curve expressed in terms of α and β, where R denotes the radius of the sphere:

$$(11.4) \qquad l = \int_{T_0}^{T_1} \left[R^2 \left(\frac{d\alpha(t)}{dt} \right)^2 + R^2 \cos^2 \alpha(t) \left(\frac{d\beta(t)}{dt} \right)^2 \right]^{1/2} dt$$

or, in its infinitesimal formulation,

$$ds^2 = R^2 (d\alpha)^2 + R^2 \cos^2\alpha(d\beta)^2$$

where ds represents the same element of length as before but expressed in terms

FIGURE 11.12

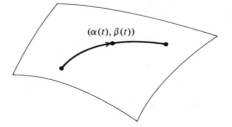

$(\alpha(t), \beta(t))$

of different coordinates, that is,

(11.5) $(ds)^2 = (dx)^2 + (dy)^2 + (dz)^2 = R^2(d\alpha)^2 + R^2\cos^2\alpha(d\beta)^2$

for curves that are restricted to this particular sphere of radius R.

Equation 11.5 can easily be derived from elementary calculus. We need the following formulae, which are found in any calculus text. If $f(t)$, $g(t)$ are two functions of t, then

$$\frac{d}{dt}(f \cdot g) = \frac{df}{dt} \cdot g + f \cdot \frac{dg}{dt}$$

$$\frac{d}{dt}(\cos f(t)) = -\sin f(t) \cdot \frac{df(t)}{dt}$$

$$\frac{d}{dt}(\sin f(t)) = \cos f(t) \cdot \frac{df(t)}{dt}$$

Using the diagram in Figure 11.13 we have from elementary trigonometry the relation

FIGURE 11.13

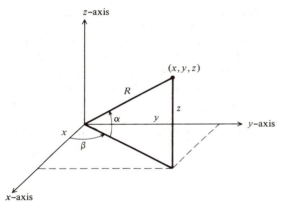

Relation between Cartesian and spherical coordinates.

between the coordinates:

$$x = R \cos \alpha \sin \beta$$
$$y = R \cos \alpha \cos \beta$$
$$z = R \sin \alpha$$

If $\alpha(t)$ and $\beta(t)$ describe a curve on the sphere of radius R, then the curve represented in the coordinates $(x(t), y(t), z(t))$ is given by

$$x(t) = R \cos \alpha(t) \sin \beta(t)$$
$$y(t) = R \cos \alpha(t) \cos \beta(t)$$
$$z(t) = R \sin \alpha(t)$$

as in Figure 11.14. Therefore, computing dx/dt, dy/dt, and dz/dt by using the differentiation formula given (and suppressing the dependence of the function on the variable t), we find

FIGURE 11.14

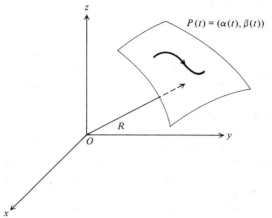

Portion of the surface of a sphere of radius R centered at O.

$$\frac{dx}{dt} = R \cos \alpha \cos \beta \frac{d\beta}{dt} - R \sin \alpha \frac{d\alpha}{dt} \sin \beta$$

$$\frac{dy}{dt} = -R \cos \alpha \sin \beta \frac{d\beta}{dt} - R \sin \alpha \frac{d\alpha}{dt} \cos \beta$$

$$\frac{dz}{dt} = R \cos \alpha \frac{d\alpha}{dt}$$

Collecting terms and using the fact that $\sin^2 + \cos^2 = 1$, we have, after some convenient cancellation of terms,

$$\left(\frac{dx}{dt}\right)^2 + \left(\frac{dy}{dt}\right)^2 + \left(\frac{dz}{dt}\right)^2 = R^2 \left[\left(\frac{d\alpha}{dt}\right)^2 + \cos^2 \alpha \left(\frac{d\beta}{dt}\right)^2\right]$$

Hence (11.4) holds and corresponds to the *infinitesimal form* (11.5).

Thus to measure the length of a curve Γ on the sphere S we need only compute the integral

$$\int_\Gamma \sqrt{R^2 \left[(d\alpha)^2 + \cos^2\alpha \, (d\beta)^2\right]}$$

which is a symbolic representation for (11.4). The symbolic expression

$$ds^2 = R^2 \left[d\alpha^2 + \cos^2\alpha \, d\beta^2\right]$$

is called the *metric form* (*infinitesimal metric*) for the sphere S of radius R. The coefficients of the symbols $d\alpha^2$ and $d\beta^2$ are functions that vary over the surface of the sphere. We have seen that by substituting for the symbols $d\alpha^2$ and $d\beta^2$ the derivatives $(d\alpha/dt)^2$ and $(d\beta/dt)^2$ for a given curve $(\alpha(t), \beta(t))$ on S we can integrate this function of t and obtain the length of the curve; thus the length of the curve on S is determined by the *form* of the expression

$$ds^2 = R^2\,d\alpha^2 + R^2\cos^2\alpha\,d\beta^2$$

and the relation between α and β which defines the curve.

Now we can define the *metric* on S. Consider two points P_1 and P_2 on S and let Γ be any curve joining P_1 to P_2, as in Figure 11.15, with the additional property that the derivatives of the functions $\alpha(t)$, $\beta(t)$ defining the curve exist for all t so that (11.4) will make sense. The length of Γ can be computed as above, that is,

$$l(\Gamma) = \int_\Gamma ds$$

where ds is given by (11.5). Define the *distance between P_1 and P_2 on S* to be the *minimum value* of the lengths of all such curves connecting P_1 and P_2. Intuitively such a minimum should exist and indeed it can be shown that it does by appealing to some basic properties of real numbers. Therefore this method does define a *metric* on the sphere S, and it is the one we are interested in. Knowing this metric is the same thing as knowing the *metric form ds^2*.

FIGURE 11.15

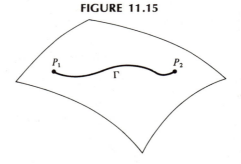

If we consider a sphere flattened at its poles (a better approximation to the shape of the earth's surface), we would have a similar situation (Figure 11.16). The same angles α and β could be used as coordinates, but the expression for the

FIGURE 11.16

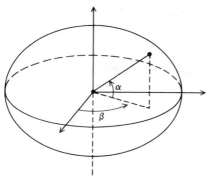

Latitude and longitude coordinates on an ellipsoid.

metric form would be different; it would be given by

$$(11.6) \qquad ds^2 = E(\alpha, \beta)\, d\alpha^2 + 2F(\alpha, \beta)\, d\alpha \cdot d\beta + G(\alpha, \beta)\, d\beta^2$$

where E, F, and G are functions of the latitude and longitude coordinates. For the sphere of radius R we had

$$E(\beta) = R^2 \cos^2 \beta$$
$$F(\alpha, \beta) = 0$$
$$G(\alpha, \beta) = R^2$$

which made things simple. For a given surface S, such as an ellipsoid with coordinates α and β, there would be a corresponding metric form that would look like (11.6); the functions E, F, and G would depend on how the surface S is contained in E^3, as, for instance, in Figure 11.17. The lengths of curves Γ *on S* could be de-

FIGURE 11.17

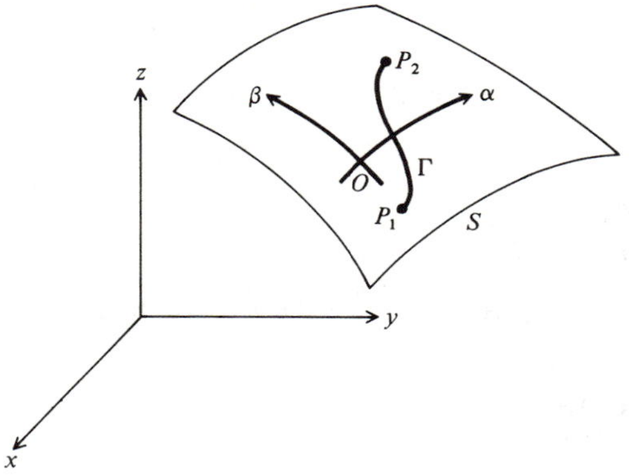

Coordinate system on a surface in three-dimensional Euclidean space.

termined by using this fundamental metric form if the coefficient functions E, F, and G were known for the surface S. The *distance* between two points P_1 and P_2 on S could be determined by finding the minimum length of all curves that lie on S and join P_1 to P_2. Thus the *metric form* determines the *metric*. It would be interesting to know what the metric form for the surface of the earth is, since it would determine the geodesics, those curves on the surface (of the earth) that have minimal length. For a sphere they are known to be the great circles, which played an important role in Chapters 5 and 6 and are generally adequate approximations to the actual geodesics of the earth. More generally, they would be more complicated curves, depending on the three functions E, F, and G mentioned above. How they depend on E, F, and G is a basic problem of differential geometry that can be solved by applying the differential calculus to this problem of minimization. It is a much more difficult problem than the simple minimization and maximization problems discussed in Chapter 9 but it is the same kind. Problems of this sort were studied by Leonhard Euler (1707–1783), Joseph Louis Lagrange (1736–1813), and various members of the Bernoulli family (see the dynastic chart in Figure 10.3). The result is that a *geodesic curve* $(\alpha(t), \beta(t))$ on a surface S with metric form

$$ds^2 = E \, d\alpha^2 + 2F \, d\alpha \, d\beta + G \, d\beta^2$$

must satisfy two *differential equations* (in the "time" variable t) of the form

(11.7)
$$\frac{d^2\alpha}{dt^2} + f_1\left(\frac{d\alpha}{dt}\right)^2 + g_1\left(\frac{d\alpha}{dt}\right)\left(\frac{d\beta}{dt}\right) + h_1\left(\frac{d\beta}{dt}\right)^2 = 0$$
$$\frac{d^2\beta}{dt^2} + f_2\left(\frac{d\alpha}{dt}\right)^2 + g_2\left(\frac{d\alpha}{dt}\right)\left(\frac{d\beta}{dt}\right) + h_2\left(\frac{d\beta}{dt}\right)^2 = 0$$

where f_1, f_2, g_1, g_2, h_1, h_2 are functions of the coordinates α, β. These functions $f_1(\alpha, \beta)$, $g_1(\alpha, \beta)$, and so on, are, in fact, certain combinations of *derivatives* of the functions E, F, and G. If $E(\alpha, \beta)$, $F(\alpha, \beta)$, $G(\alpha, \beta)$ are constants, then the coefficients $f_1(\alpha, \beta)$, $g_1(\alpha, \beta)$, etc., are all *zero*, and the equations (11.7) reduce to the simpler forms

$$\frac{d^2\alpha}{dt^2} + 0 + 0 + 0 = 0$$

$$\frac{d^2\beta}{dt^2} + 0 + 0 + 0 = 0$$

that is, to

(11.8)
$$\frac{d^2\alpha}{dt^2} = \frac{d^2\beta}{dt^2} = 0$$

If (11.8) is valid, then the coordinates (α, β) on the surface must have the form

(11.9)
$$\alpha(t) = at + b$$
$$\beta(t) = ct + d$$

where a, b, c, and d are constants; if the coordinates (α, β) are ordinary Cartesian coordinates, (11.9) describes a *straight line;* it can describe a more complicated curve if S is not a plane. For instance (Figure 11.18), if S is a cylinder, (α, β) are coordinates on this cylinder, and the metric form satisfies $f_1 = g_1 = \cdot \cdot \cdot = 0$, then the geodesics [which are described by (11.9)] are *circles* cut out by the intersection of the cylinder with planes perpendicular to its axis, *straight lines* cut out by planes parallel to its axis, and *loxodromes* (which are not cut out by planes at all); *cp.* the discussion of the Mercator projection in Chapter 6.

Recall that when S is the surface of a sphere we have

$$E = R^2 \cos^2 \beta \neq \text{constant}$$
$$F = 0 \qquad\qquad = \text{constant}$$
$$G = R^2 \qquad\qquad = \text{constant}$$

For these values of E, F, and G the metric defined by (11.6) is relatively simple, since the derivatives of F and G will be zero (but not the derivatives of E!). The solutions of (11.7) are the familiar *great circles* of the sphere of radius R.

In the discussion of relativity in Chapter 12 we show that a *geodesic* in a *certain* space with a *particular* metric can be interpreted as the path of motion of a "par-

FIGURE 11.18

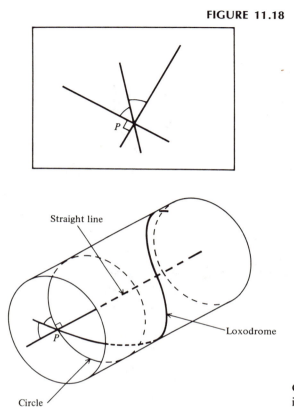

Straight line

Loxodrome

Circle

Geodesics in the plane and their image on a cylinder.

ticle" such as a spaceship, planet, or star in the mutually interacting gravitational fields of the material constituents of the universe. The "space" appropriate to cosmological geometry will have more than the *two* dimensions of the surfaces we have been using as examples, but the *ideas* and *geometric intuition* are exactly the same, and the fundamental notions are still those of metric, geodesic, and curvature, which provide the building blocks for Einstein's purely geometrical model of the universe.

§11.5. To define the *curvature of the surface S at a point P* in analogy with the curvature of a curve at a point we proceed by investigating the properties of some basic examples to decide what characteristics curvature should have. There are two essentially different local surface shapes that are not "flat" at the point in question:

(a)   those that are "round" like the surface of a sphere near one of its points;
(b)   those that are shaped like a *saddle* near one of its points.

Both shapes are illustrated in Figure 11.19*a* and *b*. The plane represents the plane

FIGURE 11.19

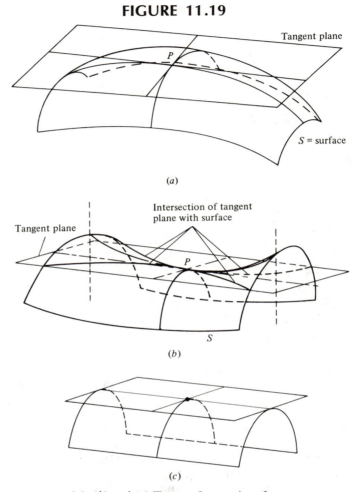

(a), (b) and (c) Types of curved surfaces.

in E^3 which is tangent to the surface at P. Note that in Figure 11.19a the tangent plane intersects the surface only at one point near P, whereas in Figure 11.19b the intersection consists of a pair of curved lines. An intermediate case is illustrated by Figure 11.19c, which shows a cylindrically shaped surface. A desirable definition of curvature should distinguish amongst these three different types of curved surface. Since there are three kinds of nonflat surface shapes (those illustrated in Figure 11.19) but only one kind of nonflat curve in the plane, it should be clear that an attempt to approximate these surfaces with spheres would not suffice to characterize the different surface types, in contrast with the approximation of curves near one of their points by circles. Therefore we shall proceed in a completely different manner. Consider any of the surfaces in Figure 11.19 and consider any plane H in E^3 that contains the straight line N which is perpendicular to the tangent plane T at point P, as shown in Figure 11.20 (we call H a "perpendicular plane"). Let Γ be the curve that is the intersection of S with H. The curve Γ is

FIGURE 11.20

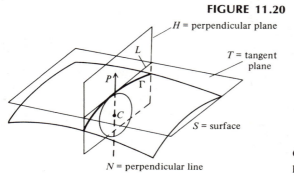

H = perpendicular plane

L

T = tangent plane

P

Γ

C

S = surface

N = perpendicular line

Curvature of the intersection of a perpendicular plane and a surface.

a curve in the plane H with a curvature $K_P(\Gamma)$ at point P, as discussed in §11.3 (that is, $K_P(\Gamma)$ is the reciprocal of the radius of the best approximating circle in H at P). Let L be the line which is the intersection of the plane H with the plane T. Let us suppose that we have *fixed* a direction on the line N as indicated by the arrow in the diagram. We define the *curvature of the surface S in the "direction" L* to be

(a) $K_P(\Gamma)$ if the center C of the approximating circle defining $K_P(\Gamma)$ lies on the *opposite* side of the line L from the direction arrow N (see Figure 11.20).

(b) $-K_P(\Gamma)$ if the opposite situation occurs.

Let us denote this number by $K_P(S, L)$, since it depends on the *surface S*, the *line L*, and the *point P*. In our discussion S, P, and the perpendicular direction N will be fixed, but we can let the line L vary (the plane H rotates around the "axis" N). Thus for each L we get a number $K_P(S, L)$ which represents the "curvature in the direction L." It should be intuitively clear that $K_P(S, L)$ has a *largest* and a *smallest* value as the direction L varies, while P and S remain fixed. The reason for introducing the $\pm$ signs is to distinguish between surfaces of the kind shown in Figure 11.19a, b. In Figure 11.21a we see that $K_P(S, L_1)$ and $K_P(S, L_2)$ are both *positive*, whereas $K_P(S, L_1)$ is positive in Figure 11.21b but $K_P(S, L_2)$ is negative. Note that in the latter case the curve Γ_2 is curving up in the same direction as the perpendicular N. This is just the *opposite* of the situation illustrated in Figure 11.20, which corresponds to the case of positive sign. In Figure 11.21c we see that Γ_2 has zero curvature in the plane H_2 and $K_P(S, L_2) = 0$.

The totality of the numbers $K_P(S, L)$ for all lines L in the tangent plane T represents as much information about the curvature as we might need, but we want a *single* number to represent curvature for surfaces, just as for curves. A simple-minded approach would be to take the *average* of all these numbers, if this were possible, to arrive at a general measure of the curvature, but this is not a fruitful

FIGURE 11.21

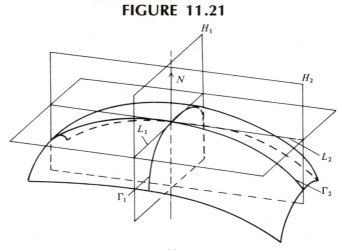

(a)

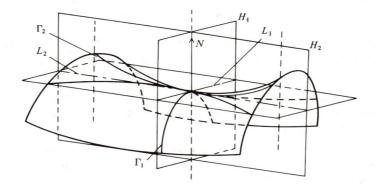

(b)

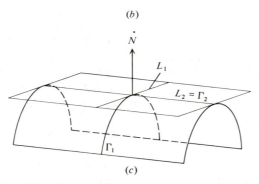

(c)

(a), (b) and (c) Curves of maximum and minimum curvature
on the types of curved surfaces.

approach. Leonard Euler proved that there are two directions L_M and L_m for which $K_P(S, L)$ achieves its *maximum* and *minimum* values. Call these numbers K_M and K_m. The *curvature of the surface S at the point P* is better defined by

$$K_P(S) = K_M \cdot K_m$$

From Figure 11.21 we see that

$$\begin{array}{ll} K_P(S_a) > 0 & \text{(Figure 11.21}a) \\ K_P(S_b) > 0 & \text{(Figure 11.21}b) \\ K_P(S_c) = 0, & \text{(Figure 11.21}c) \end{array}$$

Therefore the number $K_P(S)$ clearly distinguishes these surfaces. Also, if S is a sphere of radius R, it is easy to see from Figure 11.22 that for any point P on S we have

$$K_P(S) = \frac{1}{R^2}$$

Indeed, the curvature in *any* direction L is given by $K_P(S, L) = 1/R$, whence the maximum $K_M =$ minimum $K_m = 1/R$ and the result follows from the definition $K_P(S) = K_M \cdot K_m$.

FIGURE 11.22

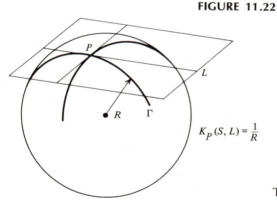

$$K_P(S, L) = \frac{1}{R}$$

The curvature of a sphere is $1/R^2$.

Now we come to a famous and remarkable result due to Gauss. Suppose we take a curve Γ in the plane and bend it as if it were a piece of string, without stretching it; the curvature obviously changes as we bend the curve. Suppose we do the same thing to a surface S, that is, deform it into another surface S' *without stretching* the surface. (Think of bending a piece of paper without folding it.) Such deformations are shown in Figure 11.23.

Then, as Gauss proved, the surface curvature $K_P(S)$ is the *same* as the curvature $K_{P'}(S')$, where P' is the point of S' that corresponds to the point P after the deformation.

Note that in the deformation of a surface the local coordinates (α, β) on the surface still identify points and the lengths of curves on the surface are unchanged

FIGURE 11.23

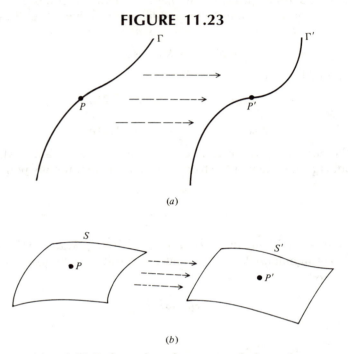

(a)

(b)

(a) and *(b)* Deformation of a curve and of a surface.

(which is what is meant by *not stretching* the surface during the deformation). So the functions $E(\alpha, \beta)$, $F(\alpha, \beta)$, $G(\alpha, \beta)$, which appear in the metric form, are unchanged by deformation, since they are functions only of the coordinates (α, β). This is shown in Figure 11.24. Gauss was able to show that the curvature $K_P(S)$ can be computed in terms of the first and second derivatives of the functions $E(\alpha, \beta)$, $F(\alpha, \beta)$, $G(\alpha, \beta)$ which appear in the metric form of S; since the functions do not change under the deformation, neither does the curvature, which proves the theorem.

FIGURE 11.24

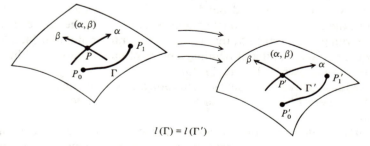

$$l(\Gamma) = l(\Gamma')$$

Deformation of a surface.

This result appeared in a famous monograph, *Disquisitiones Generales Circa Superficies Curvas* ("General investigations concerning curved surfaces"), published in 1824 ([36]). Two pages of this influential work are reproduced as Figure 11.25. Note the equation at the bottom of the first page. Gauss is using p and q as we use α and β, and indeed our choice of notation E, F, and G stems from this original paper and is now standard in all texts on differential geometry. Gauss's statement of his theorem is made on the second page of this figure.

§11.6. Gauss's monograph gave a fundamental impetus to the embryonic subject of differential geometry. In 1854 his student Bernhard Riemann presented a paper at Göttingen, with Gauss in the audience (and one year before the latter's death), which was entitled "On the hypotheses which lie at the foundation of geometry." This paper laid the groundwork for the next 50 years of research, the results of which Einstein applied to the construction of his relativistic theory of gravitation.

We can ask whether other mathematicians might have written Gauss's paper at the time he did or whether it was indeed a singular effort. Using one genius to evaluate another, we quote the following statement made by Einstein ([29], p. 350):

> The best that Gauss has given us was likewise an exclusive production. If he had not created his geometry of surfaces, which served Riemann as a basis, it is scarcely conceivable that anyone else would have discovered it.

Einstein also said of Gauss ([29], p. 349):

> The importance of C. F. Gauss for the development of modern physical theory and especially for the mathematical fundament of the theory of relativity is overwhelming indeed. . . . In my opinion it is impossible to achieve a coherent objective picture of the world on the basis of concepts which are taken more or less from inner psychological experience.

The last sentence refers in part to the manner in which Gauss was motivated to study the curvature of surfaces. Almost simultaneous with the appearance of his abstract memoir on curved surfaces (written in Latin) he published two long technical papers (in German) which dealt with the measurement of the metric form for the surface of the earth. They included many geodetic measurements he had made in the region of the small university town of Göttingen in the kingdom of Hanover, where he carried out his research. Gauss was responsible for the geodetic survey of the entire kingdom, a project that lasted some five years and immensely influenced the geometric questions he asked (and answered!) at an abstract mathematical level.

Gauss's contributions had an enormous eventual impact on the world. His predecessors studied the properties of curved lines and surfaces and applied the fast developing techniques of the calculus to solve numerous extremal problems, illustrative of which is the title of a famous book written by Gaspard Monge (1746–1818): *Applications de l'analyse à la géométrie* ("the applications of calculus to geometry"). Gauss, however, invented a new concept when he showed that surfaces in E^3 could be deformed in E^3 without stretching and still preserve

FIGURE 11.25

DISQUISITIONES GENERALES

Combinatio huius aequationis cum aequatione (10) producit

$$DD''-D'D' = (\alpha\alpha''+\mathfrak{b}\mathfrak{b}''+\gamma\gamma''-\alpha'\alpha'-\mathfrak{b}'\mathfrak{b}'-\gamma'\gamma')\Delta$$
$$+ E(n'n'-nn'')+F(nm''-2m'n'+mn'')+G(m'm'-mm'')$$

Iam patet esse

$$\frac{dE}{dp}=2m,\quad \frac{dE}{dq}=2m',\quad \frac{dF}{dp}=m'+n,\quad \frac{dF}{dq}=m''+n',\quad \frac{dG}{dp}=2n',\quad \frac{dG}{dq}=2n''$$

sive

$$m=\tfrac{1}{2}\frac{dE}{dp},\qquad m'=\tfrac{1}{2}\frac{dE}{dq},\qquad m''=\frac{dF}{dq}-\tfrac{1}{2}\frac{dG}{dp}$$
$$n=\frac{dF}{dp}-\tfrac{1}{2}\frac{dE}{dq},\qquad n'=\tfrac{1}{2}\frac{dG}{dp},\qquad n''=\tfrac{1}{2}\frac{dG}{dq}$$

Porro facile confirmatur, haberi

$$\alpha\alpha''+\mathfrak{b}\mathfrak{b}''+\gamma\gamma''-\alpha'\alpha'-\mathfrak{b}'\mathfrak{b}'-\gamma'\gamma' = \frac{dn}{dq}-\frac{dn'}{dp}=\frac{dm''}{dp}-\frac{dm'}{dq}$$
$$=-\tfrac{1}{2}\cdot\frac{ddE}{dq^2}+\frac{ddF}{dp.dq}-\tfrac{1}{2}\cdot\frac{ddG}{dp^2}$$

Quodsi iam has expressiones diversas in formula pro mensura curvaturae in fine art. praec. eruta substituimus, pervenimus ad formulam sequentem, e solis quantitatibus E, F, G atque earum quotientibus differentialibus primi et secundi ordinis concinnatam:

$$4(EG-FF)^2k = E\left(\frac{dE}{dq}\cdot\frac{dG}{dq}-2\frac{dF}{dp}\cdot\frac{dG}{dq}+\left(\frac{dG}{dp}\right)^2\right)$$
$$+F\left(\frac{dE}{dp}\cdot\frac{dG}{dq}-\frac{dE}{dq}\cdot\frac{dG}{dp}-2\frac{dE}{dq}\cdot\frac{dF}{dq}+4\frac{dF}{dp}\cdot\frac{dF}{dq}-2\frac{dF}{dp}\cdot\frac{dG}{dp}\right)$$
$$+G\left(\frac{dE}{dp}\cdot\frac{dG}{dp}-2\cdot\frac{dE}{dp}\cdot\frac{dF}{dq}+\left(\frac{dE}{dq}\right)^2\right)$$
$$-2(EG-FF)\left(\frac{ddE}{dq^2}-2\frac{ddF}{dp.dq}+\frac{ddG}{dp^2}\right)$$

12.

Quum indefinite habeatur

$$dx^2+dy^2+dz^2 = Edp^2+2Fdp.dq+Gdq^2$$

patet, $\sqrt{(Edp^2+2Fdp.dq+Gdq^2)}$ esse expressionem generalem elementi linearis in superficie curva. Docet itaque analysis in art. praec. explicata, ad inveniendam mensuram curvaturae haud opus esse formulis finitis, quae coordina-

Two pages from Gauss's celebrated memoir of 1824, which inaugurated differential geometry.

FIGURE 11.25 (*Continued*)

tas x, y, z tamquam functiones indeterminatarum p, q exhibeant, sed sufficere expressionem generalem pro magnitudine cuiusvis elementi linearis.   Progrediamur ad aliquot applicationes huius gravissimi theorematis.

Supponamus, superficiem nostram curvam explicari posse in aliam superficiem, curvam seu planam, ita ut cuivis puncto prioris superficiei per coordinatas x. y, z determinato respondeat punctum determinatum superficiei posterioris, cuius coordinatae sint x', y', z'.   Manifesto itaque x', y', z' quoque considerari possunt tamquam functiones indeterminatarum p, q, unde pro elemento $\sqrt{(dx'^2 + dy'^2 + dz'^2)}$ prodibit expressio talis

$$\sqrt{(E'dp^2 + 2F'dp \cdot dq + G'dq^2)}$$

denotantibus etiam E', F', G' functiones ipsarum p, q.   At per ipsam notionem *explicationis* superficiei in superficiem patet, elementa in utraque superficie correspondentia necessario aequalia esse, adeoque identice fieri

$$E = E', \quad F = F', \quad G = G'$$

Formula itaque art. praec. sponte perducit ad egregium

THEOREMA. *Si superficies curva in quamcunque aliam superficiem explicatur, mensura curvaturae in singulis punctis invariata manet*

Manifesto quoque *quaevis pars finita superficiei curvae post explicationem in aliam superficiem eandem curvaturam integram retinebit.*

Casum specialem, ad quem geometrae hactenus investigationes suas restrinxerunt, sistunt superficies in planum explicabiles.   Theoria nostra sponte docet, talium superficierum mensuram curvaturae in quovis puncto fieri $= 0$, quocirca, si earum indoles secundum modum tertium exprimitur, ubique erit

$$\frac{d\,dz}{dx^2} \cdot \frac{d\,dz}{dy^2} - \left(\frac{d\,dz}{dx \cdot dy}\right)^2 = 0$$

quod criterium, dudum quidem notum, plerumque nostro saltem iudicio haud eo rigore qui desiderari posset demonstratur.

13.

Quae in art. praec. exposuimus, cohaerent cum modo peculiari superficies considerandi, summopere digno, qui a geometris diligenter excolatur.   Scilicet quatenus superficies consideratur non tamquam limes solidi, sed tamquam soli-

certain of their original properties, such as length of curves and, most surprisingly, the curvature at a point, which by its very definition certainly seems to depend on the way the surface is contained in E^3. These properties, which do not vary as the surface is deformed, are called *intrinsic*. The proof that curvature is intrinsic stimulated the search for other intrinsic properties and for their physical interpretation which were to prove fundamental to the application of differential geometry to the description of physical reality. It is impossible to improve on Gauss's own description ([36], p. 344):

> *Diese Sätze führen dahin, die Theorie der krummen Flächen aus einem neuen Gesichtspunkte zu betrachten, wo sich der Untersuchung ein weites noch ganz unangebautes Feld öffnet. Wenn man die Flächen nicht als Grenzen von Körpern, sondern als Körper, deren eine Dimension verschwindet, und zugleich als biegsam, aber nicht als dehnbar betrachtet, so begreift man, dass zweierlei wesentlich verschiedene Relationen zu unterscheiden sind, theile nehmlich solche, die eine bestimmte Form der Fläche im Raume voraussetzen, theile solche, welche von den verschiedenen Formen, die die Fläche annehmen kann, unabhängig sind. Die letztern sind es, wovon hier die Rede ist: nach dem, was vorhin bemerkt ist, gehört dazu das Krümmungsmaass; man sieht aber leicht, dass eben dahin die Betrachtung der auf der Fläche construirten Figuren, ihrer Winkel, ihres Flächeninhalts und ihrer Totalkrümmung, die Verbindung der Punkte durch kürzeste Linien u. dgl. gehört.*

Translation:

> These theorems lead one to consider the theory of curved surfaces from a new viewpoint, whereby the investigation opens up a broad still undeveloped field. If one considers surfaces not as boundaries of bodies, but as bodies one of whose dimensions vanishes, and at the same time are bendable, but not stretchable, then one understands that there are two basically different types of relations to distinguish, namely those which presuppose that the surface takes on a particular form in space, and those which are independent of the various forms a surface can assume. This last type is what is being discussed here: according to the above remarks, the curvature at a point is of this nature; one sees easily that the consideration on the surface of constructed figures, their angles, their area, and their total curvature, the connection of points by curves of shortest length, etc. are also of this type.

What Gauss is saying here is that one can do geometry of a certain type *on* the surface and forget about the ambient three-dimensional space. It is this abstracted "geometry on a surface" that was Gauss's fundamental contribution to geometry.

From this point of view Bernhard Riemann (1826–1866) proceeded as follows: consider a geometrical space of *any* number of dimensions (say, four). The space consists of *points,* and near any given point we can describe this space, in this case, by four *coordinates,* that is, an ordered set of four real numbers (x_1, x_2, x_3, x_4). Associated with the coordinates is a *metric form* that in the simplest cases might look like

$$ds^2 = f_1\,dx_1^2 + f_2\,dx_2^2 + f_3\,dx_3^2 + f_4\,dx_4^2$$

Using the metric form, we can measure length of curves, as in §11.2 and §11.3. And so Riemann proceeded. It was a radical departure from tradition, but, working analogically from Gauss's developments for two-dimensional surfaces, he constructed a far-reaching theory of the geometry of space of arbitrary dimension. This theory is now taught to advanced undergraduates in our universities, but 100 years ago it was the province of only a handful of first-class mathematicians. It is beyond the scope of this book to discuss his generalization in detail, but throughout it exhibits a strong analogy to the case of a surface with a given metric form, including a well-defined concept of *curvature of the higher dimensional space* that is analogous to the curvature of the surface. This will be discussed more thoroughly in Chapter 12, in which it is shown that the universe in which we live can most conveniently be considered as a *four-dimensional space:* near any *point* in the four-dimensional universe we can consider rectangular *space coordinates* and a *time coordinate*. Then we ask: Is there an *intrinsic metric form* on this four-dimensinal space? How does the *curvature* vary throughout this space? What are the *geodesics*?

EXERCISES

11.1. Compute the distance between the following pairs of points
(a) in E^2
  (i) $(1, 0)$ and $(0, 1)$
  (ii) $(\pi, 1)$ and $(0, 0)$
(b) in E^3
  (i) $(0, 0, 1)$ and $(0, 0, 2)$
  (ii) $(1, 0, 1)$ and $(0, 1, 0)$

11.2. Suppose (x_0, y_0, z_0) is a fixed point in E^3. Let S be the set of points (x, y, z) in E^3 that satisfies

$$(x - x_0)^2 + (y - y_0)^2 + (z - z_0)^2 = 1$$

Describe the set S in geometric terms.

11.3. Suppose (x_0, y_0, z_0) is a fixed point in E^3. Let S be the set of points (x, y, z) in E^3 that satisfies

$$(x - x_0)^2 + (y - y_0)^2 = 1$$

Describe the set S in geometric terms.

11.4. Consider the curve Γ defined by the equation $y = x^2$ in the (x, y)-plane. Find the point or points on Γ where (a) the curvature is maximal, (b) the curvature is minimal, if these curvatures exist. Draw a graph indicating the location of the points in (a) and/or (b).

11.5. What types of curve in E^2 have constant curvature at each of their points?

11.6. Suppose $P(t)$ is a variable point in E^3 given by

$$P(t) = (x(t), y(t), z(t)),$$

where

$$x(t) = t$$
$$y(t) = t$$
$$z(t) = t$$

for $0 \leqslant t \leqslant 1$.
(a) Show that the path of $P(t)$ as t varies represents a straight-line segment in E^3 and draw a diagram to illustrate this line segment.
(b) Use formula (11.1) to compute the length of this straight-line segment and compare the result with that obtained by using the formula for distance given in Figure 11.2. (*Hint:* To compute the integral use the Fundamental Theorem of Calculus.)

11.7. Proceed as in Problem 11.6 for the curve defined by

$$x(t) = 1 + t$$
$$y(t) = 2$$
$$z(t) = 3$$

for $0 \leqslant t \leqslant 2$.

11.8. Consider a curve Γ on the sphere of radius R in E^3 defined by

$$\alpha(t) = 0$$

$$\beta(t) = t \qquad 0 \leqslant t \leqslant \pi$$

in which α and β are the latitude and longitude coordinates, as in Figure 11.12.
(a) Describe this curve geometrically.
(b) Compute the length of Γ by
    (i) elementary geometry,
    (ii) using formula (11.4).

11.9. Consider a curve Γ on a sphere of radius R in E^3 defined by

$$\alpha(t) = t \qquad 0 \leqslant t \leqslant \frac{\pi}{4}$$

$$\beta(t) = 0$$

(a) Describe this curve geometrically.
(b) Find the length of this curve by using Formula (11.4) and compare it with the result obtained by elementary methods.

11.10. Let Γ be a curve on a sphere of radius R described by

$$\alpha(t) = \frac{\pi}{4} \qquad 0 \leqslant t \leqslant \frac{\pi}{4}$$

$$\beta(t) = t$$

(a) Describe this curve geometrically.
(b) Find the length of the curve (i) by elementary geometry and trigonometry, (ii) by using Formula (11.4).

11.11. Is the curve Γ in Problem 11.10 a geodesic? Why?

11.12. Consider the surface S in E^3 defined by the equation $z = x^2 + y^2$. Consider the point $P = (0, 0, 0)$ and the line L_0 in E^3 passing through P, which is given by $y = 0$, $z = 0$ (this is the x-axis).
(a) Draw a diagram to illustrate the plane tangent to surface S at point P, the line L_0, and the plane H through P perpendicular to the tangent plane (see Figure 11.20).
(b) Compute the curvature $K_P(S, L)$ for this choice of S, L, and P.

11.13. Consider the surface S defined in Problem 11.12.
(a) How will $K_P(S, L)$ vary as L varies over all directions emanating from P?
(b) Compute the (Gaussian) curvature of S at P.

11.14. Compute the Gaussian curvature at $P = (0, 0, 0)$ for the surface S defined by (a) $z = x^2 - y^2$, (b) $z = 3x^2 + y^2$. (*Hint:* Draw a schematic diagram and intuitively determine the maximum and minimum values of $K_P(S, L)$.)

11.15. Show that a cylinder in E^3 (whose metric is defined by calculating the lengths of curves in E^3 that lie on the cylinder) has Gaussian curvature 0 at all of its points by using Gauss's theorem. (*Hint:* Note that a piece of paper can be rolled up into a cylinder. That is the idea.) Determine the Gaussian curvature from the maximum and minimum curvatures of curves passing through an arbitrary point of the cylinder.

11.16. If a surface S is deformed in E^3 without stretching (that is, without changing the concept of distance on S), it follows that geodesics remain geodesics. Describe the geodesics on a cylinder (which is a nonstretched rolled-up portion of a plane) by using this fact and your knowledge of what the geodesics in a plane are.

11.17. Suppose you indent a rubber ball with your finger. Are you stretching the surface? (*Hint:* Use Gauss's theorem and determine whether the indented ball has the same curvature at all points as it had in its original form.)

11.18. Describe in your own words what the concept "four-dimensional space" means.

11.19. Describe in your own words what the concept "n-dimensional space" means, in which n is any positive integer.

CHAPTER 12

Models of the Universe

This final chapter is devoted to relativity theory. With the machinery already introduced it is possible to solve a special case of the Einstein equations which, together with measurements of the Hubble red shift, *enable us to estimate the* age of the universe *for the* model of the universe *under consideration. Mathematical models are discussed more generally. The chapter ends with a geometrical space-time description of the path of a spaceship traveling from earth to moon to show how it is related to the problems of navigation and cartography discussed in the earlier chapters. The questions remain ever the same, but they are asked about different and more complex geometrical objects.*

§12.1. Models are used in everyday life for a variety of purposes. Scale models of ships and planes are constructed by designers to illustrate the appearance and configuration of the proposed product and also for experimental use in towing tanks and wind-tunnel tests. There is certainly a difference between a real airplane and a wind-tunnel model of one, but as far as their aeronautical properties are concerned they are nearly equivalent; the correspondence between their reactions is close. The model accurately represents these properties of the "real" airplane, although it does not represent many other properties at all.

The construction of a physical or an *abstract model* of a part of reality serves to focus attention on a restricted class of properties while ignoring or minimizing those that are not of interest at the moment. To the extent that it *accurately* isolates the aspects of interest it is a successful model.

For millenia man has concerned himself with the succession of the seasons, the prediction of eclipses and comets, and the position of the moon and planets, and in the course of history many different models of parts of the universe have been created to focus attention on some of these aspects of reality. The earliest accurate astronomical models were invented by the Babylonians to predict eclipses and the angular positions of the moon and planets on the celestial sphere. Their models were arithmetical: computational procedures were invented that permitted the accurate calculation of these positions but no geometrical model was developed to relate the various motions in a natural and meaningful way (cp. §3.5 and [50]). Their arithmetical model was not concerned with explaining how eclipses occurred, nor with the geometrical shapes of the heavenly bodies or their mutual spatial positions; only certain aspects of their apparent motion against the background of the celestial sphere were represented in the model and they were done quite well.

Aristarchus and his contemporaries were more interested in geometrical questions: the distances between the earth and other astronomical bodies; their shape; their size. These questions demand a different, more geometrical, model. First of all, if, say, the moon and the sun are at fixed distances from the earth (so that the first question will make sense), then, because of their apparent motion as seen from the earth, they must revolve about it in circular orbits. This geometrical *assumption* not only makes the question of calculating their distances from the earth sensible, but also suggests a geometrical model for eclipses: a solar eclipse occurs when the moon is positioned between the earth and sun, whereas a lunar eclipse results when the earth lies between the sun and moon (*cp.* §4.2).

This reasonable geometrical model is fundamentally different from the Babylonian model, which might correspond to a geometrical model in which the earth is the center of the celestial sphere and all the other astronomical bodies move on it. Such a model, however, cannot provide an interpretation of eclipses and also suggests that all celestial objects lie at the same distance from the earth, which is not a fruitful implication.

In addition to his interest in astronomical distances, Aristarchus was also concerned about the size of celestial bodies. Therefore in his model they are represented by balls. As we know today, this is not a completely accurate correspondence, since rotating bodies are flattened at their poles, but for the purpose of measuring astronomical distances the ball is a sufficiently accurate geometrical model. It will not do for other purposes.

A more subtle feature of Aristarchus' model is that it is really only two-dimensional. Since the earth, moon, and sun travel in what appears to be nearly the same plane in space, no information is lost regarding the size or mutual distance of the balls if they are replaced by disks in the plane of their joint orbit. The most subtle aspect of the model lies in the interpretation of distance. Aristarchus assumed that the plane of the orbit is the usual Euclidean plane equipped with the usual distance formula. In his time there was no direct evidence to show that space is Euclidean for distances larger than a few miles. It might, for instance, actually be the three-dimensional surface of an enormous four-dimensional ball; locally, it would appear to be like ordinary Euclidean space, but on the scale of the solar system it might require three-dimensional spherical geometry to interpret observations correctly, just as navigational observations on the earth demand that its surface be considered as a two-dimensional sphere, not a Euclidean plane. Today we know that the neighborhood of the earth is not quite Euclidean; space is curved or *bent*, but for most purposes the departure from Euclidean flatness is negligible.

§12.2. When later astronomers tried to incorporate all the planets into Aristarchus' model and to determine their mutual motions and distances, it was found necessary to develop new model representations for their orbits in place of simple circles concentric with the earth. In addition to these circular motions, simultaneous circular motions about other centers were introduced; orbits compounded from such simultaneous motions are called *epicycles*. This newly complexified model achieved its final expression in the hands of Ptolemy and is now commonly referred to by his name. It lasted until the sixteenth century when it was displaced by the Copernican-Keplerian-Newtonian heliocentric model.[1] Astronomers held

[1] *Ptolemy assumed that the earth was fixed and that the other celestial bodies revolved around it, a viewpoint that was predominant until the time of Copernicus. Some earlier Greeks, however, notably Aristarchus, thought that the earth revolved around the sun.*

onto Ptolemy's geocentric model as long as they could, but in order to bring it into agreement with new observational evidence, various modifications were made which introduced aspects of the ultimate Copernican heliocentric model. Such a hybrid intermediate version, the work of the great Danish astronomer Tycho Brahe (1546–1601), is illustrated in Figure 12.1 (from Flammarion [82]).

FIGURE 12.1

NOVA MVNDANI SYSTEMATIS HYPOTYPOSIS AB
AUTHORE NUPER ADINUENTA, QUA TUM VETUS ILLA
PTOLEMAICA REDUNDANTIA & INCONCINNITAS,
TUM ETIAM RECENS COPERNIANA IN MOTU
TERRÆ PHYSICA ABSURDITAS, EXCLU-
DUNTUR, OMNIAQUE APPAREN-
TIIS CŒLESTIBUS APTISSIME
CORRESPONDENT.

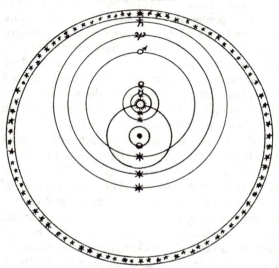

Tycho Brahe's model of the Solar system.

Trigonometry is enough to describe and compute all the relationships between circles and between points moving on epicycles with constant speed on each component circle. Consequently the mathematics available to Ptolemy in +150 was sufficient for the study and use of the Ptolemaic model. The exploitation of the more subtle Copernican heliocentric model had to wait for the development of the calculus by Newton and Leibniz, since Copernican-Keplerian orbits are compounded from ellipses. Recent models of the universe, which are the subject of this chapter, rely on the mathematics of differential geometry, developed in the nineteenth century, only about 50 years before Einstein's first model. It is the purpose of this chapter to describe the Einstein models but also to indicate the reasons for passing from Newton's simple and effective viewpoint to the much more subtle and technically difficult models developed by Einstein. To further our discussion of the differences between these models a brief description of

Newton's, in a form convenient for our later needs but rather different from the way it is usually introduced in a freshman physics course, is given next.

§**12.3.**   Three hundred years after its invention, Newton's model of the universe still underlies all basic physics courses. Following Aristarchus and Ptolemy, Newton assumed that the fabric of space could be accurately represented by three-dimensional Euclidean geometry and that time is an external parameter with no direct geometrical representation. Aristarchus had not concerned himself with the motions of heavenly bodies, and Ptolemy included their orbits in his model by assuming that they could be represented by epicycles: both models provided *geometrical* interpretations of all phenomena of interest to their creators. Newton's model is only partly geometrical. In effect, he assumed that under "normal" (that is, "force-free") circumstances the path of a body in space would be a geodesic and that *time* could be used to measure the *distance* traveled by the body by multiplying the time of travel by a constant. This amounts to choosing an appropriate scale of measure for time. Since the geodesics of Euclidean space are straight lines, Newton's assumptions imply that if no "forces" act on an object it will move in a straight line; furthermore, the distance traveled in a time t, say $s(t)$, will be given by $s(t) = vt$, where v is a constant. This means that the body travels along a straight line with constant speed v. Most interesting bodies do not travel in straight lines with constant speed. For instance, a massive object dropped from a high tower falls in a straight line but with ever-increasing speed (*cp.* §9.9); a projectile fired at an angle with the vertical travels in a parabolic path with a speed that varies according to its position on the path; the orbit of the earth is an ellipse, a closed curve quite unlike a straight line in three-dimensional space. To account for such motions in his geometrical model Newton introduced certain *nongeometrical* concepts, foremost of which are *forces*. Forces are the hypothetical causes of departures from straight-line constant-speed motion. They do not correspond to any natural geometrical structures in three-dimensional Euclidean space but are fundamentally similar to the arithmetical aspects of Babylonian models, although they are much more subtle and have far-reaching implications. In common with the Babylonian arithmetical models, the properties and nature of forces must be determined for certain important classes of motion in a way that has nothing to do with the geometrical parts of the model; once this is done, the effects of forces can be interpreted in geometrical terms, such as the orbit of the body under investigation, by using a *differential equation* discovered by Newton that relates the nongeometrical forces to the equations of the orbit. It works this way (recall a similar discussion in §9.9): choose a Cartesian coordinate system in space so that the coordinates of a typical point are (x, y, z). Suppose that a moving body in space is represented by a moving point in the model (intuitively, this point corresponds to the center of mass of the body). If the point (body) is acted on by a force F_x in the direction of the x-axis, a force F_y in the direction of the y-axis, and a force F_z in the direction of the z-axis, then the coordinates (x, y, z), which describe the position of the point, will change with time. The equations that determine this change are

$$\text{(12.1)} \qquad Newton's\ equations \quad \begin{cases} m\,\dfrac{d^2x}{dt^2} = F_x \\[2mm] m\,\dfrac{d^2y}{dt^2} = F_y \\[2mm] m\,\dfrac{d^2z}{dt^2} = F_z \end{cases}$$

Here m is a constant that depends only on the nature of the body; it is called its *mass*.

If these equations can be solved, then it may be possible to eliminate the time t from the solutions, which will lead to two equations relating x, y, and z. These equations represent a curve in three-dimensional space that corresponds to the orbit of the moving body.

For example, if, as in §9.9, the force is that of gravity near the surface of the earth, it is known (from experiments) to point downward (in the direction of the negative z-axis, say) and is proportional to the mass m of the body. This means that

$$F_x = 0 \qquad F_y = 0 \qquad F_z = -mg$$

where g is the constant of proportionality (the gravitational constant), in which case Newton's equations simplify to

$$\text{(12.2)} \qquad \begin{aligned} \frac{d^2x}{dt^2} &= 0 \\[2mm] \frac{d^2y}{dt^2} &= 0 \\[2mm] \frac{d^2z}{dt^2} &= -g \end{aligned}$$

In §9.4 we learned how to differentiate polynomials; in particular, it turned out that the second derivative of a first-degree polynomial is zero and the second derivative of a second-degree polynomial is a constant. These are just the facts needed to solve the equations in (12.2); they imply that the solutions are

(12.3a)

(12.3b)

(12.3c)

$$x = at + b$$
$$y = ct + d$$
$$z = -\frac{g}{2}\,t^2 + et + f$$

where $a, b, ..., f$ are certain constants whose values must be determined from the way in which the body was initially released to the action of the gravitational force; we need not worry about them here.

To find the geometrical orbit and thus bring this solution back into the realm of the geometrical model, the time, which is not a geometric quantity (that is, which has no representation in Euclidean three-dimensional space), must be eliminated from the three equations in (12.3). From (12.3a) find $t = (x - b)/a$; substitute it in (12.3b) to find, after a slight simplification,

(12.4a) $$ay - cx = ad - bc$$

the equation of a plane parallel to the z-axis (see Figure 12.2). Also substitute for t in (12.3c) to find

(12.4b) $$z = -\frac{m}{2}\left(\frac{x-b}{a}\right)^2 + e\left(\frac{x-b}{a}\right) + d$$

which is the equation of a parabolic cylinder with its axis parallel to the y-axis. The orbit of the body consists of those points that lie on both surfaces, that is, the curve formed by the intersection of the two surfaces. This parabola is shown in Figure 12.2.

FIGURE 12.2

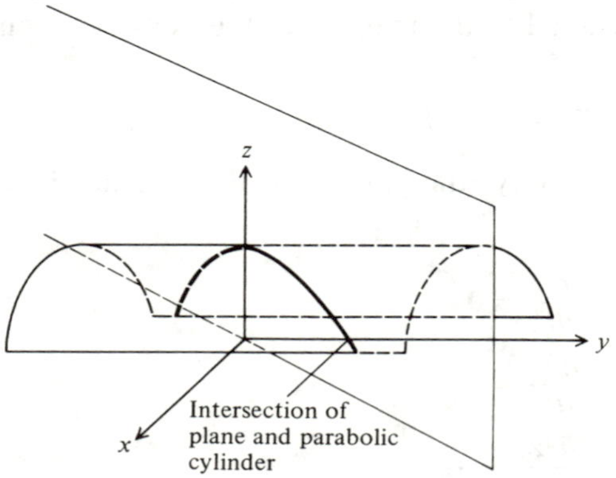

Intersection of plane and parabolic cylinder

Parabola realized as the intersection of a plane and a parabolic cylinder.

The significance of these calculations is that the geometrical orbit can be found by solving Newton's equations and eliminating the time. Thus, through the medium of the nongeometrical force and time concepts, coupled to the geometry by equations involving derivatives, a geometrical representation of motion in three-dimensional Euclidean space is possible.

§**12.4.**   The orbits that occur in Newton's model bear no intrinsic relation to the underlying Euclidean space. Indeed the only *natural* curves on a surface or in a space are its geodesics, the curves that trace the shortest paths between points in the space. No other curve is related to the geometry of the surface in such an immediate and meaningful way. This means that *any model of the universe that represents physical space by three-dimensional Euclidean space must introduce nongeometrical concepts such as force if it is to describe the motions of planets and other astronomical bodies.* Thus no minor adjustments of Newton's model will suffice to generalize the force concept in a geometrical way, for there is no natural geometrical object in Euclidean three-dimensional space that could fill such a role.

To create a *completely geometrical model* of the motions of astronomical bodies moving under the influence of their mutual gravitation three-dimensional Euclidean space must be given up. We give it up in two distinct stages to show how Newton's model will lead to Einstein's model if we try to geometrize Newton's force concept in a natural way.

First of all we must agree on what we require of a completely geometrical model. The fundamental demand is that the *paths of astronomical bodies be geodesics of the model space.* If, for instance, the orbit of the earth about the sun is considered in a Euclidean three-dimensional model space, it is clearly far from a geodesic, that is, a straight line, in this model space; this suggests that Euclidean three-dimensional space must be given up, as noted above. It must be replaced by a space whose geodesics agree closely with the observed orbits of astronomical bodies like the earth. There are two ways to do this, which are not mutually exclusive. First, increase the number of dimensions of the space; this will change the relationship between the space and its geodesics and, after the new dimensions have been interpreted in physical terms so that the geometrical quantities in this new space will have interpretations, it may turn out that the geodesics do correspond closely to actually observed orbits. Second, bend the Euclidean space so that its geodesics are no longer straight lines. Both modifications of the original Euclidean space can be made; the result is a higher dimensional, non-Euclidean model space. This is just what Einstein did to construct his model, which is four-dimensional and non-Euclidean.

The meaning of "dimension" was briefly discussed in Chapter 11; here it will be enough if the reader will suspend his disbelief long enough to agree that whatever other properties it may have four-dimensional *Euclidean* space is "flat" and its geodesics are straight lines, just as in two- and three-dimensional Euclidean spaces. Granted these facts, let us examine the appearance of the path of the earth about the sun as it would be represented in a four-dimensional Euclidean model. Three of the dimensions are used to represent ordinary physical space, whereas the fourth coordinate axis, called the *t-axis,* represents the variation of time. This simply means that each point in four-dimensional Euclidean space will be represented by an ordered set of four numbers (x, y, z, t), with the first three numbers referring to the position in physical space expressed relative to a convenient system of Cartesian coordinates and the fourth referring to the time associated with the event under consideration.

Time might be measured in years and distance in miles. How can such measurements be compared? What is a mile's equivalent of time or a year's equivalent of miles? Experimental physicists have provided the answer to this question. They have discovered that the speed of light is a constant, independent of the environment through which the light may pass, equal to about 186,000 miles per second (nearly 3×10^{10} centimeters/second). Thus it takes light a fixed time to travel a fixed distance, and therefore we might as well agree that a fixed distance can be measured by the time of flight of light traversing it. For instance, in a year light will travel about six trillion miles; this distance, called a *light year,* is a standard unit of astronomical measurement of *distance*. The earth is about 92 million miles from the sun, so light requires about 8 minutes to travel from there to here; it can be said that the earth is *8 light minutes distant from the sun.*

If it is agreed to measure distances in terms of time this way, and if the plane of the earth's orbit about the sun is chosen as the *x-y* plane, with the origin of coordinates at the center of the earth's spatial orbit (which, for simplicity, we assume to be a circle), then the orbit of the earth about the sun represented in the *four-dimensional* model is a *helix* (the thread of a machine bolt, the curve on a barber pole) whose axis is the time axis. Figure 12.3 illustrates this path; the *z*-axis has been suppressed, just as Aristarchus suppressed the unnecessary space axis in his model, since it contributes no information. Time increases in the upward direction on the *t*-axis. After the passage of a year the point representing the earth lies over the space position (represented in the *x-y* plane) that it occupied at the same time exactly one year before. That is, the helix has wound around the time axis exactly once.

FIGURE 12.3

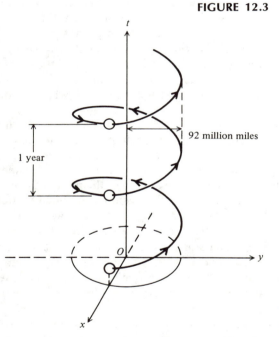

Space-time diagram of the earth's motion relative to the sun.

Figure 12.3 does not do justice to physical reality because the scales of space and time measurement are not shown in the same units, which makes it impossible to form a good idea of the pitch of the helix, that is, of the relative height along the *t*-axis required for the helix to make one complete loop about the *t*-axis. Figure 12.4 uses the measurement of a distance in terms of the time required for light to traverse it to provide a more accurate representation, but even in Figure 12.4 the helical orbit of the earth has been distorted so that it can be distinguished from the time axis. The pitch of the helix is 1 year/8 minutes = 525,600 minutes/8 minutes

FIGURE 12.4

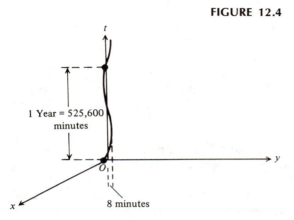

1 Year = 525,600 minutes

8 minutes

Figure 12.3 drawn on a more accurate scale.

= 65,700; the helix climbs so steeply that it is virtually indistinguishable from the time axis in any diagram.

In this four-dimensional model the path of the earth about the sun is indeed nearly a straight line; that is, it is nearly a geodesic of the four-dimensional Euclidean space. This suggests that just a slight deformation of four-dimensional Euclidean space might produce a new, bent, four-dimensional surface whose geodesics would include the helical earth orbit. In such a model the curvature of the surface would be the "cause" of the curved geodesics and would thereby provide a geometrical interpretation of Newton's gravitational "force." In such a model there would be no need for equations relating geometrical quantities to external nongeometrical concepts.

In 1916 Albert Einstein (1879–1955) created such a model. In addition to a qualitative description similar to the one given, such a model requires a mathematical prescription that will determine, or at least limit, the exact nature of the geodesics from point to point on the surface that represents the universe. Since the geodesics, which are the paths of bodies, are determined by the curvature, it is plausible that the mathematical prescription will be a set of equations that restricts the possible values of the curvature at the various points of the four-dimensional surface. Einstein's celebrated *gravitational* equations, which are the heart of his contribution, do just that; they are differential equations relating values of the curvature to values of its derivatives (that is, rates of change of the curvature) for each point on the four-dimensional space-time surface which is the model representation of the universe; *cp.* Figure 12.5.

If the four-dimensional surface is nearly flat in some region, its geodesics will be nearly straight-line segments there, in which case Einstein's complicated equations simplify to the form of Newton's equations, but in place of Newton's force term there is a geometrical quantity. This means that Einstein's model, when it is a nearly flat four-dimensional space, is equivalent to Newton's model (including his nongeometrical gravitational force concept) in three-dimensional Euclidean

FIGURE 12.5

$$g_{ik;s} = 0, \quad \Gamma_i = 0$$
$$+$$
$$R_{ik} = 0, \quad R_{ik,l} + R_{kl,i} + R_{li,k} = 0$$

Einstein's unified field equations as they appeared on the front page of *The New York Times* on Monday, March 30, 1953.

space. No one is surprised by this because Newton's model works very well for the description of planetary motions. For instance, the (four-dimensional) orbit of the earth about the sun is nearly a straight line and it lies in a region that is nearly flat; therefore Einstein's model should be equivalent to Newton's for this case, and indeed astronomers are not yet able to detect any difference between the predictions of the two models for the orbit of the earth.

§12.5. When considering the earth's orbit, the only possible advantage of Einstein's model is a conceptual one. Many physicists and mathematicians think that it makes observed motions and properties more intuitive and easier to explain. There are other astronomical phenomena for which the two models (Einstein's and Newton's) predict different results; comparison of their predictions with reality is difficult because all the differences now known are slight and observations of the phenomena must be made with great care and specially designed apparatus. All experimental tests that have been made since the introduction of Einstein's model in 1916 appear to confirm that it provides a more accurate representation of the observed phenomena than any other known model.

There are three principal experimental tests of Einstein's model. One concerns an anomaly in the motion of the planet Mercury which was known to nineteenth-century astronomers but could not be represented by Newton's model; the other two were new phenomena predicted by Einstein's model but not previously known to exist. Experiments show that they definitely do exist; this qualitative support for the model is strengthened by the quantitative agreement of the experimental observations with the predictions made from the model.

The three tests are these:

1. The shift of the perihelion of Mercury.
2. The bending of light in a gravitational field.
3. The gravitational red shift.

In the remainder of this section each of these phenomena is described; in the next section a model of Einstein's model which provides an intuitively reasonable interpretation of each of them is presented.

1. *The Shift of the Perihelion of Mercury.* The early Greeks thought that the planets revolved in circular orbits. Kepler's exacting study of their orbits together with the heliocentric model of Copernicus showed that they actually revolve in

ellipses. One focus[2] of an elliptical orbit lies very nearly at the center of mass of the rotating planet and the sun, but for our purposes it will do to suppose that the focus actually coincides with the sun's center. There are two geometrically distinguished points on an elliptical orbit: one at which the planet is as close as it ever is to the sun and one at which it is as far as possible from it. The latter is an *aphelion* point; the former a *perihelion* point.

Newton's theory predicts that a planet will move in a perfectly elliptical orbit. Suppose that a planet moving in a perfectly elliptical orbit starts at some time from a perihelion point; after exactly one of its years it will return to that precise point and its annual motion will repeat itself. If the orbit is not perfectly elliptical, then after a year's motion the planet may return to nearly the point of perihelion from which it began its motion. A measure of the departure from perfect ellipticity (hence also from Newton's theory) is the angle θ formed by the line joining the sun and the perihelion point and the sun and the position of the planet after one revolution; this angle is shown in Figure 12.6. With each revolution about the sun the orbit rotates by an angle θ; more picturesquely stated, the elliptical Newtonian orbit *precesses* by the angle θ with each revolution.

FIGURE 12.6

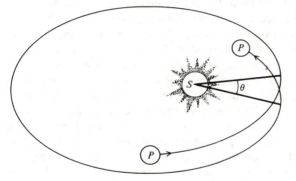

Perihelion precession for one revolution.

Careful measurements show that the amount of annual precession is extremely small for planets in our solar system. For the earth, Landau and Lifschitz [46] give

(12.5a) $$\theta_{\oplus} = 0.046'' \pm 0.027''$$

The precession is greatest for Mercury

(12.5b) $$\theta_{\female} = 0.426'' \pm 0.009''$$

and had already been detected by nineteenth-century astronomers. Recall that a *second of arc*, abbreviated by the ''sign, is 1/3600 degree; therefore, the deviation

---

[2] *The* foci *of an ellipse are two points* F_1 *and* F_2 *interior to the ellipse with the property that the sum of the distances* $PF_1 + PF_2$ *is constant as P ranges over the points of the ellipse. This property was used by the Greeks for a* synthetic *definition of an ellipse. When* F_1 *and* F_2 *coincide, the ellipse becomes a circle with center* $F_1 = F_2$.

$\theta_{\yen}$ is approximately 1/100,000 of a right angle, a measurement of astounding accuracy!

Einstein's model predicts the precession of planetary orbits; the numerical values calculated from the model for earth and Mercury are

(12.6a)   (model) $\theta_{\oplus} = 0.038''$

and

(12.6b)   (model) $\theta_{\yen} = 0.430''$

in excellent agreement with the observations in (12.5a, b).

2. *The Bending of Light in a Gravitational Field.* In certain circumstances light behaves as if it were a wavelike disturbance in a hypothetical medium called *aether*, but otherwise it behaves like an assemblage of particles. When considering the behavior of light in a gravitational field, the particle viewpoint is the appropriate one. All terrestrial experiments appear to show that light traveling in a vacuum follows a straight-line path, and it was natural to assume that this would be true of light passing through any region in which there is a gravitational force. Furthermore, since the force of gravity was supposed to affect the paths of material bodies by acting on their mass and since experiments showed that light particles could not be thought of as having mass (at least in the usual sense), it seemed that there was no way for the gravitational force to exert its influence on light and that light therefore would travel along the straight-line geodesics of Newton's model. The fundamental idea underlying Einstein's model, that the force of gravity acts by creating curvature in space and that the natural paths of bodies are geodesics of this deformed space, suggests that the paths of light rays should be deformed from straight lines wherever the geodesics due to gravity are not straight. In particular, this model suggests that the path of a light particle should bend toward the more curved regions (in physical terms, toward the *sources* of gravitational force). For a light ray just grazing the limb of the sun the Einstein model predicts that the direction of the ray will be bent toward the sun by an angle of 1.75'', as shown in Figure 12.7. This prediction has been verified by observations of the paths of light rays from stars near the limb of the sun during solar eclipses and comparison of their apparent direction with their direction when the earth stands between the stars and the sun.

FIGURE 12.7

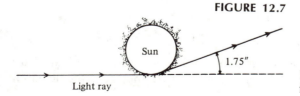

Light ray

Path of a light ray in a gravitational field.

3. *The Gravitational Red Shift.* Another novel and unexpected but experimentally confirmed prediction of Einstein's model is that light traveling from a region of greater curvature to a region of lesser should change its color, becoming redder. This phenomenon has been verified by observing the light emitted by

hydrogen in extremely dense "blue dwarf" stars, which are sources of intense gravitational forces and consequent curvature in the model. The color of the emitted light is compared with the color of corresponding light emitted by hydrogen atoms here on earth; the difference is a measure of the strength of the gravitational force near the surface of the observed star.

§12.6. A simple "model of Einstein's model" can be conceived, which helps to place the complicated four-dimensional geometrical properties implied by Einstein's curvature equations into an intuitive perspective. The general nature of the orbits of planets and double-star systems, the bending of light, and the gravitational red shift all become reasonable expectations after a few moments of thought about this heuristic simplified form. Although it cannot be used to calculate the values of any of the interesting quantities, it is useful because it places matters in the proper qualitative perspective.

Let a smooth, thick layer of rubber be placed on a flat horizontal tabletop. Its surface will represent two of the three spatial dimensions of the four-dimensional model of the universe. Suppose that a number of balls of various radii and density are available, some composed of a dense substance such as lead, others hollowed out so as to have a very low density. These balls will play the role of stars, planets, and other astronomical objects. Furthermore, suppose that particles of light are represented by exceedingly small balls of virtually no weight whatever; and that if any of these balls is rolled across the rubber surface there will be no friction to impede its motion.

Place a large dense ball on the rubber surface; it will cause a depression to be formed in the surface. The denser the ball, the deeper the depression for balls of a fixed radius; this also means that the curvature of the rubber surface will be greater for the denser balls, at least in their immediate vicinity. The curved rubber surface represents the gravitational force of the object represented by the ball. Now place a small, less dense ball on the surface at some distance from the large one (we call the latter a "star" and the former a "planet"). The small planet ball will form a small depression in the rubber, but it will already be situated in the depression of the star and will tend to roll downhill into the star's depression. In the same way the small depression made by the planet will be neither very curved nor very deep but it will nevertheless extend over the entire tabletop so that the star will be situated on part of this small depression and will tend to roll downhill toward the planet, although at a much slower rate than that of the planet rolling toward the star. In this way we see that any two bodies attract each other; this is the force of universal gravitation, as represented in the model. See Figures 12.8 and 12.9.

It is easy to see how a planet could orbit about a star. If left alone, the planet will simply roll downhill and finally collide with the star. If, however, the planet is rolling along at a great but constant speed (which is not diminished by friction, since we have assumed that there is no friction), it is easy to convince ourselves that its path will start out as a straight line but will dip toward the star somewhat as the planet progresses into the star's depression and will (because the speed is assumed great enough) climb back up out of the depression, headed off on a new

FIGURE 12.8

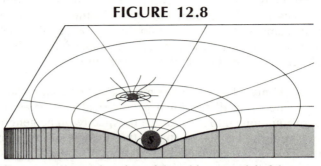

Cutaway perspective view of the tabletop model of the curvature of space.

straight-line course. The diagram that describes this path is the same as that for the bending of light (Figure 12.7) but the angle of deflection, of course, will vary. *If* the direction and speed of motion of the planet are chosen properly, it can be arranged that the effect of the curvature striving to pull the planet toward the depression will be exactly counterbalanced by the planet's tendency, due to its initial speed, to fly away from the star. In intuitive terms, if a planet with no initial speed will fall into the star, but a planet with great initial speed will fly away from it, then there must be some intermediate speed and direction that will balance the planet perfectly between falling in and flying off; this balanced state is the bound orbit. The planet circulates around the star down in the depression, as shown in Figure 12.10.

In the same way it is possible to represent the motion of two stars that form a double-star system; each "falls into" the other's depression with just the right initial direction and speed to avoid both collision and ultimate separation.

This simple tabletop model does not make it clear whether the precession of the perihelion of a planet can be represented; that would require a complicated calculation based on the theory of elasticity of materials and is more trouble than solving Einstein's equations for the real four-dimensional model. The other two experimental tests of Einstein's model can be interpreted easily in the tabletop model.

The bending of light rays near the sun is a consequence of the remarks already made. A light particle does not form a significant depression of its own, since it has virtually no weight in the model, but its path, if it is given a push, will be a geodesic

FIGURE 12.9

Cutaway sectional view of the configuration in Figure 12.8.

FIGURE 12.10

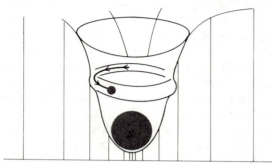

Exaggerated view of a planet orbiting a star in the tabletop model.

of the deformed rubber surface. If the particle of light is rolled so that it passes close to the star, it will fall into the star's depression, but, because the density of a typical star is relatively low and the speed of light is the greatest speed possible, it will not be captured by the star as a planet, nor will it fall into the star; its path will be deflected toward the star ball by the depression, as Einstein's model predicts and as experiments confirm.

To interpret the gravitational red shift in the tabletop model we need an experimental fact. At the beginning of the twentieth century it was discovered that the energy carried by a light particle was not related to its speed, which is constant, but rather to its *color*. "Color" is the common word for the physicist's term "frequency." Max Planck (1858–1947) discovered that the energy of a light particle is proportional to its frequency; the constant of proportionality is now called *Planck's constant* in honor of this important contribution. Light near the blue end of the spectrum has greater energy than light near the red end. This means that a shift toward the red corresponds to a loss of energy. With this in mind, consider what happens when a light particle travels from a region of greater to one of lesser gravitational force. In terms of the tabletop model, the ball representing the light particle is climbing out of a depression in the rubber surface. To climb out it must give up some energy; if it were a planet, it would lose speed as it climbed out, but the speed of light is constant. The only way light can lose energy is to change its color by becoming redder. This is the gravitational red-shift phenomenon.

§12.7. The tabletop model gives a reliable picture of Einstein's model in the vicinity of a star, but it appears to imply that at great distances from the star the four-dimensional surface representing the universe is flat, like Euclidean space. This is probably not so, and is certainly not implied by Einstein's model; therefore a few words have to be said about the global shape (or *shape in the large*) of the universe. A model of the complete universe, expressing its global shape, is called a *cosmological model*. Models of the universe are constructed to agree with the observed evidence concerning the large-scale nature of the physical universe; they also are limited to maintain some degree of mathematical simplicity so that the equations that arise can be treated by currently available mathematical techniques.

One of the most successful current cosmological models assumes that the spatial universe would appear the same in its large-scale properties to observers located anywhere in it at a given instant of time. This notion corresponds to the idea that the surface of the earth appears spherical in the neighborhood of any observer; that is, by smoothing the mountains and filling the seas the earth would appear to be spherical at each of its points. Similarly, smoothing the concentrations of matter that constitute the stars and considering the large-scale average properties of the spatial universe, it is reasonable, although not certain, that it would appear to be the same to all observers. Geometrically, this means that it must be a three-dimensional space of constant curvature. From the study of differential geometry it is known that there are only three possibilities (locally): the space must have constant *positive* curvature like a sphere, or *zero* curvature like Euclidean space, or constant *negative* curvature. There is no completely natural model that can be used to prompt intuition in this last case, but the surface of a hyperboloid of one sheet (Figure 12.11) is a two-dimensional space of nonconstant negative curvature that may help to form your intuition (*cp.* also Figure 5.15). In any event, notice how the observational assumption is translated into a geometrical model. Now, according to Einstein, any geometrical model of the universe must satisfy certain differential equations involving the curvature, but the curvature is the curvature of *space-time,* a four-dimensional geometrical space not to be confused with three-dimensional physical space. This means that although the spatial curvature of physical three-dimensional space may be constant the curvature of space-time will be governed by Einstein's equations.

FIGURE 12.11

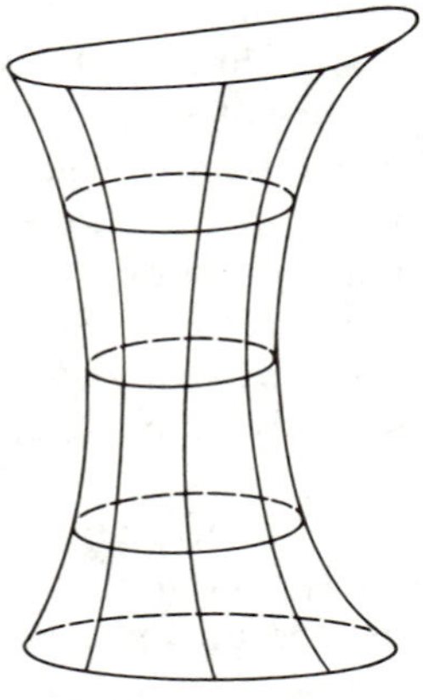

Hyperboloid of one sheet: a surface of negative curvature.

So far nothing has been said about how the spatial universe may vary with time. If, for instance, space has constant positive curvature, then it can be imagined as a three-dimensional sphere and, as time passes, the radius of this sphere may change. If the radius increases with time, it makes sense to speak of an *expanding universe;* this will have certain observable physical consequences that are discussed below. The way in which the radius of the universe (or analogous quantities if the spatial universe has zero or negative curvature) will change with time is determined by the Einstein equations.

We can use Einstein's equations in conjunction with the measurement of the *observable* expansion of the universe to determine the *age* of the universe. A brief outline of this derivation is given in the following paragraphs.

Einstein's equations reduce to two simple equations for this (constant curvature) model. Denote the radius of the universe by $r(t)$ (for the case of positive curvature; the analogous geometrical quantity for the other cases will not be described here); it may depend on the time t. Suppose that the density of matter in the spatial universe is denoted by ρ ($\rho =$ mass/volume) and that g denotes the universal constant of gravitation (a number whose value we do not need to know). Finally, let ϵ be $+1$, 0, -1, depending, respectively, on whether the model we choose has a three-dimensional physical space of constant positive, zero, or negative curvature. The Einstein equations then reduce to

$$(12.7a) \qquad \frac{\epsilon}{r(t)^2} + \frac{r'(t)^2}{r(t)^2} + 2\,\frac{r''(t)}{r(t)} = 0$$

$$(12.7b) \qquad \frac{\epsilon}{r(t)^2} + \frac{r'(t)^2}{r(t)^2} - \frac{1}{3}\,g\rho = 0$$

In these equations $r'(t)$ denotes the derivative of $r(t)$ with respect to time t and $r''(t)$ denotes the second derivative with respect to t.

Suppose that we set $r'(t)/r(t) = h(t)$ by definition. This will simplify the form of the equations (12.7a,b), but more importantly $h(t)$ has an interesting physical meaning. It represents the rate of change of the radius per unit of radial distance. To understand what this means let us suppose that the spatial universe is the surface of a balloon (that is, two-dimensional instead of three) and that the stars are points attached to the surface in fixed positions. If the balloon were to expand, its radius would increase with time and the stars on its surface would move farther and farther away from one another. An observer located on or near one of the stars would conclude that all the other stars were receding from him and that his universe was *expanding;* he clearly would not be justified in concluding that his star or planet was distinguished by being at the center of the universe in any sense, for all observers in the universe would observe the same phenomenon. On the other hand, it seems clear that the stars nearby would appear to be retreating at one rate, whereas those far off would appear to be retreating at much greater rates.

Indeed, the distance $D(t)$ between stars A and B in Figure 12.12 is just given by $D(t) = r(t)\theta$, where θ is the central angle subtended by the stars. Therefore the speed of recession of B from A is the derivative, $dD/dt;$ since θ does not depend on time, $dD/dt = \theta\, dr/dt = \theta\, r'(t)$. This states that the recession speed is proportional to the angle θ; thus the stars that are initially more distant from star A seem to recede more rapidly. To eliminate the dependence of this result on the locations of particular stars, which enters the equations through the angle θ, we divide the equation by $D(t)$. This gives $D'(t)/D(t) = \theta\, r'(t)/D(t) = r'(t)/r(t)$ since $D(t) = \theta\, r(t)$, and shows that the ratio D'/D is the same as the ratio r'/r, which we called $h(t)$; hence the meaning of $h(t)$ is that it *measures the mutual recession of stars in terms of the radius of the universe.* The quantity D'/D, which is the rate of recession of the stars divided by their distance from us, is known as *Hubble's expansion;* it was discovered and measured by the astronomer Edwin Hubble about 1936.

Keeping in mind that Hubble's expansion $h(t)$ has been measured experimentally, we express Einstein's equations (12.7a,b) in terms of this quantity to see what they imply

FIGURE 12.12

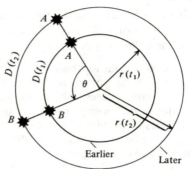

A sectional view of two instants in the life of an expanding spherical universe.

about the global properties of the universe, that is, about the cosmological models we are studying. After some elementary simplifications (12.7a,b) become:

(12.8a)
$$\frac{\epsilon}{r^2} + 2h'(t) + 3h(t)^2 = 0$$

(12.8b)
$$\frac{3\epsilon}{r^2} + 3h(t)^2 = g\rho$$

Our further study of these equations will be greatly simplified if we suppose that $\epsilon = 0$ first, for the leftmost term in each of the equations in (12.4) vanishes, leaving

(12.9)
$$2h'(t) + 3h(t)^2 = 0$$

(12.10)
$$3h(t)^2 = g\rho$$

You can easily check that the solution of the differential equation (12.9) is just

(12.11)
$$h(t) = \frac{2}{3(t - t_0)}$$

where t_0 is a constant that corresponds to a choice of the origin of time measurement. If we take $t_0 = 0$, then (12.11) becomes

(12.12)
$$h(t) = \frac{2}{3t}$$

Time in this equation is a measure of the *age of the universe*, for when it is zero in (12.12), $h(0)$ makes no sense, and therefore physical time cannot be extended back to negative values. Equation 12.12 states that as time passes the ratio represented by Hubble's expansion decreases. Of much more interest is the fact that since Hubble's expansion is known from observational measurements for the present time, by substituting its observed value in (12.12) we find a completely explicit and specific estimate of the age of the universe; it turns out to be about 18 billion years (see [67] for a detailed discussion of various age estimates and [83] for the most recent result).

It is an amazing consequence of the model — a purely geometrical one — that it should tie the age of the universe to a simple measurable quantity like the Hubble expansion; it is also an excellent example of the power of abstract geometrical models of physical systems. But before becoming overjoyed by this result it will be useful to compare it with other ways of estimating the age of the universe. We also must note that the initial geometrical and observational assumptions that we made to obtain this result cannot have been valid during the early life of the universe; there is no reason to suppose that it would have appeared smooth and the same to all observers just after its "creation," that is, for very small values of time. What the phrase "age of the universe" must mean in our context is the amount of time that the universe has endured in a condition of spatial symmetry such that all observers *do* obtain the same view of its spatial aspect. One other remark that makes matters simpler: to obtain the age given above (18 billion years) it was assumed that space was Euclidean: $\epsilon = 0$. This simplified the equations, but even if ϵ is not zero it is possible to work out $h(t)$ as a function of t (perhaps by using a computer for $\epsilon = -1$). We then find an age that is still about 18 billion years, although not precisely the same as that which followed from the Euclidean assumption.

There are two other ways to estimate the age of the universe. Actually, they provide lower bounds for the age. One technique, of dubious validity, requires calculating how long it would have taken the galaxies to assume their present general relative configuration in space based on certain assumptions about their configuration at the beginning of the expansion of the universe. This intricate scheme leads to an age in the same general range as that determined above from the cosmological model, but the details are much too complex to be discussed here. The second method concerns the decay of uranium. From studies of uranium found on earth its law of decay is precisely known. Furthermore, studies of the spectral properties of the light from stars show that uranium is found in them, and the relative amounts can be determined for stars of various types. The suggestion is that uranium was produced at the same time as the other elements; this provides a means for estimating the age of uranium and therewith of the universe constituted essentially as we now know it. The result lies between 7 and 15 billion years, which also agrees in general with the cosmological model's results.

It is interesting to recall that geological estimates of the earth's age fall between 2 and 4 billion years; the earth is not a newcomer on the celestial stage.

§12.8. In December 1968, man freed himself for the first time from the gravitational dominance of the earth when the Apollo 8 spaceship circumnavigated the moon. Spaceflight is in a primitive stage of development. The energy required to move freely in the gravitational currents of space is currently stored in chemical fuels that are bulky and massive; therefore only a limited amount of powered flight is possible, and for most of its space journey the spaceship must ride the geodesics of space, which do not require the expenditure of energy. This means that the ability to navigate with great precision is a precious skill; if the geodesic from the point of departure to the point of arrival can be precisely calculated, the amount of fuel that must be used can be minimized, which makes longer and more complex journeys possible. For instance, the flight of Apollo 8 consisted of five parts. First, energy had to be expended to lift the spacecraft from the surface of the earth into earth orbit; second, additional energy was expended to propel the orbiting ship into a space-time geodesic that passed close to the surface of the moon. While on this leg of the flight no energy had to be expended, just as no energy was required to maintain motion along the earth-orbit path, which also is a geodesic. On reaching the vicinity of the moon, a small amount of energy was used to change the path of motion from the earth-to-moon geodesic to a new geodesic which orbited the moon. To return to earth this sequence of events was repeated in inverse order: from moon-orbit geodesic to moon-to-earth geodesic to earth-orbit geodesic, thence back to the earth's surface.

Since geodesics in the neighborhood of the sun are nearly straight, but earth and moon orbits are, in three-dimensional space, almost circular, and even in slightly deformed three-dimensional space would not be geodesics, it is clear that the simplest way to view spaceflight is from the four-dimensional space-time aspect. The situation is similar to that shown in Figures 12.8 and 12.10, but now, since the mass of the moon is a significant fraction of the mass of the earth, we have to decide just how the coordinate axes of the graphic representation should be chosen so that the resulting "map" is simple. One convenient way is to suppose that the time axis coincides with the path of the more massive body — the earth. Seen from the earth, the moon appears to orbit about it; hence its path in the diagram will be a helix about the time axis, as in Figure 12.13. We could calculate the pitch and radius of the moon's helix easily enough; the radius is the same as the distance from the earth to the moon (for the sake of simplicity suppose it is constant; that is, suppose that the moon travels in a circle as seen from the earth) and so was already known to Aristarchus, at least approximately; but space-time coordinates require that distances and times be measured in compatible units; therefore this distance measurement should be converted into terms of the time light would require to pass between the earth and the moon, which works out to about 1.29 seconds. The time required for the moon to traverse one full loop of the helix is a lunar month — about 28 days. Hence the pitch of the moon's helix about the earth is about (28 days)/(1.29 seconds) = approximately 2,390,000. On p.

FIGURE 12.13

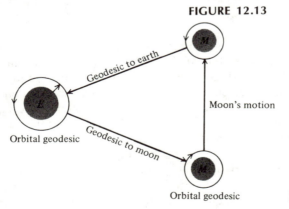

Diagram of Apollo 8 voyage.

332 our computation showed that the pitch of the helix of the earth as it moves about the sun is only about 65,700; this means that the helix corresponding to the moon in motion about the earth is much more nearly a straight line than the helix of the motion of the earth about the sun. The reason is simple: the gravitational field of the sun is much stronger than that of the earth, and therefore it distorts the space-time manifold more than the earth can, which in turn makes the geodesics less similar to straight lines.

To travel from earth orbit to moon orbit without expending energy along the way we must find a geodesic that passes through the earth-orbit path and the moon-orbit path; then enough energy must be expended to transfer the spacecraft from the orbital geodesic to that passing from one body to the other. To simplify the rest of the discussion assume that we are seeking a geodesic that passes through the center of the earth and the center of the moon; forget about the orbits for the moment. Then what is wanted is a geodesic that passes from the time axis, which is the path of the earth in space-time, to the nearly straight helix which is the path of the moon in space-time; this situation is described in Figure 12.14 in which a possible geodesic links the two bodies. If the calculations used to determine the geodesic are not completely accurate, it will not be the right one; instead of passing to the moon, it might only come near it, and a *mid-course correction* must be made; that is, some additional energy must be expended to move the spacecraft from the nearly correct geodesic to the correct one.

FIGURE 12.14

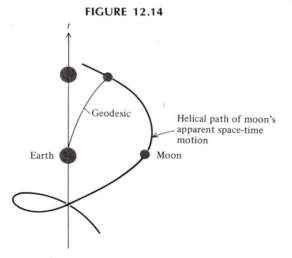

Notice that the problem of finding the proper geodesic is really a task for differential geometry. Looked at from another point of view, it is just the same as the traditional problem of navigating a ship from one point on the earth's surface to another in a tolerably efficient manner. Navigation and differential geometry are inextricably intertwined. Moreover, the space-time diagrams we have been drawing are really *maps* of the geometrical situation designed to show the important navigational features in a convenient representation. Man has not traveled enough in space to have found the map that is most convenient; our space-time diagrams are neither as simple nor as efficient as the Mercator projection was for its purpose. Nor, of course, are the diagrams we have drawn entirely accurate; they are gross representations that do not take into account the subsidiary effects of the sun and other planets. More accurate maps will undoubtedly result from surveying activities as travel in space becomes popular and systematic.

The analogy between fifteenth-century exploration of the earth's surface by seamen navigating the vast reaches of the oceans and twentieth-century exploration of the solar system by astronauts navigating the still more immense reaches of space-time is surprisingly accurate in its technical aspects. It also suggests what the effect on civilization of this enormous and inspiring effort may be. The major consequences of fifteenth-century explorations of the earth were not the technological improvements that filtered down to "the household kitchen" nor were they the discovery of natural resources such as gold and spices.

§**12.9.**   Einstein's model of the universe is surely not the last significant step in our understanding of the geometrical nature of space, but it is an important one in the evolution of man's mathematical mastery of the universe. From Babylonian arithmetical models, through the synthetic geometrical models of Aristarchus and Ptolemy, to the analytical calculus-based model of Newton, and on to the differential geometrical model of Einstein, the ideas of mathematical scientists have influenced the entire civilized world as the consequences of their models have been studied and explored in intricate detail. The next significant step awaits the efforts of some unknown young scientist, perhaps a reader of this book, whose imagination and daring will add yet another name to the roll of *memorable mathematicians.*

EXERCISES

12.1. What is a "mathematical model" of the physical world?

12.2. Describe briefly three different models of the universe, indicating their nature, origin, and utility.

12.3. Why doesn't the Babylonian model of the universe adequately explain an eclipse, even though it can accurately predict its time of occurrence and duration?

12.4. Let $P(t) = (x(t), y(t), z(t))$ be the path of a particle in a Newtonian model of the universe. Suppose that the coordinates are measured in meters and no forces are acting on the particle. Moreover, suppose that at time $t = 0$ the particle is at the origin $(0, 0, 0)$ and moving along the x-axis with a velocity of 2 meters/second.
(a) What equation must $(x(t), y(t), z(t))$ satisfy?
(b) Using the data above, write down the solution of the equations. [*Hint.* The general solution will be of a form similar to that given in (12.3) without, however, a force of gravity acting.] The specific value of the constants will be determined by the data above; that is,

$$x(0) = y(0) = z(0) = 0$$

$$\frac{dx}{dt}(0) = 2 \qquad \frac{dy}{dt}(0) = 0 \qquad \frac{dz}{dt}(0) = 0$$

(c) How far will the particle have gone in 10 seconds?

12.5. (a) Is the situation described in Problem 12.4 a realistic one for particles in the everyday world? Why?
(b) Is the situation described in the text [(12.2) and (12.3)] realistic? Why?

12.6. Suppose a 10-pound ball is dropped from a tower 100 feet high. Given that the gravitational constant $g = 32$ feet/second^2, how long will it take the ball to reach the ground? [*Hint.* Use the solution of Newton's equations given by (12.3).]

12.7. In Problems 12.4 and 12.6 the path of the moving object is a straight line, which is a geodesic in the Newtonian model. Yet in one case a force is acting and in the other it is not. How do these motions differ?

12.8. In Problem 12.6 suppose that the ball had been thrown out from the tower in the direction of the x-axis at a speed of 10 feet/second.
(a) Along what kind of a curve would the ball travel?
(b) How long would it take to reach the ground?

12.9. In Einstein's model of the universe what plays the role of Newton's forces that causes particles (or astronomical bodies) to move as they do?

12.10.   Give examples of situations in which it is more practical to use the
        (a) Babylonian model of the universe
        (b) Newtonian model of the universe
        (c) Einsteinian model of the universe
        Give brief reasons for your choices.

12.11.   Why do the three tests of Einstein's theory show that his relativity model is a better model of the universe than Newton's?

12.12.   In the "tabletop" model of a relativistic universe describe the motion of a particle approaching the sun from a great distance. (*Hint.* There are several distinct possibilities.) Is it possible that such a particle can go into a closed orbit about the sun? Give examples of astronomical objects that behave in the manner you describe in this exercise.

12.13.   Newton's model has been used to describe the motion of astronomical objects in the solar system for more than 250 years. Its results are known to be generally accurate. What does this imply about the size of the curvature of the four-dimensional universe in the vicinity of the solar system?

12.14.   Do you think the universe is *finite* and *bounded* (like the surface of a ball)? Why? What are the alternatives?

SUPPLEMENT
Twentieth-Century Mathematics

§S.1.   The development and growth of mathematics has continued unabated into the twentieth century, and its impact on civilization has become increasingly profound. It takes little reflection to recognize the role played by calculation in the establishment of flight paths for Apollo moon explorers and the still more complicated Mariner and Voyager spacecraft journeys to Mars and the outer planets. Yet these examples of the use of mathematics by civilization are mere superficial reflections of a more subtle and pervasive mathematical presence which interweaves our comprehension of the structure of the physical universe, of that peculiarly human creation, language, and of the nature of thought itself.

The two main strands of mathematical progress which were used to tie the accomplishments of mathematicians to the needs of civilization in the previous periods covered by this book retain their significance for twentieth-century developments. Twentieth-century studies of the *geometrical nature of space* include, amongst other advances, the creation of a comprehensive understanding of *infinite dimensional spaces* which, in their particular realization as *Hilbert spaces*, provide the mathematical foundation upon which *quantum theory* was built. This greater understanding has led to the command of physical reality so clearly exemplified by the mushrooming development of high-speed microelectronic computers whose physical operation depends upon processes which occur during time intervals and within spatial volumes characteristic of quantum phenomena, effects totally unsuspected during the previous century. The microelectronic computer, the laser and the nuclear power reactor are three striking examples of the symbiotic growth of mathematics and our ability to adapt nature to our desires during the twentieth century.

Although the physical operation of modern computers owes everything to the mathematical study of infinite dimensional spaces which underlies quantum theory, the logical organization of the physical components which constitute the computer and the applications to which the computer is set are practical consequences of a far-reaching advance along the second of our two major themes: the development of the *ability to compute*. This advance has roots in Aristotle's attempt to formalize the processes of reason and in the concepts of proof and axiomatic method first introduced by the Greek mathematicians and embodied in Euclid's *Elements of Geometry*, Greek antiquity's authentic intellectual monument. Attempts to formalize proof and reason continued through centuries of fitful development until, at the end of the nineteenth century, the problem of the nature of proof was transformed and reformulated in such a way that the twentieth-century mathematical logicians, especially Kurt Gödel, were able to make decisive and shocking advances of which their predecessors in previous times had not a glimmer of anticipation.

The word "shocking" is rarely found in a book about mathematics, yet it seems to be the only one that corresponds to the situation. Pre-twentieth-century science developed and became increasingly mathematical. As technology became able to translate the new understanding of the nature of physical reality into ever more complex adjuncts to daily life, the belief in the power of reason and in mathematics as the epitome of reason realized as a structured discipline, continued to grow. It seemed to many that if only one could accumulate all the observations, the future of the mechanical universe could be predicted by what amounted to an (admittedly impractically large) calculation. It was thought that, in principle at least, the course of the physical universe was completely understandable, even mostly understood, and that what remained was merely to collect more data and mathematically work out the details. In particular, it was generally accepted that every true mathematical statement could be proved, although the proof of some statements might require considerably more effort and ingenuity than others.

These views were consequences of the great success of Newton's theory of physical phenomena embodied in his famous three laws of motion and law of universal gravitation. They were given substance by the mathematical methods of calculus, a tool which grew so powerful as to be apparently capable of the solution of any physical problem. This was the predominating viewpoint until the last quarter of the nineteenth century when two quite different lines of investigation emerged, each of which signaled that neither physical theories of nature nor mathematical methods could any longer be accepted as complete descriptions. Another fifty years were to pass before, in the middle third of the present century, it was fully understood that neither the corporeal reality of nature nor the abstract reality of mathematics can be completely known — there are physical quantities whose values cannot be measured and there are true mathematical statements which cannot be proved. In natural philosophy this essential incompleteness of knowledge is usually called the *Uncertainty Principle* of Heisenberg; in mathematics it is known as the *Incompleteness Theorem* of Gödel.

These facts carry the connotation that there are veils through which our gaze shall never pass, mysteries which must remain forever unsolved — that mathematics and the disciplines it supports have an essential and unremediable limitation, which some may think is a vital flaw. These are indeed shocking statements. Yet we are simultaneously exposed to an apparent contradiction, for in the years which followed recognition of the limits of physical science and mathematical methods, mathematics and science advanced at an even more hectic pace, and our ability to predict and control the course of nature and to resolve the open problems of mathematics grew far more powerful and effective than before! Order out of uncertainty and precision from incompleteness characterize the mathematical advances of the twentieth century.

§S.2. We are going to give a brief description of some of the main discoveries of twentieth-century mathematics and discuss their interconnections, concentrating first on the geometrical theme. The concept of an *infinite dimensional space* was considered during the nineteenth century, but these spaces and transformations of them were first systematically and successfully investigated during the early decades of the present century, and are still actively studied today. We can picture two-dimensional space as the usual Euclidean plane equipped with coordinates (x, y) of its points. Points in three-dimensional Euclidean space have corresponding coordinates (x, y, z). In the discussion of general relativity in Chapter 12, it was useful to introduce four-dimensional space, whose points have coordinates (x, y, z, t) or (x_1, x_2, x_3, x_4), where x_1, x_2, x_3, x_4 are independent real numbers. Thus *ordered sets of numbers* can be used to *define* the concept of point for higher dimensional spaces. The *vector* from the origin of a coordinate system to a point can be identified with that point and hence with its coordinates. Vectors can be added by the parallelogram rule illustrated in Figure S.1.

FIGURE S.1

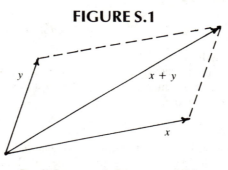

Parallelogram rule for vector addition.

Because vectors can be identified with points, addition of them is equivalent to addition of the corresponding coordinates of the vector summands. A vector can also be multiplied by a number, thereby changing its length and/or direction (cp. Figure S.2).

FIGURE S.2

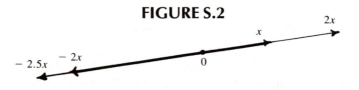

Multiplication of a vector by various numbers.

Now consider the set of all *sequences* of real numbers, $V = \{(x_1, x_2, \ldots, x_n, \ldots)\}$. We will call V a *space* and each sequence in it a *point*. V is an example of an *infinite-dimensional space*. Two sequences in V can be added by adding corresponding entries, thus

$$(x_1, x_2, \ldots, x_n, \ldots) + (y_1, y_2, \ldots, y_n, \ldots) =$$
$$(x_1 + y_1, x_2 + y_2, \ldots, x_n + y_n, \ldots)$$

and multiplied by scalars (i.e., by numbers)

$$a(x_1, x_2, \ldots, x_n, \ldots) = (ax_1, ax_2, \ldots, ax_n, \ldots)$$

Any set of mathematical objects which possesses a *vector addition* and *multiplication by scalars* satisfying certain natural conditions which generalize those which are valid for vectors in the usual two- or three-dimensional Euclidean space is called a *vector space*. The space will be infinite-dimensional if there are an "infinite number of degrees of freedom," a concept which was given precise meaning around the turn of the century.

The Euclidean *scalar product* of two vectors in three-dimensional space $\mathbf{R}^3$, given by the formula

$$\mathbf{x} \cdot \mathbf{y} = (x_1, x_2, x_3) \cdot (y_1, y_2, y_3) = x_1 y_1 + x_2 y_2 + x_3 y_3,$$

includes as a special case the square of the Euclidean *length* of vector $\mathbf{x}$: set $\mathbf{y} = \mathbf{x}$ to obtain

$$\mathbf{x} \cdot \mathbf{x} = x_1^2 + x_2^2 + x_3^2, \quad \text{usually denoted by } \|\mathbf{x}\|^2.$$

This formula leads to a definition of the length of a vector in an infinite-dimensional space. For the infinite-dimensional space V of sequences introduced above, consider the subset of all sequences for which the sum of the squares of the components of a vector converges, that is, consider

$$l^2 = \{(x_1, x_2, \ldots, x_n, \ldots) : x_1^2 + x_2^2 + \ldots + x_n^2 + \ldots < \infty\}$$

It turns out that this subset of V is itself an infinite-dimensional vector space which has a concept of length defined for each of its vectors. It is called a *Hilbert space* after the German mathematician David Hilbert (1862–1943). The scalar product of two vectors in l^2 can be defined by

$$\mathbf{x} \cdot \mathbf{y} = x_1 y_1 + x_2 y_2 + \ldots + x_n y_n + \ldots$$

If $\mathbf{x} = (x_1, x_2, \ldots, x_n, \ldots)$ and $\mathbf{y} = (y_1, y_2, \ldots, y_n, \ldots)$, and the length of $\mathbf{x}$ is

$$\|\mathbf{x}\|^2 = x_1^2 + x_2^2 + \ldots + x_n^2 + \ldots$$

the sums converge to finite numbers.

Another important example of a Hilbert space is the set L^2 of square integrable functions defined on the unit interval $[0, 1]$, i.e., those functions f which satisfy

$$\int_0^1 (f(x))^2 \, dx < \infty$$

Unbounded functions will belong to this space, but only those whose square encloses a finite area under its graph. An example of such a function is $f(x) = x^{-1/4}$, since

$$\int_0^1 (f(x))^2 \, dx = \int_0^1 x^{-1/2} \, dx = \lim_{\epsilon \to 0} \int_\epsilon^1 x^{-1/2} \, dx = \lim_{\epsilon \to 0} 2x^{1/2} \Big|_\epsilon^1 = 2$$

If f and g belong to L^2, then their scalar product is defined by

$$f \cdot g = \int_0^1 f(x)\, g(x)\, dx$$

A function can be multiplied by a scalar (i.e., a number), and two functions can be added in the usual way. Hence, L^2 is an infinite-dimensional vector space with a scalar product, that is, a Hilbert space. To make these conclusions entirely rigorous one must employ a theory of *integration of functions* created by the French mathematician Henri Lebesgue (1875–1941) in 1900. The theory is more comprehensive than the one initiated by Archimedes and formalized in the mid-nineteenth century by Riemann (*cp.* Chapters 4 and 5).

Recall from Chapter 8 that a quadratic function of two real variables x and y of the form

$$Q(x, y) = ax^2 + bxy + cy^2$$

is the algebraic equivalent of a geometrical conic section, i.e., it corresponds to an ellipse, a hyperbola or a parabola. Pierre de Fermat (1601–1665) discovered an important property of this quadratic function called the *Principal Axis Theorem*, which asserts that there are vectors **u** and **v** in $\mathbf{R}^2$ which point along the principal axes of the conic section, and there are real numbers α and β which represent the elongation of the curve in these directions (Figure S.3); in formulae,

FIGURE S.3

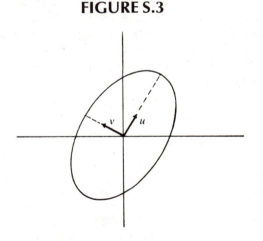

Principal axes of an ellipse.

$$Q(\mathbf{u}, \mathbf{u}) = \alpha \|\mathbf{u}\|^2, \quad Q(\mathbf{v}, \mathbf{v}) = \beta \|\mathbf{v}\|^2$$

Here α and β represent the amount of stretching or compression of the vectors **u** and **v** respectively along the principal axes, and $\|\mathbf{u}\|$ and $\|\mathbf{v}\|$ denote the length respectively of **u** and **v**. α and β are called *characteristic values* or *eigenvalues* of the quadratic function Q, and $\mathbf{u}$ and $\mathbf{v}$ are called its *characteristic vectors* or *eigenvectors*. The set of eigenvalues of the quadratic form is its *spectrum*. We see that the spectrum of a quadratic function Q of two variables consists of precisely two real numbers, as shown by our example. The far-reaching generalization of these simple ideas to quadratic functions of vectors which belong to infinite-dimensional Hilbert spaces has led to a remarkably rich and complex *spectral theory* which describes their eigenvalues and eigenvectors. This theory was primarily the work of Hilbert and his school during the second and third decades of this century.

Although the spectrum of a quadratic function on a Hilbert space can in general be very complicated, the best-behaved examples mimic the behavior of the spectrum of the quadratic function of two variables described above: there is an infinite dimensional version of the Principal Axis Theorem which involves an infinite number of eigenvectors $\mathbf{v}_1, \mathbf{v}_2, \ldots, \mathbf{v}_n, \ldots$ and an infinite number of discrete eigenvalues $\alpha_1, \alpha_2, \ldots, \alpha_n, \ldots$ which satisfy the infinite collection of equations

$$Q(\mathbf{v}_1, \mathbf{v}_1) = \alpha_1 \|\mathbf{v}_1\|^2, \quad \ldots, \quad Q(\mathbf{v}_n, \mathbf{v}_n) = \alpha_n \|\mathbf{v}_n\|^2, \quad \ldots$$

where $\|\mathbf{v}\|^2 = \mathbf{v} \cdot \mathbf{v}$ is the squared length of the vector $\mathbf{v}$ in the infinite-dimensional space. In 1925–26 Werner Heisenberg (1901–1976) and Erwin Schrödinger (1887–1961) independently found different specific Hilbert spaces and associated quadratic functions Q which enabled them to explain the experimentally verified fact that the possible states of an electron in an atom (which, in the present context, can be thought of as electron orbits about the central atomic nucleus) are restricted to a well-defined discrete set characterized by specific energy levels. The levels turned out to correspond to the eigenvalues of the quadratic function and the differences of pairs of eigenvalues to the experimentally observed frequencies of the spectral lines of light radiated by thermally excited atoms (Figure S.4). This remarkable discovery opened the path in the decade after 1926 to the development of the highly successful theory of atomic processes called *quantum mechanics*.

FIGURE S.4

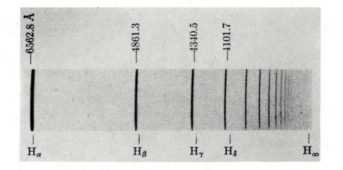

Spectrum of the hydrogen atom (Balmer series).

In order to simplify the discussion let us suppose that the physical system of interest is free to vary only along the x-axis. In Schrödinger's representation of quantum mechanics the state of the physical system (say an electron bound to the nucleus of an atom) is described by a function of position ψ in the sense that $\bar{\psi}\psi dx$ is the probability that the electron will be found in the "infinitesimal volume" dx (cp. Chapter 9) centered at x (here, as always, $\bar{\psi}$ is the complex conjugate of ψ). The possible state functions ψ belong to L^2 (but with the functions permitted to take on values which are *complex numbers*). The quadratic function $Q(\psi)$ is

(S.1)
$$Q(\psi) = \int \bar{\psi}\left(-\frac{\hbar^2}{2m}\frac{d^2}{dx^2} + V(x)\right)\psi\, dx$$

where $\hbar$ is a universal constant ("Planck's constant divided by 2π," approximately equal to 1.05×10^{-27} erg sec) and $V(x)$ is the *potential function* associated with the system (not to be confused with the vector space V). For an electron bound to an atomic nucleus, the potential function represents the electrostatic attraction between the nuclear protons and the orbiting electron.

§S.3.
Another feature of the quantum mechanics is that all observable quantities correspond to linear operators which carry state functions in L^2 onto other vectors. In particular, the function in (S.1) above corresponds to the observable *energy* and leads to a linear operator which transforms one state function into another as follows: the numerical valued quadratic function can be expressed in the form

$$Q(\psi) = \int \bar{\psi} Q\psi\, dx$$

where Q is an *operator* which transforms one vector into another and generally is an expression which involves differentiation operators and multiplication by ordinary functions. In the particular example above, the operator Q which corresponds to the quadratic function $Q(\psi)$ defined by (S.1) is

$$Q = -\frac{\hbar^2}{2m}\frac{d^2}{dx^2} + V(x)$$

and Q transforms the vector ψ into the vector

$$Q\psi = -\frac{\hbar^2}{2m}\frac{d^2\psi}{dx^2} + V(x)\psi$$

The eigenvalues α and eigenvectors ψ of the quadratic function $Q(\psi)$ are precisely those numbers and vectors which satisfy the operator equation

$$Q\psi = \alpha\psi$$

and α is an observable and measurable value of the physical variable represented by Q.

If it is possible to simultaneously measure the values of the physical variables represented by operators, say Q_1 and Q_2, no matter what the state of the given physical system, then according to what has been said above, $Q_1\psi = \alpha_1\psi$ and $Q_2\psi = \alpha_2\psi$, with α_1 and α_2 the respective eigenvalues. Hence

$$Q_1Q_2\psi = Q_1(\alpha_2\psi) = \alpha_2(Q_1\psi) = \alpha_2\alpha_1\psi$$

and

$$Q_2Q_1 = Q_2(\alpha_1\psi) = \alpha_1(Q_2\psi) = \alpha_1\alpha_2\psi$$

so $Q_1Q_2\psi = Q_2Q_1\psi$, from which it follows that

(S.2) $$Q_1Q_2 - Q_2Q_1 = 0$$

Operators which satisfy this relation are said to *commute*.

Conversely, if (S.2) holds, then it can be shown that the physical variables corresponding to Q_1 and Q_2 can be simultaneously measured. This is the content of the *Heisenberg Uncertainty Principle:* physical states which correspond to noncommuting operators cannot be simultaneously observed. An example of a pair of physical variables whose values cannot be simultaneously measured are the *position* and *momentum* of a particle. The operator which corresponds to the former multiplies a state vector by $x: \psi \to x\psi = Q_1\psi$, and the operator which corresponds to momentum is proportional to the operator of differentiation: $\psi \to d\psi/dx$. These operators do not commute because

$$(Q_2Q_1 - Q_1Q_2)\psi = \frac{d(x\psi)}{dx} - x\frac{d\psi}{dx} = \psi$$

i.e., as operators,

$$Q_2Q_1 - Q_1Q_2 = 1 \text{ (multiplication by 1)}.$$

This astonishing feature of quantum mechanics is the reason that algebras of noncommuting operators play an important role in physics.

§S.4.   The rule which associates to the pair of operators (Q_2, Q_1) the new operator $Q_2Q_1 - Q_1Q_2$ can be thought of as a kind of multiplication rule for operators. This product (which is usually denoted by $[Q_2, Q_1]$) together with the usual rule for addition of operators turns the set of operators into a particular example of a *Lie algebra* (named in honor of the Norwegian mathematician Marius Sophus Lie, 1842–1899). The operators which correspond to the observable features of physical systems belong to certain Lie algebras whose structure reflects symmetries of the corresponding physical system. From a mathematical standpoint, the quantum mechanical description of a particular physical system amounts to the prescription of a certain Lie algebra whose structure embodies "laws of nature," and the further prescription of a certain infinite dimensional Hilbert space whose vectors are the various possible state vectors of the physical system. The problem of determining the observable consequences of the laws of nature is converted, in this mathematical model, into a pair of purely mathematical problems. The first is to determine all possible explicit realizations (technically called *representations*) of the Lie algebra of the physical system as a Lie algebra of operators which act on the vectors of some Hilbert space. Each realization (or representation) corresponds to a possible physical system whose behavior is governed by the natural laws embodied in the Lie algebra structure, and it can be built up from a collection of particular representations which are said to be *irreducible*. The irreducible representations are the building blocks from which the various possible systems are composed. Remembering that physical variables correspond to operators, the observable values of physical variables turn out to be the eigenvalues of the operators which correspond to them in the various representations.

The study of the representations of Lie algebras is primarily a twentieth-century accomplishment although its roots can be traced to the last quarter of the nineteenth century. Élie Cartan (1869–1951) discovered the close connection between differential geometry and Lie algebras, but he would no doubt have been astonished to learn that his classification of the possible structures of Lie algebras has been used by physicists to describe the relationships among particular collections of elementary particles. The work of Hermann Weyl (1885–1955) clarified the connection between symmetry in physical systems and Lie algebra structures and showed how the various representations can be constructed in a systematic way. Both the mathematical investigations of the representations of Lie algebras and their application to the description of nature continue to flourish at the present moment.

The complex and subtle mathematical theories of infinite-dimensional spaces and representations of Lie algebras provide the machinery in terms of which contemporary theories of atomic and nuclear physics are expressed. Their influence on ordinary life is daily increasing as machines based on quantum phenomena become universal companions. In electronic calculators and watches, in home computers and laser video playback entertainment systems, and soon, no doubt, in superconducting electrical power transmission systems, this mathematical knowledge finds physical realization and demonstrates the bond between the practical and the abstract.

FIGURE S.5

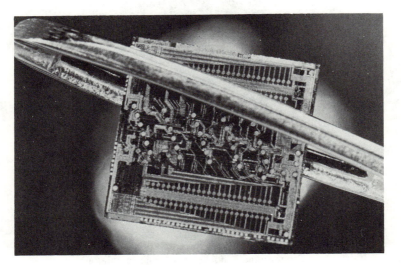

Large-scale integrated (LSI) electronic circuit chip.

§S.5. Now we turn to our second theme, the *ability to compute*. In a few years the pocket electronic calculator has developed from a machine only able to perform the four elementary arithmetic operations and the evaluation of the elementary transcendental functions (that is, the trigonometric and exponential functions and their inverses) to a much more powerful and general device which can be programmed to perform a vast variety of calculations requiring large memory stores and including decision-making operations such as branching instructions. This is visible and ubiquitous evidence that, in the common sense of the phrase, our ability to compute has taken a great leap forward. But this view is misleading, for apart from their capability of being programmed, pocket calculators compute the same functions, and by means of essentially the same algorithms, that were previously known and used. In many cases the computational methods were already known in the eighteenth century. The calculating machines merely compute much faster and with less error than a person can. The feature of programmability, however, reflects more recent and subtle mathematical achievements whose implications are novel and important and suggest ideas about the nature of thought itself.

FIGURE S.6

Programmable pocket electronic calculator.

What is a *calculation*? It can be thought of as a particular kind of *proof* and its result as a *theorem*. A proof is merely a sequence of statements, each of which is either a hypothesis (that is, a statement which is assumed to be true) or a theorem, or is derived from the previous statements of the proof by means of *rules of logic*. These rules are formal statements of the usual reasoning processes used by all rational people. For instance, one of them (number 2 in Figure S.7) states that if a statement or proposition P is true, and if the proposition "P implies Q" is true, then the statement Q is necessarily true. Since each step in a calculation follows from those that precede it by using arithmetic and some statements which are already known to be theorems, it is clear that a calculation is really a special kind of proof.

FIGURE S.7

(Adapted from S. C. Kleene,
Introduction to Metamathematics, D. van Nostrand, 1950)

HYPOTHESES FOR THE PROPOSITIONAL CALCULUS

1a. $P \Rightarrow (Q \Rightarrow P)$ 1b. $(P \Rightarrow Q) \Rightarrow ((P \Rightarrow (Q \Rightarrow R)) \Rightarrow (P \Rightarrow R))$

2.   If P and $P \Rightarrow Q$, then Q
3. $P \Rightarrow (Q \Rightarrow P \wedge Q)$
4a. $P \wedge Q \Rightarrow P$ 4b. $P \wedge Q \Rightarrow Q$
5a. $P \Rightarrow P \vee Q$ 5b. $Q \Rightarrow P \vee Q$
6. $(P \Rightarrow R) \Rightarrow ((Q \Rightarrow R) \Rightarrow (P \vee Q \Rightarrow R))$
7. $(P \Rightarrow Q) \Rightarrow ((P \Rightarrow \sim Q) \Rightarrow \sim P)$
8. $\sim \sim P \Rightarrow P$

ADDITIONAL HYPOTHESES FOR THE PREDICATE CALCULUS

"x" denotes a variable and "c" denotes a constant. "$P(x)$" denotes a proposition which depends on the variable x.
"$\forall x$" denotes "for all x;"
"$\exists x$" denotes "there exists an x such that."

9.   If $R \Rightarrow P(x)$, then $R \Rightarrow (\forall x)P(x)$
10. $(\forall x)P(x) \Rightarrow P(c)$
11. $P(c) \Rightarrow (\exists x)P(x)$
12. If $P(x) \Rightarrow Q$, then $(\exists x)P(x) \Rightarrow Q$

Only relatively recently have mathematicians and logicians become aware that every proof can also be considered as a kind of calculation, so that there is really no difference between the two concepts. Indeed, all of the symbols which might be used in the logical statements that occur within a proof can be coded by numbers, say sequences of zeros and ones, as is commonly done when symbols such as alphabetic letters are put into computers. Furthermore, the operations which correspond to applications of the rules of logic can be realized as calculations which involve the numbers associated with the logical statements. George Boole (1815–1864) appears to have been one of the first to reduce a proof of certain types of propositions to calculations in essentially this way. The procedures he conceived are now in common use by electrical engineers who design computing machinery circuits.

Boole's idea can be described in the following way. Consider any propositional statement P composed of other statements $P_1, P_2, \ldots, P_n$ connected (in any way consistent with the rules of logic) by the logical operations "and," "or," "not" and "implies." These operations are usually denoted by special symbols, thus: "$P \wedge Q$" for "P and Q," "$P \vee Q$" for "P or Q," "$\sim P$" for "not P," and "$P \Rightarrow Q$" for "P implies Q." If we know whether each of the propositions $P_1, P_2, \ldots, P_n$ is true, we will be able to determine whether the combined statement P is true by applying the rules of logic, i.e., we will be able to determine whether P is a theorem given the hypotheses $P_1, P_2, \ldots, P_n$. In order to convert this process into a calculation, let us define a numerical valued *truth function* t such that for any proposition P, $t(P) = 1$ if P is a true statement, otherwise $t(P) = 0$. Then it is easy to check that for any propositions P and Q,

$$t(P \wedge Q) = t(P)t(Q)$$

and

$$t(\sim P) = \frac{1 + (-1)^{t(P)}}{2}$$

As an application of this calculational scheme, observe that the statement $P \vee Q$ can be expressed by the formula

$$P \vee Q = \sim ((\sim P) \wedge (\sim Q))$$

so $t(P \vee Q)$ can be determined by successively applying the two formulae above, which leads to the result

$$t(P \vee Q) = \frac{1}{2}\left(1 + (-1)^{\left(1 + (-1)^{t(P)} + (-1)^{t(Q)} + (-1)^{t(P)+t(Q)}\right)/4}\right)$$

This just means that $t(P \vee Q) = 0$ if $t(P) = t(Q) = 0$ (i.e., if P and Q are both false), and $t(P \vee Q) = 1$ otherwise. In a similar way we can calculate $t(P \Rightarrow Q)$ from $t(P)$ and $t(Q)$, and indeed $t(P)$ from $t(P_1), t(P_2), \ldots, t(P_n)$, so the truth of statements which only involve the logical operations specified above, the so-called statements of the *propositional calculus*, is reduced to numerical calculations.

The truth of statements belonging to the *predicate calculus* can also be reduced to certain numerical calculations. We can think of a predicate calculus statement as having the form $P = P_1 \wedge P_2 \wedge P_3 \wedge \ldots \wedge P_n \wedge \ldots$; thus, P may consist of an infinite number of propositions concatenated by conjunctions. In order to determine whether such a statement is true we must calculate the value of the product $t(P) = t(P_1)t(P_2)t(P_3) \ldots t(P_n)$ $\ldots$. Is it possible to calculate the value of this product, which may involve an infinite number of factors? There are certainly many cases for which it can be done, but to settle the general question we will have to understand the nature of calculation and be able to answer the fundamental question of which functions can be explicitly computed. The Austrian mathematical logician Kurt Gödel (1906–1974) and the English mathematician Alan Turing (1912–1954) approached this problem from different but complementary viewpoints.

§S.6. In 1931 Gödel showed that there are mathematical statements which cannot be proved or disproved. This means that such an *undecidable statement* cannot occur as the last line of any proof. It implies that the concept of mathematical truth is different from that of mathematical provability. Particular statements which have since been shown to be undecidable are generally rather complicated and do not lend themselves to nontechnical discussion. There is one which, although it has not yet been proved, many mathematicians believe to be undecidable and can be easily discussed. We refer to the celebrated Last Theorem of Fermat which asserts that the equation $x^n + y^n = z^n$ has no solution in positive integers x, y, z and n if n is greater than 2. Extensive machine calculations have failed to find a solution, which of course encourages the belief. Nevertheless, we can readily conceive what it means for Fermat's statement to be true or false. It may be *impossible* to determine this. There are more complicated assertions which one can *prove* are undecidable.

The question whether a statement can be proved or disproved using the rules of logic and hypotheses about properties of numbers in addition to the predicate calculus hypotheses contained in Figure S.7 addresses the strength of the system of formal reasoning and whether it is sufficiently powerful to identify all that is in fact true. The theorem of Gödel shows that it is not.

If our system of reasoning contains the undecidable statement U, can we not evade the difficulty by including U as one of the hypotheses in the system? Another theorem of Gödel dashes this hope, for it asserts that if the undecidable statement U is adjoined to the hypotheses, then there will be another proposition which can be formulated in the newly extended system which will be undecidable. It is the essence of complex systems of formal reasoning that they cannot be strong enough to prove all that is true. This in turn means that there can be knowledge which is not the result of formal reason.

In 1936 Turing analyzed the process of calculation and decomposed it into its most elementary constituents. In this way he was led to conceive the *Turing Machine*, an idealized general calculating machine which was the conceptual ancestor of the real computing machines constructed a decade later. A Turing machine consists of a two-way infinite sequence, called a tape, of memory locations. Each of the memory locations can hold just one bit of information, that is, the binary digit 0 or 1. The tape can be moved past the read/write station of a control unit location by location. When a memory location is positioned at the read/write head, the content of the location can be changed or the tape moved one location in either direction as determined by the bit it contains and the current instruction. After execution of the instruction the control unit selects the next instruction. The capabilities of the Turing machine seem to correspond to our intuitive notions of computability, and any specific finite calculation can be reorganized and coded so that it can be executed by such a machine.

A function f can be said to be computable if, whenever an x for which the function is defined is read into the Turing machine, it calculates and writes onto the tape the value $f(x)$. If an x and a program or rule for the calculation of f are given, we may ask whether the Turing machine will actually be able to calculate $f(x)$ and stop in a finite number of steps. Turing proved that there exist functions which are not computable in this sense. If we accept the identification of calculations with proofs, then Gödel's theorem shows that not every function can be computed. Let us number all possible mathematical statements with the natural numbers 1, 2, . . ., and consider the function f defined so that $f(x) = 1$ if the statement labeled x is true, but $f(x) = 0$ otherwise. If there were a program for a Turing machine which could calculate the values of f, then for each x we could translate the calculation of $f(x)$ into a proof of the statement whose label is x.

This would imply, however, that every true statement is provable, contradicting Gödel's theorem. In terms of the Turing machine, when an x which labels an unprovable but true statement is read in, the calculation of $f(x)$ cannot be completed, so the machine never stops.

If we imagine that the human brain and central nervous system constitute just a complex biological computer which, in principle at least, does not differ from electronic digital computers in any essential way, then the results of Gödel and Turing must apply to thought; there must be true statements which are in principle not accessible to the processes of reason. It is not yet clear what role random events might play in either human thought processes or in the theory of computation, but if they do play a role, it might be to provide a means for drawing inferences about truth which extend beyond reason.

Although the theories of computability and provability have shown the essential identity of these two concepts, many working mathematicians still consider them to be very different in practice. Thus a proof which is phrased as a calculation will generally be considered less beautiful and less revealing than a "noncomputational" proof. It seems that these viewpoints really express matters of taste rather than substance. The recent proof of the famous four-color-map conjecture (which asserts that four colors suffice for any map) involved traditional techniques of proof interwoven with a vast machine calculation quite beyond human powers. It no doubt signals just the beginning of an era in which the speed and accuracy of calculating machines will serve to augment, and to some extent replace, human formal reasoning processes.

§S.7.   Thus far we have distinguished computable functions, whose values can be calculated by a Turing machine in a finite time, from uncomputable functions, for which the Turing machine may never come to a halt. The class of computable functions can be further refined into subclasses according to the *computational complexity* of the various computable functions. There are various ways to measure computational complexity. One that is suitable for the present purpose is to measure complexity in terms of the time that an ideal computer programmed in the most efficient possible way would require to calculate the function value $f(x)$, and then to consider the greatest time as x varies through the domain of definition of the function f. Denote this time by $T(f)$. If $T(f) = \infty$, then f is uncomputable in the sense of the discussion above, for this means that there is an x such that the computer or Turing machine will never halt in an attempt to calculate $f(x)$. In these terms the results of Gödel and Turing show that there are functions such that $T(f) = \infty$.

Among the computable functions f, for which $T(f) < \infty$, it certainly makes a practical difference whether, say, $T(f) = 1$ second or $T(f) = 10^{100}$ seconds. In the former case the function values can be calculated in a practical period of time whereas the latter is much longer than the estimated duration of the universe. It was recently proved that no matter how large the number N is, there are functions f such that $T(f) > N$, that is, there are functions of finite but arbitrarily great computational complexity. This and similar results have motivated mathematicians and computer scientists to try to characterize families of functions and problems according to their computational complexity in order to obtain some insight into which kinds of problems could actually be solved in practice rather than only in principle. Using the equivalence between proofs and calculations, these investigations amount to a study of the classes of theorems which could be proved by machines in a sufficiently short time to be of assistance to the creative mathematician.

It is not yet clear to what extent research on computational complexity will lead to profound and powerful mathematical results. Surprisingly, it has already had stimulating and potentially important practical consequences, especially in applications to cryptography. The problem of cryptography is to encipher a message so that only the intended recipient who holds the "key" can decipher it. A message can be thought of as a sequence of symbols drawn from some fixed inventory, which, without loss of generality, may be taken to be the digits 0 and 1. Thus a message can be identified with the binary expansion of a number between, say, 0 and 1. Encryption of a message amounts to replacing one such number by another. Since, in order to be of practical value, an encryption algorithm must be applicable to all possible messages, it follows that each number between 0 and 1 will be carried over to another of the same sort, so the encryption rule is simply a function, say f. Moreover, since a practical rule for encryption must permit the ciphered message to be unraveled by its intended recipient, it must be possible to calculate the message x from the encrypted version $f(x)$; this means that the function f has an inverse f^{-1}. Let us suppose that, without knowledge of the key, the function f^{-1} is computationally more complex than f. This would mean that even if the encryption function f were made known, without the key it would be effectively impossible, from a practical point of view, to calculate f^{-1} and decipher messages enciphered by f.

Certain functions and processes which arise in the theory of numbers appear to share these properties. It is, for instance, believed by many mathematicians that the computational complexity of calculating the prime factors of an integer is substantially greater than the complexity of calculating the products of integers. Based on this belief, encryption algorithms have been proposed which begin with preselected large prime numbers and utilize an encryption function f whose definition involves their product, say P. For these particular functions, calculation of the inverse function entails knowledge of the prime factors of P, which will not be practically computable if the complexity of factorization algorithms is indeed very great. The person who created the particular function will, of course, have initially selected and thus know the prime factors of P, which will be the "key" making decipherment possible. Problems of this type are currently the subject of active research and have focused attention on factorization and related fundamental aspects of synthetic number theory, which are among the most ancient in pure mathematics. Now they have found their role in applications as well.

§S.8.   Our themes of the *geometrical nature of space* and the *ability to compute* are, finally, fully interwoven and realized within ourselves. The biological realization of the abstract computing machine that is our brain appears as the common meeting ground of the Incompleteness Theorem and the Uncertainty Principle, the great indeterminacies of physical reality and abstract thought.

SOLUTIONS TO SELECTED EXERCISES

Chapter 1

1.1. (a) $346 = 2^8 + 2^6 + 2^4 + 2^3 + 2^1 = (101011010)_2$
$346 = 5 \times 8^2 + 3 \times 8^1 + 2^1 = (532)_8$
(b) $64 = 2^6 = (1000000)_2$
$64 = 8^2 = (100)_8$
(c) $512 = 2^9 = (1000000000)_2$
$512 = 8^3 = (1000)_8$
(d) $1696 = 2^{10} + 2^9 + 2^7 + 2^5 = (11010100000)_2$
$1696 = 3 \times 8^3 + 2 \times 8^2 + 4 \times 8^1 = (3240)_8$

1.2. (a) $(10011)_2 = 2^4 + 2^1 + 2^0 = 16 + 2 + 1 = 19$
(b) $(1071)_8 = 8^3 + 7 \times 8^1 + 8^0 = 512 + 56 + 1 = 569$
(c) No; octal (base 8) representation uses digits less than 8.

1.3. (a) $(1512)_{10} = 10 \times 12^2 + 6 \times 12^1 = (T60)_{12}$
(b) (i) $(TEE)_{12} = 10 \times 12^2 + 11 \times 12^1 + 11 \times 12^0 = (1583)_{10}$
(ii) $(T0E)_{12} = 10 \times 12^2 + 11 \times 12^0 = (1451)_{10}$
(iii) $(111)_{12} = 12^2 + 12^1 + 12^0 = (157)_{10}$
(iv) $(21E3)_{12} = 2 \times 12^3 + 12^2 + 11 \times 12^1 + 3 \times 12^0 = (3735)_{10}$

1.4. (a) $(110)_2 \times (101)_2 = (11110)_2$
(b) $(123)_8 + (457)_8 = (602)_8$
(c) $(310E)_{12} - (TEE)_{12} = (2210)_{12}$
(d) $(55)_8 \div (17)_8 = (3)_8$

1.5.

	EGYPTIAN	GREEK	ROMAN
(a) $251 =$	ⲉⲉ𓎤𓎤ꞁ	$\sigma\, \upsilon\, \alpha$	CCLI
(b) $2,462 =$	𓏢𓏢ⲉⲉ𓎤𓎤𓎤	$,\beta\, \upsilon\, \varsigma\, \beta$	MMCCCCLXII
(c) $28 =$	𓎤𓎤𓏺𓏺𓏺𓏺	$\kappa\, \eta$	XXVIII
(d) $10,270 =$	𓆼ⲉ𓎤𓎤𓎤	$M\, \sigma\, o$	impractical

1.6.
(a) ⲉⲉ𓎤𓎤ꞁꞁꞁ $+$ 𓆼𓏢ⲉⲉⲉ𓎤𓎤𓎤ꞁꞁ $= 246 + 11{,}355 = 11{,}601 =$ 𓆼𓏢ⲉⲉⲉⲉⲉⲉꞁ
(b) $\chi\,\lambda\,\alpha + \omega\,\iota\,\beta = 631 + 812 = 1443 = ,\alpha\,\upsilon\,\mu\,\gamma$
(c) $\phi\,\kappa\,\beta + \sigma\,\xi = 522 + 260 = 782 = \psi\,\pi\,\beta$
(d) ⲉⲉ𓎤𓎤𓎤 $-$ ⲉꞁꞁꞁꞁꞁ $= 250 - 107 = 143 =$ ⲉ𓎤𓎤ꞁꞁꞁ

1.7. (a) 1   (b) 4   (c) 9   (d) 16   (e) 26   (f) 64
(g) 81   (h) 625   (i) $2 \times 60^3 + 18 \times 60^2 + 53 \times 60^1 + 20 \times 60^0 = 500{,}000$

1.8. (a) 𒁹 (b) 𒌋𒈫 (c) 𒈫𒐲𒈫𒈫 (d) 𒐈𒐈𒐉𒐲𒐲 (e) 𒈫𒈫𒈫𒐈𒐲𒐲𒐉𒐲𒐲𒐲𒐲
(f) 𒐉𒐈𒈫𒈫𒐈𒐲𒐲𒐲𒐉𒈫𒈫𒐲𒐲𒐲 (g) 𒐲𒐲𒐉𒐉𒈫𒐈𒐈𒐲𒐲𒐲𒐉𒐲𒐲𒐲

1.9. (a) $(621)_{10} = 10,21$ (b) $(3600)_{10} = 1,0,0$ (c) $(0.24)_{10} =$ $0;14,24$ (d) $(120.50)_{10} = 2,0;30$ (e) $\frac{1}{3} = 0;20$ (f) $\frac{1}{7} =$ $0;8,34,17,8,34,17, \ldots$

1.10. $\frac{1}{7}$ has a nonterminating sexagesimal expansion.

1.11. (a) $1,1 = 61$ (b) $0;1 = 1/60 = 0.01666\ldots$ (c) $1;0,1 = 1\frac{1}{3600} = 3601/3600$
(d) $0;1,50 = \frac{1}{60} + \frac{50}{60^2} = \frac{110}{60^2} = \frac{11}{360}$ (e) $12,20;21 = 740.35$
(f) $1;24,51,10 = 1 + \frac{24}{60} + \frac{51}{60^2} + \frac{10}{60^3} = 1.41421296\ldots$

1.12. (a) $12,2 + 15,4 = 27,6$
(b) $12,45 + 13,55 = 26,40$
(c)
$$
\begin{array}{r}
25,6;37 \\
\times \quad 1,2;3 \\
\hline
1,15;19,51 \\
50,13;14 \\
25,\;6,37 \\
\hline
25,58,\;5;33,51
\end{array}
$$
(d) $12,20 \div 2,1 = 6;6,56, \ldots$
(e) $1 \div 36 = 0;1$

1.13. Let the rational number be written as $I + \frac{m}{n}$, where I is its integer part and m/n is its fractional part. We must show that m/n has an ultimately repeating binary expansion. The binary expansion of m/n is produced by long division of m by n expressed in binary notation. As was shown on pp. 31–32 for decimal expansions, division by n can produce only the n distinct remainders $0, 1, 2, \ldots,$ $n - 1$ (expressed in binary notation), so after at most n steps of the long division process, a remainder must repeat and the digits of the binary expansion must begin to repeat also.

*1.14. Same as above, with "binary" replaced by "base b" for any fixed base b.

1.15. (a) Egyptian vs. Roman: developed procedure for calculating with fractions; Greek vs. Egyptian: more compact;
Babylonian vs. Greek: positional notation permits expression of any real number, no matter how large or small, using only a finite number of different symbols;
decimal vs. Babylonian: fewer multiplications to memorize;
binary vs. decimal: least number of different symbols and smallest multiplication table;
decimal vs. binary: more compact representation for large numbers.
(b) Babylonian, decimal, binary.
(c) Positional notations are better than nonpositionals. Different bases have different advantages.

1.16. $2 + \frac{4}{3} + \frac{8}{9} + \frac{16}{27} + \cdots = 2(1 + \frac{2}{3} + (\frac{2}{3})^2 + (\frac{2}{3})^3 + \cdots)$
$$= 2\frac{1}{1 - \frac{2}{3}} = 6$$

1.17. (a) $1.212121 \ldots = 120/99$

(b) $0.121212\ldots = 12/99$

(c) $0.18591859\ldots = 1859/9999$

(d) $21.90111\ldots = 21.9 + \frac{1}{900} = \frac{19711}{900}$

1.18. (a) No.

(b) If "yes," then $\sqrt{5}$ is rational, hence $\sqrt{5} = m/n$ for some positive integers m and n without common factors other than 1. Then $5n^2 = m^2$, so 5 divides m. Say $m = 5k$. Then $5n^2 = (5k)^2$, so $n^2 = 5k^2$ and hence 5 divides n. Thus 5 is a divisor of both m and n, a contradiction.

1.19. $x = 0.10100100010\ldots$ defined so that each sequence of zeros includes one more zero than the sequence to its left. This never repeats, hence is not rational.

1.20. A one-to-one correspondence between $\mathbf{N}$ and $\mathbf{Z}$ is given by $n \mapsto n/2$ if $n \in \mathbf{N}$ is even, and $n \mapsto (1 - n)/2$ if $n \in \mathbf{N}$ is odd.

1.21. $n \mapsto n^2$ is a one-to-one correspondence.

Chapter 2

2.1. (a) ⫼ ($= 9$) occurs at the end of the third row.

ℓ ($= 100$) occurs in the fourth row.

(b) The right-hand column of calculations contains columns of numbers being added up. They translate to:

	1680	135
	200	
	100	10
sum	1980	145
	625	45
	630	61
	525	38
sum	1780	143 (note error in addition!)
	200	2

2.2. (b) In the diagram

$b = 4 - 2 = 2$

$h = 6$

$56 = V$

$a = 4$

$b = 4 - 2 = 2$
$a = 4$
$h = 6$

The Egyptian computed

$$a^2 + ab + b^2 = 4^2 + 4 \times 2 + 2^2 = 16 + 8 + 4 = 28$$

$$\frac{1}{3}h = \frac{1}{3} \times 6 = 2$$

$$(a^2 + ab + b^2) \times \frac{h}{3} = 28 \times 2 = 56$$

The computations near the diagram:

to the left: " $\begin{matrix} 28 & 1 \\ 56 & 2 \end{matrix}$ "

represent doubling 28 or multiplying $2 \times 28 = 56$.

at the bottom: "28 equals 8 $\mathcal{H}$ 16 add."
So $\mathcal{H}$ must mean "$2 \times 4 = 8$," and this represents the first addition in the main computation.

(c) To derive the formula for the volume, construct a larger pyramid of height $h + k$ as in the diagram below. Then the volume of the frustum of the

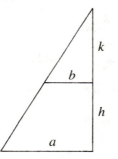

cone is given by the volume of the large pyramid minus the volume of the small pyramid at the top. Namely,

$$V = \frac{(h+k)}{3} \times a^2 - \frac{k}{3} \times b^2$$

We want a formula for V in terms of only h, a and b. Notice that by proportionality

$$\frac{h+k}{a} = \frac{k}{b}$$

and we can solve for k in terms of a, b, and h, obtaining

$$k = \frac{hb}{a-b}$$

Substituting this into the formula for V above we obtain

$$V = \frac{1}{3}\left[ha^2 + \frac{hba^2}{a-b} - \frac{hb^3}{a-b} \right]$$

$$= \frac{h}{3} \left[\frac{a^2(a - b) + ba^2 - b^3}{a - b} \right]$$

$$= \frac{h}{3} \left[\frac{a^3 - b^3}{a - b} \right]$$

$$= \frac{h}{3} \left[a^2 + ab + b^2 \right]$$

recalling that $(a - b)(a^2 + ab + b^2) = a^3 - b^3$

2.3.   Calculate: $\frac{3}{3} = 1$, $3 - 1 = 2$, let $x = 2$. Then check

$$2 \times 2 + 2 = 6$$
$$\tfrac{1}{2}(2 \times 2 + 2) = 3$$

so $(2 \times 2 + 2) - \tfrac{1}{2}(2 \times 2 + 2) = 3$ as desired. Two disadvantages of this procedure include:

(i) it does not work for all equations of this type;

(ii) you only know it does not work after trying to check the solution, which involves the majority of the computations.

By usual algebra, we find that

$$2x + x - \frac{1}{2}(2x + x) = 3$$

simplifies to

$$3x - \frac{3}{2}x = 3$$
$$\frac{3}{2}x = 3$$
$$x = \frac{6}{3} = 2$$

2.4.   (a)

so $3 \times 15 = 45$, and 3 is the desired quotient.

(b)

476 is the desired product.

(c)

so $10 \times 10 = 100$, and 10 is the desired quotient.

2.5. (a)

64	1
128	2
256	4
512	8
1024	16
2048	32'
2048 ⌣	32

(b)

64	1'
128	2'
256	4'
512	8'
1024	16'
2048	32
1984 ⌣	31

(c)

31	1'
62	2
124	4
248	8
496	16
992	32
1984	64'
2015 ⌣	65

(d)

31	1'
62	2'
124	4'
248	8'
496	16'
992	32'
1984	64
1953 ⌣	63

(e)

10	1
20	2'
40	4
80	8'
100 ⌣	10

(f)

10,000	1
20,000	2'
40,000	4
80,000	8'
100,000 ⌣	10

2.6. (a) 10100
   (b) 1100100
   (c) 1110111
   (d) 1100.10

2.7. (a) $(1.01 \ldots)_2$
   (b) $1.01 \times 1.01 = (1.1001)_2 = 1 + \frac{1}{2} + \frac{1}{16}$
   (c) The check shows only that $(1.01)^2$ is approximately 2.

2.8. (a) $\overline{6} + \overline{2}$ (b) $\overline{12} + \overline{4}$
   (c) $\overline{18} + \overline{6}$ (d) $\overline{24} + \overline{8}$
   (e) $\overline{30} + \overline{10}$ (f) $\overline{36} + \overline{18}$

2.9. $\dfrac{2}{n} = \dfrac{1}{4} \times \dfrac{1}{n} + \dfrac{7}{4} \times \dfrac{1}{n}$

2.10. (a) $2^0 + 2^1 + 2^2 + 2^3$ (b) $2^3 + 2^4$

       (c) $2^0 + 2^1 + 2^2 + 2^3$ (d) $2^0 + 2^1 + 2^2 + 2^3 + 2^4$

             $+ 2^4 + 2^5$ $+ 2^5 + 2^6 + 2^7 + 2^8 + 2^9$

       (e) $2^0 + 2^3 + 2^4 + 2^5 + 2^6$ (f) $2^0 + 2^2 + 2^4 + 2^6 + 2^8$

             $+ 2^7 + 2^8 + 2^9 + 2^{12}$

*2.11. Let N be a given positive integer. Choose n_0 to be the largest positive integer such that $2^{n_0} \leqslant N$, then let n_1 be the largest integer such that $2^{n_1} \leqslant N - 2^{n_0}$. It is clear that $n_1 < n_0$. Proceed in this manner, and choose n_k to be the largest positive integer such that $2^{n_k} \leqslant N - 2^{n_1} - 2^{n_k - 1}$, having chosen n_0, n_1, . . ., n_{k-1} in the same manner. Then eventually, we will obtain that

$$N - 2^{n_0} - 2^{n_1} - \ldots - 2^{n_k} = 0$$

for some k with $n_k \geqslant 0$, since this process will terminate. Then we have

$$N = 2^{n_0} + 2^{n_1} + 2^{n_2} + \cdots + 2^{n_k}$$

with $n_0 > n_1 > \ldots > n_k \geqslant 0$, as desired. No, there is only one expression of this kind.

Chapter 3

3.2. (a) $\frac{1,5}{4} = (1,5)(\frac{1}{4}) = (1,5)(0;15) = 16;15$

     (b) $\frac{20,1}{1,4} = (20,1)(\frac{1}{1,4}) = (20,1)(0;56,15) = 18,45;56,15$

     (c) $\frac{2}{5} = (2)(\frac{1}{5}) = (2)(0;12) = 0;24$

* 3.3. Suppose $n = 2^a 5^b$. Then $\frac{1}{n} = \frac{1}{2^a} \cdot \frac{1}{5^b} = \frac{10^a}{2^a} \cdot \frac{10^b}{5^b} \cdot 10^{-a-b} = (5^a \cdot 2^b) 10^{-a-b}$; write $M = 5^a \cdot 2^b$. Then $M 10^{-a-b}$ is a finite decimal. Conversely, if $\frac{1}{n} = \frac{M}{10^c}$ with M an integer, then $n = \frac{10^c}{M}$ is an integer by hypothesis. Since $10 = 2 \cdot 5$, $n = \frac{2^c \cdot 5^c}{M}$ is an integer. Therefore the only prime factors of M are 2 or 5. QED

* 3.4. Suppose $n = 2^a 3^b 5^c$. Then $\frac{1}{n} = \frac{1}{2^a} \cdot \frac{1}{3^b} \cdot \frac{1}{5^c} = (\frac{60^a}{2^a} \cdot \frac{60^b}{3^b} \cdot \frac{60^c}{5^c}) 60^{-a-b-c} = (30^a \cdot 20^b \cdot 12^c) 60^{-a-b-c}$; write $M = 30^a \cdot 20^b \cdot 12^c$, an integer. Then $\frac{1}{n} = \frac{M}{60^{a+b+c}}$ is a finite sexagesimal expansion. Conversely, if $\frac{1}{n} = \frac{M}{60^d}$ for some integer M, then $n = \frac{60^d}{M}$. Since $60 = 2^2 \cdot 3 \cdot 5$, the prime factors of M must be 2, 3 or 5.

3.5. $\frac{1}{n}$ such that n is a power of 2.

3.6. Probably; Plimpton 322 lists many examples which would be difficult to discover without knowledge of the Pythagorean theorem. The Babylonians left no proofs of any theorems.

3.7. $\sqrt{3}$: $a_1 = 1$, $a_2 = \frac{1 + (3/1)}{2} = 2$, $a_3 = \frac{2 + (3/2)}{2} = 1.75$, $a_4 = \frac{1.75 + (3/1.75)}{2} = 1.7321^+$. Note: $\sqrt{3} = 1.73205^+$.

3.8. (a) $\sqrt{5}$: choose $a_1 = 2$. Then $a_2 = \frac{2 + (5/2)}{2} = 2.25$, $a_3 = \frac{2.25 + (5/2.25)}{2} =$

$2.2361 \ldots$, $a_4 = \frac{2.2361 \ldots + (5/2.2361 \ldots)}{2} = 2.260 \ldots = 2.26$, correct to two decimals.

(b) $\sqrt{7}$: choose $a_1 = 2$. $a_4 = 2.6457$, $\sqrt{7} = 2.64$, correct to two decimals.

(c) $\sqrt{10}$: choose $a_1 = 3$. Then $a_3 = 3.1622 \ldots$, $\sqrt{10} = 3.16$, correct to two decimals.

3.9. (a) $\sqrt{1 - x^2}$: choose $a_1 = 1$. Then $a_2 = \frac{1 + \frac{1 - x^2}{1}}{2} = 1 - \frac{1}{2}x^2$ and $a_3 =$

$$\frac{(1 - \frac{1}{2}x^2) + \frac{1 - x^2}{(1 - \frac{1}{2}x^2)}}{2}$$

Calculate the second term in the numerator by long division:

$$
\begin{array}{r}
1 - \frac{1}{2}x^2 - \frac{1}{4}x^4 - \frac{1}{8}x^6 - \cdots \\
1 - \frac{1}{2}x^2 \overline{\smash{\big)}\, 1 - x^2 } \\
\underline{1 - \frac{1}{2}x^2 } \\
-\frac{1}{2}x^2 + 0 \\
\underline{-\frac{1}{2}x^2 + \frac{1}{4}x^4 } \\
-\frac{1}{4}x^4 + 0 \\
\underline{-\frac{1}{4}x^4 + \frac{1}{8}x^6} \\
-\frac{1}{8}x^6 + 0 \\
\cdot \\
\cdot \\
\cdot
\end{array}
$$

so $a_3 = \frac{1}{2}\{(1 - \frac{1}{2}x^2) + (1 - \frac{1}{2}x^2 - \frac{1}{4}x^4 - \frac{1}{8}x^6 \ldots)\} = 1 - \frac{1}{2}x^2 - \frac{1}{8}x^4 \ldots$.

Continue this way to calculate a_4, a_5, etc. The binomial theorem produces

$$\sqrt{1 - x^2} = (1 - x^2)^{1/2} = 1 - \frac{1}{2}x^2 + \underbrace{\frac{(\frac{1}{2})(-\frac{1}{2})}{1\cdot2}x^4}_{} - \underbrace{\frac{(\frac{1}{2})(-\frac{1}{2})(-\frac{3}{2})}{1\cdot2\cdot3}x^6}_{} + \cdots,$$

which agrees with a_n up through its nth term.

3.10. The number below the diagonal and the number above and to the left of the square must have sexagesimal points placed so that their ratio approximates $\sqrt{2} = 1.414 \ldots$; the number above the diagonal is an approximation of $\sqrt{2}$, so it must have the sexagesimal point to the right of the first (leftmost) cuneiform symbol, thus: 1;24,51,10.

3.11. p, q positive. Set $h = p^2 - q^2$, $b = 2pq$, $d = p^2 + q^2$. Then $h^2 + b^2 = (p^2 - q^2)^2 + (2pq)^2 = (p^4 - 2p^2q^2 + q^4) + (4p^2q^2) = p^4 + 2p^2q^2 + q^4 = (p^2 + q^2)^2 = d^2$. Consider $(120, 119, 169)$. Since only one of these three numbers is even, it must be of the form $2pq$. So $120 = 2pq$, whence $60 = pq$. Since $p^2 + q^2 > p^2 - q^2$, then $p^2 + q^2 = 169$ and $p^2 - q^2 = 119$. Add these equations to obtain $2p^2 = 169 + 119 = 288$, so $p^2 = 114$ and $p = 12$. Hence $60 = pq = 12q$ yields $q = 5$.

3.12. (a) $xy = 720 = P$, $x + y = 72 = S$. Let $x = \frac{S}{2} + E = 36 + E$. From p. 81, $E^2 = (\frac{S}{2})^2 - P = 36^2 - 720 = 576$, so $E = \pm 24$. Then $x = 36 + 24 = 60$, $y = 36 - 24 = 12$.

(b) $xy = 1 = P$, $x + y = a = S$, $a > 0$. As above, $E^2 = (\frac{S}{2})^2 - P = \frac{a^2}{4} - 1$, so $E = \sqrt{\frac{a^2}{4} - 1}$ and $x = \frac{a}{2} + \sqrt{\frac{a^2}{4} - 1}$,

$y = \frac{S}{2} - E = \frac{a}{2} - \sqrt{\frac{a^2}{4} - 1}$.

3.13. (a) $P > 0$ and $S > 0$

(b) $S^2 - 4P > 0$

Chapter 4

4.3. $\dfrac{\sin .017}{.017} \cong \dfrac{.017}{.017} = 1.00$

$\dfrac{\sin .175}{.175} \cong \dfrac{.174}{.175} \cong .99$

etc., these are decreasing.

4.4. $\dfrac{\tan .017}{.017} \cong \dfrac{.017}{.017} = 1.000$

$\dfrac{\tan .175}{.175} \cong \dfrac{.176}{.175} \cong 1.005$

etc., these are increasing.

4.5.   We have:

$$\frac{\sin \frac{\pi}{3}}{\frac{\pi}{3}} < \frac{\sin \frac{\pi}{4}}{\frac{\pi}{4}} = \frac{\sqrt{2}}{2} \cdot \frac{4}{\pi}$$

which implies that

$$\sin \frac{\pi}{3} < \frac{2}{3} \cdot \sqrt{2} < \frac{2}{3}(1.415) = .94$$

The exact value of $\sin \dfrac{\pi}{3} = \dfrac{\sqrt{3}}{2} \cong .866 \ldots$

4.6. $\dfrac{\sin 2y}{\sin y} < \dfrac{2y}{y} \Rightarrow \sin 2y < 2 \sin y$,

$\dfrac{2y}{y} < \dfrac{\tan 2y}{\tan y} \Rightarrow 2 \tan y < \tan 2y$

4.10. $4\left(\dfrac{8}{9}\right)^2 = $ ⦀ ⚲ ⦀ ⦀ ⦀ ꩜꩜꩜꩜ ꩜꩜꩜꩜ ⎮ $\left(= 4 + \dfrac{2}{3} + \dfrac{1}{9} + \dfrac{1}{81}\right)$

4.11. $\left(\dfrac{17}{12} - \sqrt{2}\right)^{-1} \approx 1.4, \; a_1 = 1,$

$$\approx -0.0833, \; a_2 = \frac{3}{2},$$

$$\approx \infty, \quad a_3 = \frac{17}{12}$$

$$144\left(\frac{17}{12} + \sqrt{2}\right) \approx 348, \quad a_1 = 1,$$

$$\approx 420, \quad a_2 = \frac{3}{2},$$

$$\approx 408, \quad a_3 = \frac{17}{12}$$

4.12. $(\sqrt{2} - 1)^{-1} \approx \infty, \quad a_1 = 1,$

$$\approx 2, \quad a_2 = \frac{3}{2},$$

$$\approx 2.4, \quad a_3 = \frac{17}{12}$$

$\sqrt{2} + 1 \approx 2, \quad a_1 = 1,$

$$\approx 2.5, \quad a_2 = \frac{3}{2},$$

$$\approx 2.416, \quad a_3 = \frac{17}{12}$$

4.13. We have $1 < \sqrt{2} < \frac{17}{12} < \frac{3}{2}$, which implies that:

(a) $-0.0833 < 1.4 < \left(\frac{17}{12} - \sqrt{2}\right)^{-1} < \infty$

$$348 < 144\left(\frac{17}{12} + \sqrt{2}\right) < 408 < 420$$

(b) $2 < 2.4 < (\sqrt{2} - 1)^{-1} < \infty$

$2 < \sqrt{2} + 1 < 2.416 < 2.5$

4.21. (a) 2.449   (b) 3.080
          2.872   3.127
          2.972   3.136
          3.034   3.138

4.22. A rigorous proof of this uses mathematical induction, but one can see this also by observing that if we consider

$$1 + 2 + \ldots + n = N$$
$$n + (n - 1) + \ldots + 1 = N$$

and add each term, we get

$$(n + 1) + (n + 1) + \ldots + (n + 1) = 2N$$

and since there are n terms we see that

$$n(n + 1) = 2N,$$

which gives us that $N = \dfrac{n(n + 1)}{2}$, as desired.

4.23. $A = \dfrac{1}{3}$.

4.24. A formula for
$$1^m + 2^m + 3^m + \ldots + n^m$$

*4.25. For instance:
$$1.414138010010001 \ldots.$$

4.28. $\dfrac{h}{2}(b_1 + b_2)$

Chapter 5

5.4. See Figure 5.4. We are given δ to determine α. But $\delta \cong \delta'$, and $\delta' + \gamma = 90°$, and $\alpha + \gamma = 90°$, so $\delta \cong \alpha$, and we conclude that the altitude is $45°2'$.

5.5. The variation of Polaris from the true north is approximately $1\frac{1}{2}°$, and the error in computing the latitude using the altitude of Polaris would be at most this amount.

5.6. If the altitude was measured in a northward direction at the time that Sirius was in the plane formed by the zenith of the observer, the celestial north pole and the origin, then the altitude can be determined. If γ is the angle between the zenith Z and S ($=$ Sirius), and L is the southern latitude of the observer, then one finds $L - \gamma = 16°31'$, using $16°31'S$ as the latitude of Sirius in the celestial sphere. Also, $59° + \gamma = 90°$, so we find that, by adding these two relations, $L + 59° = 90° + 16°31'$, or $L = 47°31'S$. A similar argument holds if one makes a southward measurement.

5.8. No. It will have the same altitude at any longitude.

5.9. Choosing the star near $45°$ in latitude will minimize the error since the difference between the horizon and the celestial horizon will be minimized when the star is essentially vertically overhead of the observer, as discussed at the top of p. 140 in the case where the observer is at the North Pole. Both $0°N$ and the celestial north pole would have larger errors of the same magnitude since they are approximately equidistant from the zenith of the observer.

5.12. (a) $270°$
(b) Sum of angles of planar triangle $= 180° < 270°$

5.13. Yes.

Chapter 6

6.4. (a) 57.28 feet high by 1 foot wide (cylindrical);
(b) 8.113 feet high by 1 foot wide (Mercator).

Chapter 7

7.1. $\cos(x + y) = \cos x \cos y - \sin x \sin y$, $\cos(x - y) = \cos x \cos(-y) - \sin x \sin(-y)$, $\cos(-x) = \cos x$, and $\sin(-x) = -\sin x$ imply $\cos(x + y) + \cos(x - y) = 2 \cos x \cos y$.

7.2. (a) $1576 \times 998 = \frac{1576}{10,000} \times \frac{998}{1,000} \times 10^7 = (0.1576 \times 0.998) \times 10^7$. From Table 4 (or a calculator) find $0.1576 = \cos(1.414)$, $0.998 = \cos(0.070)$. Hence $x = 1.414$, $y = 0.070$; $x + y = 1.484$, $x - y = 1.344$. From Table 4 $\cos(1.484) = 0.087$, $\cos(1.344) = 0.225$, whence $\cos(1.484) + \cos(1.344) = 0.312$ and $1576 \times 998 \cong \frac{0.312}{2} \times 10^7 = 1.56 \times 10^6$.

(b) Similarly, $242 \times 358 = (0.242 \times 0.358) \times 10^6$. Find $x = 1.326$, $y = 1.204$, $x + y = 2.530$, $x - y = 0.122$, $\cos(2.530) = -\cos(\pi - 2.530) = -\cos(0.614)$ so $\frac{1}{2}\{\cos(x + y) + \cos(x - y)\} = 0.087$, $242 \times 358 \cong (0.087) \times 10^6 = 8.7 \times 10^4$.

(c) $242 \times 0.358 = (0.242 \times 0.358) \times 10^3 = (0.087) \times 10^3 = 87$ (by part b).

7.4. Prosthaphaeretic rule: $a \times b \to (0.a \times 0.b) \times 10^N$; $0.a \to \cos x$, $0.b \to \cos y$ by table look-up; $x + y$, $x - y$ by addition, $\cos(x + y)$, $\cos(x - y)$ by table look-up; $\frac{1}{2}(\cos(x + y) + \cos(x - y))$ by addition and division by 2. Thus four table look-ups, three additions, one division by 2, two normalizations by powers of 10. Logarithms: $x \to \log x$, $y \to \log y$ by table look-up. $\log x + \log y$ by addition; antilog by table look-up. Thus three table look-ups, one addition, two normalizations by power of 10.

7.5. To calculate a^b, calculate (i) $\log a$; (ii) $b \log a$; (iii) find x from table of logarithms such that $\log x = b \log a$. Example: $x = 2^5$. $\log_{10} 2 = 0.3010$; $5 \log_{10} 2 = 1.5050 = \log_{10} 32$.

7.6. If $x = y$, we can compute x^2. If $y = x^2$, we can compute x^3 by two applications of prosthaphaeretic rule. In general, x^n can be computed for *integers* n by applying the prosthaphaeretic rule $(n - 1)$ times. This method is inefficient.

7.7. (a) $e = 2.71828 \ldots$
(b) $\log_{10} e \cong 0.434 \ldots$ from Table 1

7.8. See table of logarithms (Table 1) at back of book.

7.9. We know that there must be increasingly better approximations of $\log_{10} 2$ which can be obtained by finding integers M and N such that 2^M is approximately equal to 10^N, for if the decimal expansion of $\log_{10}2$ is $\log_{10}2 = 0.a_1a_2a_3\ldots$, then

$$\log_{10}2 \cong \frac{a_1a_2\ldots a_k}{10^k}$$

is an approximation accurate to k decimals. Therefore

$$10^k\log_{10}2 \cong a_1a_2\ldots a_k$$

and so

$$2^{10k} \cong 10^{a_1a_2\ldots a_k}$$

We can choose $M = 10^k$ and $N = a_1a_2\ldots a_k$. For instance, $\log_{10}2 = 0.3010\ldots$, so $a_1 = 3$. Therefore $\log_{10}2 \cong \frac{3}{10}$, $10\log_{10}2 \cong 3$, and hence $2^{10} \cong 10^3$, which is the approximation used on p. 193. The first two digits of $\log_{10}2$ provides $\log_{10}2 \cong 0.30 = \frac{30}{10^2}$ which leads to $2^{100} \cong 10^{30}$, but this is the same as $2^{10} \cong 10^3$, as we see by taking the tenth root of each side. The next better approximation of this type arises from $\log_{10}2 = 0.301$, which leads to $2^{1000} \cong 10^{301}$! This approximation would be found "naturally" were you to successively compute powers of 2 until you arrived at a power of 2 which nearly equals a power of 10. It would require 1000 lines of calculation to find $2^{1000} \cong 10^{301}$, and cover 40 pages of typical lined paper. Notice that this calculation is beyond the capacity of current pocket calculators. This example shows that accurate values of logarithms cannot be calculated by so simple a method.

7.10.

ADVANTAGES	DISADVANTAGES
(a) Only two distinct symbols needed.	Base 2 expressions are the longest of all.
(b) Simplifies calculus formulae.	Inconvenient for arithmetic.
(c) Convenient for arithmetic.	Inconvenient for calculus.

7.12. (a) $P_H(1968 + n) = P_H(1968)(1.034)^n$, $P_{US}(1968 + n) = P_{US}(1969.5)(1.01)^{n-\frac{3}{2}}$. Populations will be equal when $P_H(1968)(1.034)^n = P_{US}(1969.5)(1.01)^{n-\frac{3}{2}}$. Insert values and calculate $\log_{10}$ to find $n\log_{10}(1.034) + \log_{10}(1,867,000) = (n - \frac{3}{2})\log_{10}(1.01) + \log_{10}(204,000,000)$. Find $n \cong 199$ years and $1968 + n = 2167$.

7.13. Yes. Multiplication can be reduced to addition/subtraction because of the identity $(x + y)^2 = x^2 + 2xy + y^2$ in which xy appears. To obtain x/y we can use $(x + \frac{1}{y})^2 = x^2 + 2\frac{x}{y} + \frac{1}{y^2}$. Hence, if a table of squares and a table of reciprocals of squares are available (recall the Babylonian table of reciprocals, Figure 3.1 on pp. 70–71), the quotient x/y can be calculated.

7.14. $S(x) = x^2/4$. Then $x^4 = 4(\frac{x^2 \cdot x^2}{4}) = 4S(x^2)$. But $x^2 = 4S(x)$, so finally $x^4 = 4S(4S(x))$.

Chapter 8

8.1. (a) $(x - 1)(x + 1) = x^2 - 1$
(b) $(x - 1)(x^2 + x + 1) = x^3 - 1$

8.2. (a) $(x - 1)(x^3 + x^2 + x + 1) = x^4 - 1$
(b) $(x - 1)(x^4 + x^3 + x^2 + x + 1) = x^5 - 1$
(c) $(x - 1)(x^5 + x^4 + x^3 + x^2 + x + 1) = x^6 - 1$

8.3. $(x - 1)(x^n + x^{n-1} + \ldots + x + 1) = x^{n+1} - 1$.
To see this is the case, simply multiply this out, term by term, obtaining
$(x^{n+1} + x^n + x^{n-1} + \ldots + x) - (x^n + x^{n-1} + \ldots + x + 1)$,
then canceling the terms of the form $x^n - x^n$, etc., we obtain $x^{n+1} - 1$.

8.5. $(x - 1)\left(\sum_{j=0}^{n} x^j\right) = x^{n+1} - 1$.

8.8. For instance, $(y - x^2)(x^2 + y^2 + 1) = 0$.

8.11. No.

8.12. Let the coordinate axes be the two straight lines, i.e., $L_1 = \{(x, y) : x = 0\}$, $L_2 = \{(x, y) : y = 0\}$, and then $L_1 \cup L_2 = \{(x, y) : xy = 0\}$.

8.13. $x - y - 1 = 0$

8.14. No.

8.15. $y = x^3 + x + 1$

8.16. $a = b > 0$

8.20. For x satisfying $0 < x < 1$, the quantity x^n gets closer to zero as n gets larger and larger, while for $x = 1$, $x^n = 1$ for all n and this does not go to zero.

Chapter 9

9.1. (a) $\lim_{n \to \infty} \frac{1}{2n} = 0$
(b) $\lim_{n \to \infty} n$ does not exist
(c) limit $= 0$
(d) limit does not exist
(e) $\lim_{n \to \infty} \frac{n+1}{n} = \lim_{n \to \infty} 1 + \frac{1}{n} = \lim_{n \to \infty} 1 + \lim_{n \to \infty} \frac{1}{n} = 1 + 0 = 1$

9.3. (a) Lines L_n given by $y = n$ are parallel to the x-axis through $(0, n)$ and march off to infinity as n increases through the positive integers.
(b) Lines L_n given by $y = \frac{1}{n}$ form a sequence of lines parallel to the x-axis which converge to the x-axis.

9.4. (a) Secant corresponding to pair $\{1 - \frac{1}{2n}, 1 + \frac{1}{n}\}$ contains points with coordinates $(1 - \frac{1}{2n}, (1 - \frac{1}{2n})^2)$ and $(1 + \frac{1}{n}, (1 + \frac{1}{n})^2)$. Equation of secant through these points is

$$y - \left(1 - \frac{1}{2n}\right)^2 = \left\{x - \left(1 - \frac{1}{2n}\right)\right\}\left\{\left(1 + \frac{1}{n}\right) + \left(1 - \frac{1}{2n}\right)\right\},$$

with limit as $n \to \infty$ the line with equation $y - 1 = (x - 1)(2) = 2x - 2$, i.e., $y = 2x - 1$, which passes through $(1, 1)$ and has slope 2.

  (b) Secant corresponding to $\{1 - \frac{1}{2n}, 1\}$ contains points (x, y) with coordinates $(1 - \frac{1}{2n}, (1 - \frac{1}{2n})^2)$ and $(1, 1)$. Secant equation is

$$y - \left(1 + \frac{1}{n}\right)^3 = \left\{x - \left(1 + \frac{1}{n}\right)\right\}\left\{\left(1 - \frac{1}{n}\right)^3 - \left(1 + \frac{1}{n}\right)^3\right\}/\left\{\left(1 - \frac{1}{n}\right) - \left(1 + \frac{1}{n}\right)\right\}.$$

$$y - \left(1 - \frac{1}{2n}\right)^2 = \left\{x - \left(1 - \frac{1}{2n}\right)\right\}\left\{1 + \left(1 - \frac{1}{2n}\right)\right\}.$$

Limit as $n \to \infty$ yields tangent line equation

$$y - 1 = (x - 1)(2) = 2x - 2; \; i.e, \; y = 2x - 1$$

as before.

9.5. Tangent line determined by sequence of x-coordinate pairs $\{1 - \frac{1}{n}, 1 + \frac{1}{n}\}$. Equation of secant line is

$$y - \left(1 + \frac{1}{n}\right)^3 = \left\{x - \left(1 + \frac{1}{n}\right)\right\}\left\{\left(1 - \frac{1}{n}\right)^3 - \left(1 + \frac{1}{n}\right)^3\right\}/\left\{\left(1 - \frac{1}{n}\right) - \left(1 + \frac{1}{n}\right)\right\}.$$

Limit as $n \to \infty$ yields equation of tangent line, $y = 3x - 2$. "Yes."

9.6. (a) Slope is given by

$$\lim_{n \to \infty} \frac{\left(x + \frac{1}{n}\right)^3 - x^3}{\left(x + \frac{1}{n}\right) - x} = \lim_{n \to \infty}\left\{\left(x + \frac{1}{n}\right)^2 + \left(x + \frac{1}{n}\right)x + x^2\right\} = 3x^2.$$

9.8. (a) $2 + 6x$ (b) $4\sqrt{2}\,x^3 + \dfrac{2\pi}{e}x$

  (c) $= \dfrac{d}{dx}\{acx^2 + (ad + bc)x + bd\} = 2acx + ad + bc$

  (d) $3(x - 1)^2$

9.9. (b)

$$\frac{df}{dx} = \begin{cases} 1 & \text{if } x > 0 \\ -1 & \text{if } x < 0 \\ \text{undefined} & \text{if } x = 0 \end{cases}$$

9.10. Limit sequences from the right of $x = 0$ have $\frac{df}{dx} = 1$; those from the left have $\frac{df}{dx} = -1$, so no unique limit exists at $x = 0$.

9.11. (I) A: $\frac{1}{2}(1)(1) = \frac{1}{2}$.   B: $A = \lim_{n \to \infty}\left\{\dfrac{1}{n^2} + \dfrac{2}{n^2} + \dfrac{3}{n^2} + \cdots + \dfrac{n-1}{n^2}\right\}$

$$= \lim_{n \to \infty}\frac{1}{n^2}(1 + 2 + 3 + \cdots + (n - 1)) = \lim_{n \to \infty}\frac{1}{n^2}\left\{\frac{n(n-1)}{2}\right\}$$

$$= \lim_{n \to \infty} \frac{n-1}{2n} = \lim_{n \to \infty} \left(\frac{1}{2} - \frac{1}{2n} \right) = \frac{1}{2}.$$

C: $\int_0^1 x\,dx = \frac{x^2}{2} \Big]_0^1 = \frac{1}{2}.$

(II) A: No elementary way. B: Requires knowledge of

$1^2 + 2^2 + \cdots + n^2.$ C: $\int_0^1 x^2 dx = \frac{x^3}{3} \Big]_0^1 = \frac{1}{3}.$

(III) A: No elementary way. B: Requires knowledge of

$1^5 + 2^5 + \cdots + n^5.$ C: $\int_0^1 x^5 dx = \frac{x^6}{6} \Big]_0^1 = \frac{1}{6}.$

(IV) A: Area of circle $= \pi R^2$, so area shown $= \pi/4$.

B: No elementary way. C: $\int_0^1 \sqrt{1 - x^2}\,dx$. This integral is not expressible in terms of polynomial functions.

9.13. $V =$ volume. Then $V = x(2 - 2x)^2 = x(4 - 8x + 4x^2) = 4x - 8x^2 + 4x^3$. Maximum volume occurs for $0 = \frac{dy}{dx} = 4 - 16x + 12x^2 = 4(1 - 3x)(1 - x)$, so $x = \frac{1}{3}$ or $x = 1$. If $x = 1$, volume is 0, the minimum; if $x = \frac{1}{3}$, $V = \frac{1}{3}(2 - \frac{2}{3})^2 = 16/27$, the maximum.

9.14. (a) $\int_0^1 (2x - 3x^2)\,dx = (x^2 - x^3) \Big]_0^1 = 0$

(b) $\int_2^3 (1 - x^2)\,dx = (x - \frac{x^3}{3}) \Big]_2^3 = 1 - \frac{19}{3}$

(c) $\int_0^2 \sqrt{2}\, x^2\,dx = \frac{\sqrt{2}\, x^3}{3} \Big]_0^2 = \frac{8\sqrt{2}}{3}$

(d) $\int_0^1 dx = x \Big]_0^1 = 1$

9.15. Consider a small portion of Γ whose endpoints are endpoints of secant S_k. If this portion of the curve is sufficiently small, it will be convex with respect to S_k. Let P be any point on it other than the endpoints. Then S_k will be less than the sum of the other two sides of the triangle formed by P and the endpoints of S_k. Repetition of this process approximates the curve ever more closely, and the approximation of its length increases to a limiting value.

Chapter 10

10.1. (a) $T_2 = 5 + 3x^3$ (b) No error.

10.2. $T_2 = 5 + 3x^2$, error $= 2x^3$.

10.3. (a) $(1 + x)^5 = 1 + 5x + 9x^2 + 9x^3 + 5x^4 + x^5$

10.4. $f(x) = \sqrt{4 + x}$ implies:

$\frac{df}{dx}(0) = \frac{1}{2^2}$

$$\frac{d^2f}{dx^2}(0) = -\frac{1}{2^5}$$

$$\vdots$$

$$\frac{d^nf}{dx^n}(0) = \frac{(-1)^n(1 \cdot 1 \cdot 3 \cdot 5 \cdot \ldots \cdot (2n-3)\,)}{2^{3n-1}}$$

which implies that

$$f(x) = 2 + \frac{1}{2^2}x - \frac{1}{2 \cdot 2^5}x^2 + \frac{3}{2 \cdot 3 \cdot 2^8}x^3 - \frac{3 \cdot 5}{2 \cdot 3 \cdot 4 \cdot 2^{11}}x^4 + \cdots$$

10.5. $\sqrt{2} = \sqrt{4-2}$, so we have

$$\sqrt{2} \approx 2 - \frac{1}{2^2}(2) - \frac{1}{2 \cdot 2^5}2^2 - \frac{3}{2 \cdot 3 \cdot 2^8}2^3 - \frac{3 \cdot 5}{2 \cdot 3 \cdot 4 \cdot 2^{11}}2^4 + \cdots \text{etc.,}$$

and one can use a calculator to calculate enough terms to give 4-decimal-place accuracy.

10.6. \$1.98

10.10. (a) (i) 5 terms
            (ii) 50 terms
            (iii) 500,000 terms
        (b) 3.3396
        (c) not as good

*10.12. $a + ar + ar^2 + \cdots + ar^n = \frac{a}{1-r} - \frac{ar^{n+1}}{1-r}$. Error $= \frac{ar^{n+1}}{1-r}$. Choose values of $a, r, -1 < r < 1$ to give specific examples.

Chapter 11

11.1. (a) (i) $\sqrt{2}$ (ii) $\sqrt{\pi^2 + 1}$
        (b) (i)   1   (ii) $\sqrt{3}$

11.2. S is the sphere of radius 1 centered at (x_0, y_0, z_0).

11.3. S is the cylinder whose axis the line through $(x_0, y_0, 0)$ perpendicular to the xy-plane and whose cross section, perpendicular to this axis, is the circle of radius 1.

11.4. Curvature is $K_p = \frac{\pm f''(x)}{(1 + \{f'(x)\}^2)^{3/2}}$, from p. 300. From $f(x) = x^2$, find $K_p = \frac{\pm 2}{(1 + 4x^2)^{3/2}}$. Then $\frac{dK_p}{dx} = \pm\{\frac{-24x}{(1 + 4x^2)^{5/2}}\}$. Maximum/minimum occurs where $0 = \frac{dK_p}{dx}$, i.e., at $x = 0$. At $x = 0$, $K_p = 2$. If $x \neq 0$, then denominator of K_p is > 1, so $K_p < 2$.

11.5. $K_p = \frac{\pm f''(x)}{(1 + \{f'(x)\}^2)^{3/2}}$ = constant. Intuition suggests (1) (straight) lines, (2) circles. There are no others. To prove this one must find all functions $y = f(x)$ such that $K_p = \frac{\pm f''(x)}{(1 + \{f'(x)\}^2)^{3/2}} =$ constant, say C; that is, solve a *differential equation*. (1) If $C = 0$, then $f''(x) = 0$, so $f(x) = ax + b$; a, b constants by the Fundamental Theorem of Calculus. These are the lines. (2) The differential equation with $C \neq 0$ is too difficult to solve here.

11.6. $P(t) = (x(t), y(t), z(t)) = (t, t, t), 0 \le t \le 1$.

(a) Straight line segment joining $(0, 0, 0)$ to $(1, 1, 1)$. Indeed, points satisfy the pair of equations $x(t) = y(t)$ and $x(t) = z(t)$, *i.e.*, x $=$ y and x $=$ z. These are equations of planes perpendicular, respectively, to the xy-plane and the xz-plane. The intersection of two planes is a line.

(b) $l = \int_0^1 [(\frac{dx}{dt})^2 + (\frac{dy}{dt})^2 + (\frac{dz}{dt})^2]^{1/2} dt$. $\frac{dx}{dt} = \frac{dy}{dt} = \frac{dz}{dt} = 1$, so $l = \int_0^1 \sqrt{3}\, dt = t\sqrt{3}\,]_0^1 = \sqrt{3}$.

11.7. $x(t) = 1 + t, y(t) = 2, z(t) = 3, 0 \le t \le 2$. Curve lies in line through $(0, 2, 3)$ and in fact is a line segment. Find

$$l = \int_0^2 \sqrt{1^2 + 0^2 + 0^2}\, dt = 2.$$

11.8.   Sphere of radius R. $\alpha(t) = 0, \beta(t) = t, 0 \le t \le \pi$.

(a) Portion of equator from 0° to 180° longitude.

(b) (i) Circumference of circle of radius R equals $2\pi R$ so semicircle has length πR.

(ii) $l = \int_0^\pi [R^2 \cdot 0 + R^2 \cos^2 \alpha(t) \cdot 1^2]^{1/2} dt = R \int_0^\pi dt = \pi R$.

11.9. $\alpha(t) = t, \beta(t) = 0, 0 \le t \le \pi/4$.

(a) Longitude is constant so curve is a portion of a great circle: 1/8 of circle from equation northward through longitude 0°.

(b) Elementary geometry implies $l = \frac{1}{8}(2\pi R) = \pi R/4$. Also, $l = \int_0^{\pi/4} R[1^2]^{1/2} dt = Rt\,]_0^{\pi/4} = \pi R/4$.

11.10.   Radius R; $\alpha(t) = \pi/4, \beta(t) = t, 0 \le t \le \pi/4$.

(a) Latitude constant so curve is part of a circle of latitude; indeed, $\frac{1}{8}$ of the particular latitude circle between longitude 0° and 45° at latitude 45°.

(b) (i) Elementary geometry shows $l = \pi\sqrt{2}\,R/8$. (ii) The formula provides $l = \int_0^{\pi/4} R(\cos^2 \frac{\pi}{4} \cdot 1^2)^{1/2} dt = (\frac{R}{2} \cos \frac{\pi}{4}) t\,]_0^{\pi/4} = \pi \frac{R}{8} \cos \frac{\pi}{4} = \pi R\sqrt{2}/8$ since $\cos \frac{\pi}{4} = \sqrt{2}$.

11.11.   No. Geodesics of sphere are portions of great circles, *i.e.*, of intersections of sphere with planes through its center.

11.12. $S:z = x^2 + y^2, P = (0, 0, 0), L_0: y = 0, z = 0$ (the x-axis).

(a) Paraboloidal cylinder; intersection of S with any plane through the z-axis is a parabola; with any plane perpendicular to the z-axis which meets the z-axis at $z = R^2$, a circle of radius R.

(b) $K_p(S, L) =$ curvature of $y = x^2$. By Exercise 11.4, curvature of $y = x^2$ at $(0, 0)$ is $K = 2$.

11.13.   Since paraboloid is symmetrical about z-axis, its curvature will not depend on the direction of L in the tangent plane at $P = (0, 0, 0)$. Hence (see top of p. 313) $K_M = K_m = 2$ and $K_p(S) = K_M \cdot K_m = 4$.

11.14. (a) $K = -4$

(b) Section in yz-plane $(x = 0): K_m = 2$

Section in xz-plane $(y = 0): z = 3y^2, \frac{dz}{dy} = 6y, \frac{d^2z}{dy^2} = 6, K_M = \frac{6}{(1 + (6y)^2)^{3/2}}$.

At $y = 0, = 6$. $K = K_M \cdot K_m = 6 \cdot 2 = 12$.

11.15. Gauss's theorem asserts that Gaussian curvature is an intrinsic invariant. On a cylinder take a line in the tangent plane parallel to the cylinder's axis. Curvature in this direction is 0, hence a minimum: $K_m = 0$. For a tangent line perpendicular to the first, the curvature is $K_M = \frac{1}{R}$ where R is the radius of the cylinder. Hence the Gaussian curvature is $K_M \cdot K_m = 0$.

11.16. Geodesics of the plane are straight lines so geodesic connecting P to Q on the cylinder is curve obtained by: (1) cutting cylinder parallel to its axis and rolling it flat; (2) connecting P to Q by a line segment; (3) rerolling to obtain cylinder again.

11.17. Yes. The sphere has constant Gaussian curvature $1/R^2$, which is > 0. But an indented rubber ball has points where its surface is shaped like a saddle (*cp.* Figure 11.21b on p. 312) for which Gaussian curvature is < 0. Hence deformation of a ball must stretch its surface.

Chapter 12

12.4. (a) $\frac{d^2x}{dt^2}(t) = 0, \frac{d^2y}{dt^2}(t) = 0, \frac{d^2z}{dt}(t) = 0$

(b) $x(t) = 2t, y(t) = 0, z(t) = 0$

(c) 20 meters

12.6. Solution: $x(t) = 0, y(t) = 0, z(t) = -16t^2 + 100$, so $z(t) = 0$ for some t means $16t^2 = 100$ or $t^2 = 100/16$ whence $t = 10/4 = 2.5$ seconds. (Note: the weight 10 lbs. plays no role in this solution.)

12.8. (a) parabola

(b) the same length of time, 9.16 seconds

REFERENCES

1. Oscar S. Adams: *A Study of Map Projections in General,* U.S. Department of Commerce, Coast and Geodetic Survey Special Publication No. 60, Washington, 1919.

2. American Mathematical Society: *Combined Membership List, 1966–1967,* Providence, 1967.

3. Leo Bagrow: *History of Cartography,* revised and enlarged by R. A. Skelton, Harvard University Press, Cambridge, Mass., 1964.

4. James Phinney Baxter, 3rd: *Scientists Against Time,* Little, Brown, Boston, 1946.

5. Friedrich Bleich, C. B. McCullough, Richard Rosecrans, and George S. Vincent: *The Mathematical Theory of Vibration in Suspension Bridges,* U.S. Department of Commerce, Washington, D.C., 1950.

6. Salomon Bochner: *Eclosion and Synthesis,* Benjamin, New York, 1969.

7. Emile Borel: *Space and Time,* Dover, New York, 1960.

8. Carl B. Boyer: *A History of Mathematics,* Wiley, New York, 1968.

9. ———: *History of Analytic Geometry,* Wiley, New York, 1956.

10. Harrison Brown: *The Challenge of Man's Future,* Viking, New York, 1954.

11. Harrison Brown, James Bonner, and John Weir: *The Next Hundred Years,* Viking, New York, 1957.

12. Lloyd A. Brown: *The Story of Maps,* Little, Brown and Bonanza Books, Boston, 1949.

13. Sir E. A. Wallis Budge: *Egyptian Language,* Routledge, London, 1966.

14. Roger Burlingame: *March of the Iron Men,* Scribner, New York, 1938.

15. Florian Cajori: *A History of Mathematical Notations,* 2 vols., Open Court, Chicago, 1928–1929.

16. ———: *The Early Mathematical Sciences in North and South America,* Richard G. Badger, Boston, 1928.

17. ———: History of the exponential and logarithmic concepts, *Amer. Math. Monthly* **20,** 5–14 (1913).

18. John F. Campbell: *History and Bibliography of the* New American Practical Navigator *and the* American Coast Pilot, Peabody Museum, Salem, 1964.

19. A. M. Carr-Saunders: *World Population,* Oxford University Press, Oxford, 1936.

20. Kuan-I Chen: *World Population Growth and Living Standards,* College and University Press, New Haven, 1960.

21. Edward Chiera: *They Wrote on Clay,* University of Chicago Press, Chicago, 1938.

22. Julian Lowell Coolidge: *A History of Geometrical Methods*, Oxford University Press, Oxford, 1940.

23. Giorgio De Santillana and Hertha von Dechend: *Hamlet's Mill*, Gambit, Boston, 1969.

24. Derek J. De Solla Price: *Big Science, Little Science*, Columbia University Press, New York, 1963.

25. ———: *Science Since Babylon*, Yale University Press, New Haven, 1962.

26. T. K. Derry and Trevor I. Williams: *A Short History of Technology*, Oxford University Press, Oxford, 1960.

27. R. H. Dicke: Dating the galaxy by uranium decay, *Nature* **194**, 329–330 (1962).

28. ———: The earth and cosmology, *Science* **138**, 653–664 (1962).

29. G. Waldo Dunnington: *Gauss: Titan of Science*, Hafner, New York, 1955.

30. Arthur Stanley Eddington: *The Mathematical Theory of Relativity*, Cambridge University Press, Cambridge, 1952.

31. Albert Einstein: *The Meaning of Relativity*, 5th ed., Princeton University Press, Princeton, N.J., 1955.

32. Howard Eves: *An Introduction to the History of Mathematics*, 3rd ed., Holt, Rinehart and Winston, New York, 1969.

33. Henry Faul: *Ages of Rocks, Planets, and Stars*, McGraw-Hill, New York, 1966.

34. Frantisek Fiala: *Mathematische Kartographie*, VEB Verlag Technik, Berlin, 1957.

35. James Kip Finch: *The Story of Engineering*, Doubleday, New York, 1960.

36. Carl Freidrich Gauss: *Werke*, Vol. IV, Königlichen Gesellschaft der Wissenschaften, Göttingen, 1873.

37. I. J. Gelb: *A Study of Writing*, rev. ed., University of Chicago Press, Chicago, 1963.

38. Charles Homer Haskins: *The Rise of Universities*, Cornell University Press, Ithaca, 1957.

39. Gerald S. Hawkins: *Stonehenge Decoded*, Doubleday, New York, 1965.

40. Sir Thomas Heath: *A History of Greek Mathematics*, 2 vols., Oxford University Press, Oxford, 1921.

41. ———: *Archimedes*, Macmillan, London, 1920.

42. ———: *Aristarchus of Samos*, Oxford University Press, Oxford, 1913.

43. ———: *The Works of Archimedes*, with a supplement, *The Method of Archimedes*, Dover, New York, n.d.

44. George Huxley: *The Interaction of Greek and Babylonian Astronomy,* The Queen's University, Belfast, Northern Ireland, 1954.

45. Felix Klein: *Elementary Mathematics From an Advanced Standpoint,* 2 vols., Dover, New York, 1939.

46. L. Landau and E. Lifshitz: *The Classical Theory of Fields,* Addison-Wesley, Cambridge, Mass., 1951.

47. Jose Maria Matinez-Hidalgo: *Columbus' Ships,* Barre Publishers, Barre, Mass., 1966.

48. G. C. McVittie: *Fact and Theory in Cosmology,* Macmillan, New York, 1961.

49. Karl Menninger: *Zahlwort und Ziffer,* Vandenhoeck & Ruprecht, Göttingen, Germany, 1958.

50. O. Neugebauer: *The Exact Sciences in Antiquity,* Brown University Press, Providence, R.I., 1957.

51. O. Neugebauer and A. Sachs: *Mathematical Cuneiform Texts,* American Oriental Series, Vol. 29, American Oriental Society, New Haven, Conn., 1945.

52. Leo Oppenheim: *Ancient Mesopotamia,* University of Chicago Press, Chicago, 1964.

53. J. H. Parry: *The Age of Reconnaissance,* World, Cleveland, Ohio, 1963.

54. Palmer Cosslett Putnam: *Energy in the Future,* Van Nostrand, New York, 1953.

55. *Recent Developments in General Relativity,* Pergamon, New York, 1962.

56. H. L. Resnikoff and J. L. Dolby: On the analysis of library growth, *Rice University Studies* **55** No. 4, 55–89 (1969).

57. David B. Steinman: *A Practical Treatise on Suspension Bridges,* 2nd ed., Wiley, New York, 1949.

58. David B. Steinman and Sara Ruth Watson: *Bridges and Their Builders,* Dover, New York, 1957.

59. D. J. Struik: *A Source Book in Mathematics, 1200–1800,* Harvard University Press, Cambridge, Mass., 1969.

60. ———: *A Short History of Mathematics,* 3rd ed., Dover, New York, 1968.

61. ———: *Yankee Science in the Making,* rev. ed., Collier, New York, 1962.

62. ———: *Lectures on Classical Differential Geometry,* Addison-Wesley, Cambridge, Mass., 1950.

63. Tata Institute of Fundamental Research, School of Mathematics: *World Directory of Mathematicians, 1966,* Bombay, n.d.

64. E. G. R. Taylor: *The Mathematical Practitioners of Tudor and Stuart England,* Cambridge University Press, Cambridge, 1967.

65. George B. Thomas: *Calculus and Analytic Geometry*, Addison-Wesley, Cambridge, Mass., 1953.

66. E. M. Tillyard: *The Elizabethan World Picture*, Penguin, Hammondsworth, England, 1963.

67. R. C. Tolman: The age of the universe, *Rev. Mod. Phys.* **21**, 374–378 (1949).

68. Stephen Toulmin and June Goodfield: *The Fabric of the Heavens*, Harper & Row, New York, 1961.

69. J. Tropfke: *Geschichte der Elementarmathematik in systematischer Darstellung,* second edition, seven volumes. Walter de Gruyter & Co., Berlin, 1921–1924.

70. U.S. Hydrographic Office: *Hydrographic Publication No. 9, American Practical Navigator,* originally by Nathaniel Bowditch, Washington, D.C.

71. B. L. Van Der Waerden: *Science Awakening,* Wiley, New York, 1963.

72. Kurt Vogel: *Vorgriechische Mathematik*, 2 vols., Hermann Schroedel Verlag KG, Hannover, Germany, 1959.

73. Sir Edmund Whittaker: *From Euclid to Eddington,* Dover, New York, 1958.

74. ———: *A History of the Theories of Aether and Electricity,* Nelson, London, 1951 and 1953.

75. John A. Wilson: *The Culture of Ancient Egypt,* University of Chicago Press, Chicago, 1951 (formerly titled *The Burden of Egypt*).

76. E. M. Bruins: "On Plimpton 322, Pythagorean numbers in Babylonian mathematics," *Akad. v. Wetenshappen, Amsterdam. Afdeling Natuurkunde, Proc.* **52**, 629–632 (1949).

77. ———: *Codex Constantinopolitanus,* 3 vols., E. J. Brill, Leiden, 1964.

78. T. Eric Peet, *The Rhind Mathematical Papyrus,* Liverpool Univ. Press, 1923.

79. E. T. Whittaker, and G. N. Watson: *Modern Analysis,* Cambridge, 1927.

80. G. Loria, *Spezielle algebraische und transscendente Ebene Kurven: Theorie und Geschichte;* I, II, B. G. Teubner, Leipzig and Berlin; 1910, 1911.

81. O. Neugebauer, *Vorlesungen über Geschichte der Antiken Mathematischen Wissenschaften: Vorgriechische Mathematik,* Springer Verlag, New York, Heidelberg, Berlin, 1934.

82. Camille Flammarion, *The Flammarion Book of Astronomy,* Simon and Schuster, New York, 1964.

83. William D. Metz: The decline of the Hubble constant: a new age for the universe, *Science* **178**, 600–601 (1972).

TABLES

Table 1: Common (Briggsian) Logarithms

N	0	1	2	3	4	5	6	7	8	9
10	0000	0043	0086	0128	0170	0212	0253	0294	0334	0374
11	0414	0453	0492	0531	0569	0607	0645	0682	0719	0755
12	0792	0828	0864	0899	0934	0969	1004	1038	1072	1106
13	1139	1173	1206	1239	1271	1303	1335	1367	1399	1430
14	1461	1492	1523	1553	1584	1614	1644	1673	1703	1732
15	1761	1790	1818	1847	1875	1903	1931	1959	1987	2014
16	2041	2068	2095	2122	2148	2175	2201	2227	2253	2279
17	2304	2330	2355	2380	2405	2430	2455	2480	2504	2529
18	2553	2577	2601	2625	2648	2672	2695	2718	2742	2765
19	2788	2810	2833	2856	2878	2900	2923	2945	2967	2989
20	3010	3032	3054	3075	3096	3118	3139	3160	3181	3201
21	3222	3243	3263	3284	3304	3324	3345	3365	3385	3404
22	3424	3444	3464	3483	3502	3522	3541	3560	3579	3598
23	3617	3636	3655	3674	3692	3711	3729	3747	3766	3784
24	3802	3820	3838	3856	3874	3892	3909	3927	3945	3962
25	3979	3997	4014	4031	4048	4065	4082	4099	4116	4133
26	4150	4166	4183	4200	4216	4232	4249	4265	4281	4298
27	4314	4330	4346	4362	4378	4393	4409	4425	4440	4456
28	4472	4487	4502	4518	4533	4548	4564	4579	4594	4609
29	4624	4639	4654	4669	4683	4698	4713	4728	4742	4757
30	4771	4786	4800	4814	4829	4843	4857	4871	4886	4900
31	4914	4928	4942	4955	4969	4983	4997	5011	5024	5038
32	5051	5065	5079	5092	5105	5119	5132	5145	5159	5172
33	5185	5198	5211	5224	5237	5250	5263	5276	5289	5302
34	5315	5328	5340	5353	5366	5378	5391	5403	5416	5428
35	5441	5453	5465	5478	5490	5502	5514	5527	5539	5551
36	5563	5575	5587	5599	5611	5623	5635	5647	5658	5670
37	5682	5694	5705	5717	5729	5740	5752	5763	5775	5786
38	5798	5809	5821	5832	5843	5855	5866	5877	5888	5899
39	5911	5922	5933	5944	5955	5966	5977	5988	5999	6010
40	6021	6031	6042	6053	6064	6075	6085	6096	6107	6117
41	6128	6138	6149	6160	6170	6180	6191	6201	6212	6222
42	6232	6243	6253	6263	6274	6284	6294	6304	6314	6325
43	6335	6345	6355	6365	6375	6385	6395	6405	6415	6425
44	6435	6444	6454	6464	6474	6484	6493	6503	6513	6522
45	6532	6542	6551	6561	6571	6580	6590	6599	6609	6618
46	6628	6637	6646	6656	6665	6675	6684	6693	6702	6712
47	6721	6730	6739	6749	6758	6767	6776	6785	6794	6803
48	6812	6821	6830	6839	6848	6857	6866	6875	6884	6893
49	6902	6911	6920	6928	6937	6946	6955	6964	6972	6981
50	6990	6998	7007	7016	7024	7033	7042	7050	7059	7067
51	7076	7084	7093	7101	7110	7118	7126	7135	7143	7152
52	7160	7168	7177	7185	7193	7202	7210	7218	7226	7235
53	7243	7251	7259	7267	7275	7284	7292	7300	7308	7316
54	7324	7332	7340	7348	7356	7364	7372	7380	7388	7396

N	0	1	2	3	4	5	6	7	8	9
55	7404	7412	7419	7427	7435	7443	7451	7459	7466	7474
56	7482	7490	7497	7505	7513	7520	7528	7536	7543	7551
57	7559	7566	7574	7582	7589	7597	7604	7612	7619	7628
58	7634	7642	7649	7657	7664	7672	7679	7686	7694	7701
59	7709	7716	7723	7731	7738	7745	7752	7760	7767	7774
60	7782	7789	7796	7803	7810	7818	7825	7832	7839	7846
61	7853	7860	7868	7875	7882	7889	7896	7903	7910	7917
62	7924	7931	7938	7945	7952	7959	7966	7973	7980	7987
63	7993	8000	8007	8014	8021	8028	8035	8041	8048	8055
64	8062	8069	8075	8082	8089	8096	8102	8109	8116	8122
65	8129	8136	8142	8149	8156	8162	8169	8176	8182	8189
66	8195	8202	8209	8215	8222	8228	8235	8241	8248	8254
67	8261	8267	8274	8280	8287	8293	8299	8306	8312	8319
68	8325	8331	8338	8344	8351	8357	8363	8370	8376	8382
69	8388	8395	8401	8407	8414	8420	8426	8432	8439	8445
70	8451	8457	8463	8470	8476	8482	8488	8494	8500	8506
71	8513	8519	8525	8531	8537	8543	8549	8555	8561	8567
72	8573	8579	8585	8591	8597	8603	8609	8615	8621	8627
73	8633	8639	8645	8651	8657	8663	8669	8675	8681	8686
74	8692	8698	8704	8710	8716	8722	8727	8733	8739	8745
75	8751	8756	8762	8768	8774	8779	8785	8791	8797	8802
76	8808	8814	8820	8825	8831	8837	8842	8848	8854	8859
77	8865	8871	8876	8882	8887	8893	8899	8904	8910	8915
78	8921	8927	8932	8938	8943	8949	8954	8960	8965	8971
79	8976	8982	8987	8993	8998	9004	9009	9015	9020	9025
80	9031	9036	9042	9047	9053	9058	9063	9069	9074	9079
81	9085	9090	9096	9101	9106	9112	9117	9122	9128	9133
82	9138	9143	9149	9154	9159	9165	9170	9175	9180	9186
83	9191	9196	9201	9206	9212	9217	9222	9227	9232	9238
84	9243	9248	9253	9258	9263	9269	9274	9279	9284	9289
85	9294	9299	9304	9309	9315	9320	9325	9330	9335	9340
86	9345	9350	9355	9360	9365	9370	9375	9380	9385	9390
87	9395	9400	9405	9410	9415	9420	9425	9430	9435	9440
88	9445	9450	9455	9460	9465	9469	9474	9479	9484	9489
89	9494	9499	9504	9509	9513	9518	9523	9528	9533	9538
90	9542	9547	9552	9557	9562	9566	9571	9576	9581	9586
91	9590	9595	9600	9605	9609	9614	9619	9624	9628	9633
92	9638	9643	9647	9652	9657	9661	9666	9671	9675	9680
93	9685	9689	9694	9699	9703	9708	9713	9717	9722	9727
94	9731	9736	9741	9745	9750	9754	9759	9763	9768	9773
95	9777	9782	9786	9791	9795	9800	9805	9809	9814	9818
96	9823	9827	9832	9836	9841	9845	9850	9854	9859	9863
97	9868	9872	9877	9881	9886	9890	9894	9899	9903	9908
98	9912	9917	9921	9926	9930	9934	9939	9943	9948	9952
99	9956	9961	9965	9969	9974	9978	9983	9987	9991	9996

Table 2: Natural Logarithms

n	$\log_e n$	n	$\log_e n$	n	$\log_e n$
0.0	*	4.5	1.5041	9.0	2.1972
0.1	7.6974	4.6	1.5261	9.1	2.2083
0.2	8.3906	4.7	1.5476	9.2	2.2192
0.3	8.7960	4.8	1.5686	9.3	2.2300
0.4	9.0837	4.9	1.5892	9.4	2.2407
0.5	9.3069	5.0	1.6094	9.5	2.2513
0.6	9.4892	5.1	1.6292	9.6	2.2618
0.7	9.6433	5.2	1.6487	9.7	2.2721
0.8	9.7769	5.3	1.6677	9.8	2.2824
0.9	9.8946	5.4	1.6864	9.9	2.2925
1.0	0.0000	5.5	1.7047	10	2.3026
1.1	0.0953	5.6	1.7228	11	2.3979
1.2	0.1823	5.7	1.7405	12	2.4849
1.3	0.2624	5.8	1.7579	13	2.5649
1.4	0.3365	5.9	1.7750	14	2.6391
1.5	0.4055	6.0	1.7918	15	2.7081
1.6	0.4700	6.1	1.8083	16	2.7726
1.7	0.5306	6.2	1.8245	17	2.8332
1.8	0.5878	6.3	1.8405	18	2.8904
1.9	0.6419	6.4	1.8563	19	2.9444
2.0	0.6931	6.5	1.8718	20	2.9957
2.1	0.7419	6.6	1.8871	25	3.2189
2.2	0.7885	6.7	1.9021	30	3.4012
2.3	0.8329	6.8	1.9169	35	3.5553
2.4	0.8755	6.9	1.9315	40	3.6889
2.5	0.9163	7.0	1.9459	45	3.8067
2.6	0.9555	7.1	1.9601	50	3.9120
2.7	0.9933	7.2	1.9741	55	4.0073
2.8	1.0296	7.3	1.9879	60	4.0943
2.9	1.0647	7.4	2.0015	65	4.1744
3.0	1.0986	7.5	2.0149	70	4.2485
3.1	1.1314	7.6	2.0281	75	4.3175
3.2	1.1632	7.7	2.0412	80	4.3820
3.3	1.1939	7.8	2.0541	85	4.4427
3.4	1.2238	7.9	2.0669	90	4.4998
3.5	1.2528	8.0	2.0794	95	4.5539
3.6	1.2809	8.1	2.0919	100	4.6052
3.7	1.3083	8.2	2.1041		
3.8	1.3350	8.3	2.1163		
3.9	1.3610	8.4	2.1282		
4.0	1.3863	8.5	2.1401		
4.1	1.4110	8.6	2.1518		
4.2	1.4351	8.7	2.1633		
4.3	1.4586	8.8	2.1748		
4.4	1.4816	8.9	2.1861		

* Subtract 10 from $\log_e n$ entries for $n < 1.0$.

Table 3: Exponential Functions

x	e^x	e^{-x}	x	e^x	e^{-x}
0.00	1.0000	1.0000	3.5	33.115	0.0302
0.05	1.0513	0.9512	3.6	36.598	0.0273
0.10	1.1052	0.9048	3.7	40.447	0.0247
0.15	1.1618	0.8607	3.8	44.701	0.0224
0.20	1.2214	0.8187	3.9	49.402	0.0202
0.25	1.2840	0.7788	4.0	54.598	0.0183
0.30	1.3499	0.7408	4.1	60.340	0.0166
0.35	1.4191	0.7047	4.2	66.686	0.0150
0.40	1.4918	0.6703	4.3	73.700	0.0136
0.45	1.5683	0.6376	4.4	81.451	0.0123
0.50	1.6487	0.6065	4.5	90.017	0.0111
0.55	1.7333	0.5769	4.6	99.484	0.0101
0.60	1.8221	0.5488	4.7	109.95	0.0091
0.65	1.9155	0.5220	4.8	121.51	0.0082
0.70	2.0138	0.4966	4.9	134.29	0.0074
0.75	2.1170	0.4724	5	148.41	0.0067
0.80	2.2255	0.4493	6	403.43	0.0025
0.85	2.3396	0.4274	7	1096.6	0.0009
0.90	2.4596	0.4066	8	2981.0	0.0003
0.95	2.5857	0.3867	9	8103.1	0.0001
1.0	2.7183	0.3679	10	22026	0.00005
1.1	3.0042	0.3329			
1.2	3.3201	0.3012			
1.3	3.6693	0.2725			
1.4	4.0552	0.2466			
1.5	4.4817	0.2231			
1.6	4.9530	0.2019			
1.7	5.4739	0.1827			
1.8	6.0496	0.1653			
1.9	6.6859	0.1496			
2.0	7.3891	0.1353			
2.1	8.1662	0.1225			
2.2	9.0250	0.1108			
2.3	9.9742	0.1003			
2.4	11.023	0.0907			
2.5	12.182	0.0821			
2.6	13.464	0.0743			
2.7	14.880	0.0672			
2.8	16.445	0.0608			
2.9	18.174	0.0550			
3.0	20.086	0.0498			
3.1	22.198	0.0450			
3.2	24.533	0.0408			
3.3	27.113	0.0369			
3.4	29.964	0.0334			

Table 4: Trigonometrical Functions

Degree	Radian	Sine	Cosine	Tangent	Degree	Radian	Sine	Cosine	Tangent
0°	0.000	0.000	1.000	0.000					
1°	0.017	0.017	1.000	0.017	46°	0.803	0.719	0.695	1.036
2°	0.035	0.035	0.999	0.035	47°	0.820	0.731	0.682	1.072
3°	0.052	0.052	0.999	0.052	48°	0.838	0.743	0.669	1.111
4°	0.070	0.070	0.998	0.070	49°	0.855	0.755	0.656	1.150
5°	0.087	0.087	0.996	0.087	50°	0.873	0.766	0.643	1.192
6°	0.105	0.105	0.995	0.105	51°	0.890	0.777	0.629	1.235
7°	0.122	0.122	0.993	0.123	52°	0.908	0.788	0.616	1.280
8°	0.140	0.139	0.990	0.141	53°	0.925	0.799	0.602	1.327
9°	0.157	0.156	0.988	0.158	54°	0.942	0.809	0.588	1.376
10°	0.175	0.174	0.985	0.176	55°	0.960	0.819	0.574	1.428
11°	0.192	0.191	0.982	0.194	56°	0.977	0.829	0.559	1.483
12°	0.209	0.208	0.978	0.213	57°	0.995	0.839	0.545	1.540
13°	0.227	0.225	0.974	0.231	58°	1.012	0.848	0.530	1.600
14°	0.244	0.242	0.970	0.249	59°	1.030	0.857	0.515	1.664
15°	0.262	0.259	0.966	0.268	60°	1.047	0.866	0.500	1.732
16°	0.279	0.276	0.961	0.287	61°	1.065	0.875	0.485	1.804
17°	0.297	0.292	0.956	0.306	62°	1.082	0.883	0.469	1.881
18°	0.314	0.309	0.951	0.325	63°	1.100	0.891	0.454	1.963
19°	0.332	0.326	0.946	0.344	64°	1.117	0.899	0.438	2.050
20°	0.349	0.342	0.940	0.364	65°	1.134	0.906	0.423	2.145
21°	0.367	0.358	0.934	0.384	66°	1.152	0.914	0.407	2.246
22°	0.384	0.375	0.927	0.404	67°	1.169	0.921	0.391	2.356
23°	0.401	0.391	0.921	0.424	68°	1.187	0.927	0.375	2.475
24°	0.419	0.407	0.914	0.445	69°	1.204	0.934	0.358	2.605
25°	0.436	0.423	0.906	0.466	70°	1.222	0.940	0.342	2.747
26°	0.454	0.438	0.899	0.488	71°	1.239	0.946	0.326	2.904
27°	0.471	0.454	0.891	0.510	72°	1.257	0.951	0.309	3.078
28°	0.489	0.469	0.883	0.532	73°	1.274	0.956	0.292	3.271
29°	0.506	0.485	0.875	0.554	74°	1.292	0.961	0.276	3.487
30°	0.524	0.500	0.866	0.577	75°	1.309	0.966	0.259	3.732
31°	0.541	0.515	0.857	0.601	76°	1.326	0.970	0.242	4.011
32°	0.559	0.530	0.848	0.625	77°	1.344	0.974	0.225	4.331
33°	0.576	0.545	0.839	0.649	78°	1.361	0.978	0.208	4.705
34°	0.593	0.559	0.829	0.675	79°	1.379	0.982	0.191	5.145
35°	0.611	0.574	0.819	0.700	80°	1.396	0.985	0.174	5.671
36°	0.628	0.588	0.809	0.727	81°	1.414	0.988	0.156	6.314
37°	0.646	0.602	0.799	0.754	82°	1.431	0.990	0.139	7.115
38°	0.663	0.616	0.788	0.781	83°	1.449	0.993	0.122	8.144
39°	0.681	0.629	0.777	0.810	84°	1.466	0.995	0.105	9.514
40°	0.698	0.643	0.766	0.839	85°	1.484	0.996	0.087	11.43
41°	0.716	0.656	0.755	0.869	86°	1.501	0.998	0.070	14.30
42°	0.733	0.669	0.743	0.900	87°	1.518	0.999	0.052	19.08
43°	0.750	0.682	0.731	0.933	88°	1.536	0.999	0.035	28.64
44°	0.768	0.695	0.719	0.966	89°	1.553	1.000	0.017	57.29
45°	0.785	0.707	0.707	1.000	90°	1.571	1.000	0.000	

INDEX

The page numbers which are *italicized* in this index correspond to italicized references in the text. The suffixes F, E, and T correspond to references in Figures, Exercises, and Tables, respectively.